REVOLUTIONARY EXPERIMENTS

Revolutionary Experiments

THE QUEST FOR IMMORTALITY IN BOLSHEVIK SCIENCE AND FICTION

Nikolai Krementsov

OXFORD
UNIVERSITY PRESS

OXFORD
UNIVERSITY PRESS

Oxford University Press is a department of the University of Oxford.
It furthers the University's objective of excellence in research, scholarship,
and education by publishing worldwide.

Oxford New York
Auckland Cape Town Dar es Salaam Hong Kong Karachi
Kuala Lumpur Madrid Melbourne Mexico City Nairobi
New Delhi Shanghai Taipei Toronto

With offices in
Argentina Austria Brazil Chile Czech Republic France Greece
Guatemala Hungary Italy Japan Poland Portugal Singapore
South Korea Switzerland Thailand Turkey Ukraine Vietnam

Published in the United States of America by
Oxford University Press
198 Madison Avenue, New York, NY 10016

Library of Congress Cataloging-in-Publication Data
Krementsov, N. L., author.
Revolutionary experiments : the quest for immortality in Bolshevik science and fiction / Nikolai Krementsov.
p. cm.
Includes bibliographical references and index.
ISBN 978-0-19-999298-0 (alk. paper)
1. Literature and science—Soviet Union. 2. Biology—Soviet Union—Experiments. 3. Immortality in literature. 4. Biology in literature. 5. Science fiction, Russian—20th century—History and criticism. I. Title.
PG3026.S348K74 2014
891.709'356—dc23
2013037097
9780199992980

9 8 7 6 5 4 3 2 1

Printed in the United States of America on acid-free paper

To A. E. and E. F.

The Russian revolution is [. . .] the inauguration of an age of limitless experiment.
H. G. WELLS, *Russia in the Shadows* (1920)

The biologist is the most romantic figure on earth at the present day.
J. B. S. HALDANE, *Daedalus* (1923)

Contents

List of Illustrations

List of Abbreviations

ARAN—*Arkhiv Rossiiskoi Akademii nauk*, Archive of the Russian Academy of Sciences
GARF—*Gosudarstvennyi arkhiv Rossiiskoi Federatsii*, State Archive of the Russian Federation
GINZ—*Gosudarstvennyi institut narodnogo zdravookhraneniia*, State Institute of People's Health Protection
IEB—*Institut eksperimental'noi biologii*, Institute of Experimental Biology
KN—*Krasnaia nov'* (*Red Virgin Soil*), a journal
KP—*Krasnaia panorama* (*Red Panorama*), a journal
MIM—*Muzei istorii meditsiny Pervogo Moskovskogo meditsinskogo universiteta imeni I. M. Sechenova*, Museum of the History of Medicine of the Sechenov First Moscow Medical University
Narkompros—*Narodnyi Komissariat Prosveshcheniia*, People's Commissariat of Enlightenment
Narkomzdrav—*Narodnyi Komissariat Zdravookhraneniia*, People's Commissariat of Health Protection
Narkomzem—*Narodnyi Komissariat Zemledeliia*, People's Commissariat of Agriculture
NEP—*novaia ekonomicheskaia politika*, new economic policy
NIOR RGB—*Nauchno-issledovatel'skii otdel rukopisei Rossiiskoi gosudarstvennoi biblioteki*, Manuscript Division of the Russian State Library
NKVD—*Narodnyi Komissariat Vnutrennikh Del*, People's Commissariat of Internal Affairs
PZM—*Pod znamenem marksizma* (*Under the Banner of Marxism*), a journal
RGASPI—*Rossiiskii gosudarstvennyi arkhiv sotsial'no-politicheskoi istorii*, Russian State Archive of Socio-Political History
RSFSR—*Rossiiskaia Sovetskaia Federativnaia Sotsialisticheskaia Respublika*, Russian Soviet Federated Socialist Republic
SNK—*Sovet Narodnykh Komissarov*, Council of People's Commissars
TsGALI SPb—*Tsentral'nyi gosudarstvennyi arkhiv literatury i iskusstva Sankt Peterburga*, St. Petersburg Central State Archive of Literature and Arts

TsIK—*Tsentral'nyi Ispolnitel'nyi Komitet*, Central Executive Committee

TsKUBU—*Tsentral'naia komissiia po uluchsheniiu byta uchenykh*, Central Commission to Improve Scientists' Living Conditions

VIET—*Voprosy istorii estestvoznaniia i tekhniki* (*Issues in the history of natural sciences and technology*), a journal

VM—*Vecherniaia Moskva*, (*Evening Moscow*), a newspaper

VOFEM—*Vestnik opytnoi fiziki i elementarnoi matematiki* (*Herald of experimental physics and elementary mathematics*), a journal

VSNKh—*Vysshii sovet narodnogo khoziaistva*, the Supreme Council of People's Economy

VZ—*Vestnik znaniia* (*Herald of Knowledge*), a journal

ZhRFKhO—*Zhurnal Russkogo fiziko-khimicheskogo obshchestva* (*Journal of the Russian Physico-Chemical Society*)

Note on Names, Translations, and Transliterations

DURING THE LAST hundred years, numerous cities and institutions in the territories of the Russian Empire, the USSR, and now the Russian Federation repeatedly changed their names. Thus, St. Petersburg became Petrograd in 1914, Leningrad in 1924, and regained its original name in 1991. Similarly, the St. Petersburg Imperial Academy of Sciences became the Russian Academy of Sciences in 1917, the USSR Academy of Sciences in 1925, and once again the Russian Academy of Sciences in 1991. Throughout the text I use that name for a particular locale or institution that was in use at the time I describe. Since this book deals with a series of social, cultural, and scientific revolutions, I reserve the use of "Revolution" (with the capital "R") to refer exclusively to the Bolshevik coup in October 1917. In rendering various Russian names in Latin alphabet, I use the Library of Congress' transliteration system, except for the commonly adopted spellings of well-known names, such as, for example, "St. Petersburg," "Alexander," "Leon Trotsky," and "Elie Metchnikoff" instead of "Sankt-Peterburg," "Aleksandr," "Lev Trotskii," and "Il'ia Mechnikov." I preserved in the text the original titles of various French and German publications, for their English and Russian translations often differ, giving the available Russian translations of these titles in the references. Except for the names of the country's two major newspapers, *Pravda* and *Izvestiia*, and a popular science journal, *Priroda* (Nature), I translated into English the titles of periodicals in the text, but preserved their Russian names in the references. Although some of the original Russian sources I cite are available in various English translations, all of the translations in the book are my own.

It has been said by its opponents that science divorces itself from literature; but this statement, like so many others, arises from lack of knowledge.
JOHN TYNDALL, *The Belfast Address* (1874)

The fear of death, and not only on the part of aging senators, has always been a stronger cultural force than the love of beauty or of truth.
ERWIN CHARGAFF, *Voices in the Labyrinth* (1977)

Prologue: Science and Fiction

FROM TIME IMMEMORIAL no other dream has haunted human imagination more than the dream of eternal life, perpetual youth, and immortality. Theologians, philosophers, writers, historians, anthropologists, psychologists, literary scholars, physicians, sociologists, biologists, and ethnographers have explored the origins, developments, and uses of this dream, in the process generating an enormous body of literature that covers a dizzying array of issues, settings, times, and actors.[1] This book examines a particular fascination with this dream, and the role of science and fiction in its pursuit, in one particular locale—Russia—during the decade or so following the country's political revolutions of 1917. It trespasses on the territories of five different fields: the history of science, literary studies, the history of medicine, cultural studies, and the history of Russia. I realize that it is quite presumptuous to take on such an ambitious project, but I have deeply personal reasons for such egregious transgression of disciplinary boundaries and professional conventions.

My first passion in life was reading. I grew up surrounded by books and learned to read before I turned four—my father was a professor of Russian literature (as was his father) and my mother taught the same subject at the high school level. It never occurred to me to count how many books inhabited our apartment on the outskirts of Tashkent or my grandfather's small house in the city's "old town." But never once in my childhood did I have to go to a public library when I wanted to read a particular book: I could almost always find that book somewhere on the floor-to-ceiling bookshelves lining every wall of our apartment, from the entry hall to the glassed-off terrace. And if, by chance, I could not find the book at home, I would almost certainly find it in my grandfather's house, hidden behind the glass doors of the old oak bookcases crowding his spacious "home office." By the age of twelve, I had devoured everything a boy was supposed to read about

Greek heroes and Baghdad thieves, musketeers and buccaneers, Egyptian Pharaohs and Roman Emperors, pirates and pioneers, noble knights and runaway slaves, cowboys and Indians, aliens and spaceships, archeological digs and undersea adventures, and much, much more—especially, of course, the classic works of Russian literature. I read Tolstoy and Dostoevsky, Bulgakov and Solzhenitsyn, Il'f and Petrov, and the brothers Strugatskii. But it never crossed my mind that I should follow the family trade and make the study of literature my profession.

For my second passion was science. I was born in the first year of a new era: the year of Sputnik. I still remember quite vividly that (sitting on the shoulders of my father amidst a huge crowd lining the sidewalk of our street) I got a glimpse of Iurii Gagarin, the first man to leave Earth and go to the Cosmos: he was being driven in an open car and waving his hand at ME! As did many boys of my generation, I collected pictures of cosmonauts, glued together models of spaceships, and constantly read works of popular science and science fiction. But it was not the exploration of the Cosmos that held my imagination; it was the study of Life. From about the age of ten, enthralled by Alfred Brehm's magnificent multivolume *The Life of Animals* and Jean Henri Fabre's marvelous book *La vie des insectes,* I knew that one day I would be a biologist. By the time I graduated from high school, my decision to study biology had become irrevocable. For five years at university and then three more in graduate school, I greedily learned everything I could in and about my chosen field: the study of the brain—neurophysiology. I studied surgical techniques and electronics repair, brain anatomy and statistical analysis, the basics of veterinary medicine and computer programming, patterns of animal behavior and human psychology. I conducted my experiments at the "world's first and only" Institute of Neurocybernetics in Rostov-on-Don and at the USSR Academy of Sciences Institute of Physiology in Koltushi—Ivan Pavlov's famous "capital of conditional reflexes" near Leningrad. But when the time came for me to write up and defend my dissertation on "Multi-neuronal integrative activity of visual cortex during the sleep-wake cycle of cats," I realized that neurophysiology did not inspire me anymore. I thought that I could perhaps move from laboratory experiments to more naturalistic research and spent a year trying to retool myself through a master's program in environmental studies at Leningrad University. Alas, this experience only strengthened my conviction that it was not just neurophysiology, but "doing science" in general, that no longer held my interest. I quit academia altogether and joined a company of *shabashniki*—"free carpenters" (meaning, freelancing or independent, and not to be confused with "free masons"). I began spending my summers earning a living by building log houses in remote corners of the Soviet Union, leaving my winters to enjoy the cultural treasures of my new "hometown"—Leningrad.

During one such winter in the mid-1980s, through sheer accident, I discovered the history of science, which seemed to offer a unique opportunity to combine my old passions for reading and for science by *reading and writing about* science, as opposed to "doing science." After a year of contemplation, I passed the entry exams and enrolled in a graduate program in the department of the history and theory (read—philosophy) of evolutionary biology at the USSR Academy of Sciences Institute of the History of Natural Sciences and Technology. In the Soviet Union, graduate studies did not involve any coursework. You were supposed to start at once on your dissertation research and in three to four years complete it, along the way passing exams in a foreign language, Marxist philosophy, and your chosen specialty. So, at the suggestion of my advisor, I began working on a dissertation about the history of the interactions between studies of evolution and studies of animal behavior in twentieth-century Russia, using my background in neurophysiology and ecology as a springboard.

I happily burrowed in the vast collections of Leningrad's two largest and oldest libraries: the Library of the USSR Academy of Sciences and the Saltykov-Shchedrin State Public Library, known to aficionados as "BAN" and "Publichka," respectively. But I felt somewhat uncomfortable in my newly found vocation, for I had received no formal instruction in history and knew next to nothing about historical research. I was learning my new trade the same way I had learned carpentry a few years earlier: by reading some basic texts and following the example of my advisors, senior colleagues, and fellow students. Unfortunately, all of my teachers came either from philosophy or biology and none had any background in history, either. This lack of rigorous specialized training, which always includes learning not only the technical skills, but also the important taboos, fashions, and values of the profession, probably explains the gnawing sense of dissatisfaction I felt throughout the years of my graduate studies, as well as the profound dislike for disciplinary boundaries and professional conventions that I acquired over the same period.

What bothered me most was a deep rift, and the nearly total lack of contact/interchange/dialogue, between the history of Russia and the history of Russian science. In the books I was reading—and at that time I had very limited access to Western literature on the subjects, since most of it was considered "anti-Soviet" and thus inaccessible without special "clearance"—science appeared to be something akin to a Kantian "thing in itself." General histories of twentieth-century Russia barely mentioned science, as if it were some foreign object forcefully thrust into and existing entirely outside of the mainstream of Russian life (which admittedly it had once been—in the eighteenth century!). Conversely, books on the history of twentieth-century Russian science completely omitted such critical events as the Civil War, NEP (New Economic Policy), the Great Break, collectivization, the Great Terror, or the Cold War, as if these pivotal episodes in the country's life had no effect on its science whatsoever. In time, surveying the history of my new profession in the Soviet Union, I came to understand the origins of this strange separation.[2] But recognizing its causes did not help my attempts to bridge this divide and figure out a way to write the history of science as an integral part of general history. The dawn of Mikhail Gorbachev's *perestroika*, however, began to offer some possibilities.

In the late 1980s, Daniil Alexandrov—a fellow graduate student who shared my dissatisfaction with the current state of affairs—and I initiated a series of workshops that brought together historians, sociologists, philologists, archivists, and historians of science and published a "manifesto," calling for the (re)integration of the "social" and the "cultural" into our explorations of the history of Russian science.[3] For twenty-plus years since my graduate studies, I have sought to do exactly that in my research on particular episodes in the history of Soviet biology and medicine, testing various approaches, sources, and modes of analysis, as well as styles of writing. This book is yet another attempt at bringing science back to its rightful place in the mainstream of twentieth-century Russian history: I believe that one cannot fully understand this tortuous history without acknowledging the central role science came to play in post-revolutionary Russian culture and society.

It all started nearly a decade ago, when I decided to launch a new project on the early history of experimental biology and medicine in Russia, focusing on the years between the end of the fratricidal Civil War that followed the Bolshevik Revolution and the beginning of the so-called "revolution from above" that inaugurated the establishment of Joseph Stalin's dictatorial rule in the late 1920s and early 1930s. These two revolutions—the 1917 Bolshevik coup and the imposition of a new, Stalinist regime around 1930—undoubtedly represent the most dramatic and traumatic events in the country's history, which reverberated throughout the entire world and shaped the course of twentieth-century history. Yet the short period between the two revolutions, roughly a decade, proved to be the most dynamic, polyphonic, and creative period in practically all walks of the country's life, from economics to foreign politics, architecture to jurisprudence, literature to medicine, and arts to sciences. The body of literature dealing with the history of the Russian revolutions is enormous, but it rarely mentions science and/or medicine.[4] At the same time, even though historians of Soviet science and medicine have produced several important studies of the post-revolutionary decade, they (myself included) have paid considerably more attention to the later period of Stalin's rule.[5] In some of my previous work I have touched upon the profound impact of the revolutions on the institutional structures, patronage patterns, and social practices of Russian science.[6] In this new project I wanted to explore whether the revolutions affected the research agendas and disciplinary developments in the life sciences. In other words, I wanted to tease out the connections between social and political history and the trajectory of scientific research and discovery that is enmeshed in this history.

I began by surveying specialized periodicals of the time to identify specific lines of research in particular biomedical disciplines that could make illustrative case studies and form the core of my project. Ploughing through numerous biological and medical journals of the 1920s, I was struck by the eerie familiarity of much of what I was reading. Whether it was an article on experiments with organ transplants, tissue cultures, telepathy, hormones, or isolated organs, I felt that I had read about these various experiments before. It did not take me very long to recall exactly where I had read about them: in the short stories and novels written by Alexander Beliaev, one of my favorite science fiction writers from childhood. Admittedly, it had been more than thirty years since I had last read Beliaev, and I forgot many details. Still, I remembered enough to recognize the experiments I was reading about in professional journals as important elements of Beliaev's numerous stories. I dismissed this recognition as trivial. After all, was this not what science fiction is all about: taking the advances of contemporary science and projecting their possible consequences into the future?

Yet as I continued my survey, I encountered reports on other experiments, which immediately reminded me of something that I had (re)read quite recently and remembered much better than Beliaev's stories: Mikhail Bulgakov's novellas "Fateful Eggs" and "The Dog's Heart." This gave me pause. At the time, I knew very little about Beliaev, his life, and his writings: I did not even know exactly when the stories I remembered so fondly (if a bit vaguely) had been written or where they had first been published. Not so with Bulgakov's: I knew nearly everything there was to know about the two novellas, for Bulgakov was (and still is) my all-time favorite writer.

I first encountered Bulgakov's writings around the same time I read Beliaev's. To this day, I remember my eager-anticipation-turned-deep-frustration when the literary magazine *Moscow*, which in its November 1966 issue had published the first part of Bulgakov's "sunset novel," *Master and Margarita*, failed to print the concluding part in its December issue, as well as the ultimate joy of being finally able to read the entire novel when its conclusion appeared in the next issue of the magazine in January 1967. Of course, as a child, I mostly delighted in the humorous parts of the novel, enjoying the antics of Woland's jester Behemoth the Cat and his partner Korov'ev-Fagot. But I also fell in love with the enchanting "realism" of the Ershalaim chapters, for despite the official militant atheism of my upbringing as a Soviet "young pioneer," I knew the Bible quite well: a pre-revolutionary children's edition with Gustave Dore's remarkable engravings held a place of honor on my personal bookshelf next to a similar edition of Greek myths (both treasured gifts from my grandfather).

Master and Margarita left an indelible mark on my imagination: I wanted to read everything Bulgakov had ever written, which in due course I did. In the subsequent years, I also learned as much as I could about Bulgakov's life and works and collected different editions of his books in my personal library. I read and reread all of his writings again and again, every time finding something new, unnoticed and unappreciated during my previous readings. Thus, discovering that two of his best-known novellas were not as "fantastic" as they seemed at first glance, but rather presented a quite accurate depiction of certain scientific experiments conducted at the time when Bulgakov was writing his stories, made a much bigger impression than the earlier observation regarding Beliaev's stories. My curiosity as a historian of science, and not merely as a fan of Bulgakov's, was piqued, for I was sure that there could be nothing trivial in Bulgakov's writing.

I doubted that Bulgakov and Beliaev had read the 1920s specialized medical and biological journals I was reading. How, then, did they learn about those experiments? Why did they choose these particular experiments for their stories? What do their stories actually say about science and scientists? And what could all of this mean for my project investigating the impact of the Russian revolutions on the research agendas and disciplinary developments of the life sciences? Unfettered by the professional concerns and conventions of literary studies, I decided to include fiction in my examination of what was going on in early Soviet biology and medicine. I expanded the scope of my survey to include popular-science, adventure, and literary magazines, as well as major newspapers, of the period. I examined the catalogues of my old haunts—BAN and Publichka—in search of book-length publications. In one of the few surviving second-hand bookstores in the newly reborn St. Petersburg, I also bought a copy of the eight-volume set of Beliaev's collected works and reread it from cover to cover.

I quickly discovered that Beliaev and Bulgakov were by no means the only authors using biomedical experiments in their fiction. Indeed, during the 1920s scores of writers—some familiar, some I had never heard of before—had done the same. Scientists, professors, and doctors engaged in various biomedical experiments populated dozens of novels and countless short stories. Furthermore, it seemed that no experiment conducted by Russian researchers in the 1920s escaped the notice of the literati and failed to provide one writer or another with an intricate plot, an engaging character, an illuminating metaphor, a potent symbol, or at least a telling element of the scenery.

Clearly, contemporary *scientific* experiments inspired many Russian writers to conduct their own *literary* experiments with the ideas and techniques offered by experimental biology and medicine. Some of these literary experiments were overly enthusiastic, hailing a bright future brought about by science and scientists. Others were deeply pessimistic, predicting the ultimate failure of science that would lead to unthinkable catastrophes. To be sure, some authors were frivolous and made fun of science and its practitioners, satirizing their goals and means, theories and practices. Still others were dead serious, questioning scientists' promises and premises, ideals and actions, beliefs and biases. I also found the source of many writers' knowledge of what was going on in contemporary science. As I had suspected, writers did not need to read specialized biological and medical periodicals: all of the scientific experiments that had made their way into fiction had been extensively covered by the media—popular magazines, documentary films, radio programs, and the daily newspapers of the time.

Inspired by all of these discoveries, I began to collect systematically all possible information on the intersections between literary and scientific experiments. Soon I found myself virtually drowning in the ever-growing mass of facts, names, titles, and images related to Russian biomedical experiments and their fictional counterparts. In time, in the cacophony of diverse voices, stories, and opinions, I began to discern certain leitmotifs that intertwined and resonated through, even if they did not quite harmonize, both the science and the fiction of the 1920s: the theme of control over life and death; the theme of liberation and of breaking boundaries; and the theme of creating a new kind of "advanced" human beings.

I realized that I could not possibly aspire to cover all of these grand themes in a single project and decided to limit my present study to the theme of control over life and death, often popularly interpreted as the pursuit of perpetual youth, the abolition of death, and the quest for immortality.[7] I knew that this theme had attracted both scientists' and writers' attention long before the Bolshevik Revolution, but my reading of 1920s fiction indicated that its treatment in the post-revolutionary decade differed dramatically from practically all previous renderings and was profoundly influenced by the concurrent scientific developments. The contemporary advances of experimental biology and medicine often supplied crucial elements to plots, settings, and the characterization of protagonists. In contrast to pre-revolutionary fiction, the majority of 1920s writings were openly futuristic, depicting a future in which control of life and death, and its extension—immortality—had become a reality. In fact, these writings formed a subset of a new literary genre—scientific fantasy (*nauchnaia fantastika*) as it was called in Russia, or science fiction in the English-speaking countries—which brought together issues of the future, human biology, and human destiny.[8] As elsewhere, this genre had begun to figure in Russian fiction before the Bolshevik Revolution, but it was in the 1920s that it reached a peak in popularity among both writers and readers.

Trying to understand the interrelations between scientific ("actual") and literary ("fictional") experiments, I began by adopting the two most common points of view epitomized in the popular formulae: "life imitates art" and "art imitates life." First, following a standard approach of historians of science, I viewed fiction as a reflection of particular history-of-science/medicine narratives and illustration to various trends and research directions in experimental biology and medicine. Second, following a tradition from literary studies, I looked at science as a source of inspiration, point of departure, and raw material for writers of fiction, searching, to paraphrase Anna Akhmatova, for that "rubbish" out of which "verses shamelessly grow."[9]

I quickly recognized, however, that such neighborly "cross-gazing" over the artificially erected disciplinary walls severely limited the range of interactions between the actual and the fictional that I was able to discern in my readings. Within each disciplinary approach, the subject of one discipline was given a voice and an agency, while the subject of the other discipline was reduced to silence and passivity.[10] Most important, this cross-gazing excluded from consideration the very ways in which science and literature interacted (e.g., stimulated, cooperated, ignored, questioned, reconciled, assimilated, exploited, and competed) with each other within the larger realm of Russian cultural life. It also obscured from view the broader cultural, institutional, ideological, and economic processes, which might have affected—and were expressed in—these interactions. In particular, I came to realize that the very positions of science and fiction vis-à-vis each other, as well as their relative positions vis-à-vis other constituent components of Russian culture, such as the arts, philosophy, and, especially, religion, needed to be examined.[11] Furthermore, the very fact of an active if often indirect dialogue between science and fiction during the 1920s, as well as the various ways in which this interchange unfolded, begged explanation. So, instead of cross-gazing at scientific and literary experiments, I decided to read them in parallel. My goal was to figure out why and how various writers used (often the same) particular scientific "rubbish" to accomplish their own artistic goals, while analyzing in detail why and how various scientists generated such "rubbish" in pursuit of their own objectives.

The parallel reading of actual and fictional experiments quickly brought an important insight: whether laughing at or lamenting about, lambasting or lauding *science*, practically all literary experiments I encountered focused on the *scientist*. The image of a lonely genius assisted by a single sidekick, or at most by a small group of disciples, dominated 1920s literature. Unlike scientists themselves, writers of fiction completely ignored one of the most important and successful experiments conducted jointly by biomedical scientists and their Bolshevik patrons during the 1920s: the building of "big science." This unique experiment led to the rapid institutional development of biomedical disciplines and to the transformation of the lonely genius into a *mass profession* in the new, Soviet Russia, but it found no place in numerous fictional experiments.

Since my interest lay in uncovering patterns and trends, I had to fight the temptation to include in my study every single instance of intersection between scientific and literary experiments I encountered in my indiscriminate readings. I had to force myself to be selective and to examine closely only what I came to recognize as the most telling and representative examples. Organizing and surveying the mountain of materials I had accumulated over the years, I discovered that among several dozens writers appearing in my notes, my two old favorites—Beliaev and Bulgakov—were the ones most consistent in their attention to contemporary biomedical science, most broad-ranging in covering its various trends, disciplines, and directions, and, arguably, most talented and inventive in using science in their own literary works. Unsurprisingly, Beliaev's and Bulgakov's stories came to occupy a prominent place in this book, serving as the major representatives of the literary experiments with control over life and death conducted in the 1920s.

All these discoveries guided to a considerable degree my choice of scientific experiments, which became the four key case studies included in this book: research on isolated organs (severed heads), attempts to freeze and thaw living organisms (anabiosis), studies of the products of internal secretions (hormones), and manipulations of sex gonads (rejuvenation). Each of these four research trends had begun to develop rapidly in Russia and elsewhere in the last decades of the nineteenth century, and all four flourished, albeit in different ways, during the first decades of the twentieth century. The contrast among their individual trajectories and the differing degrees of attention they attracted from both the literary and the scientific communities after the Revolution make the four cases particularly illuminating. During the 1920s, severed heads got a lot of play in both literature and science, but anabiosis, while extensively covered in fiction, generated no new research. Rejuvenation inspired a large amount of research and a huge literary response; but the study of hormones received scant attention in fiction, even though it was the most successfully institutionalized trend in experimental biology and medicine in post-revolutionary Russia: endocrinology.

My choices of particular case studies were thus guided by three main considerations. First, I picked those trends of experimental biology and medicine that had begun to develop before the Bolshevik Revolution and continued, albeit in different forms, in its aftermath, which allowed me to trace the specific changes in their trajectories induced by the Revolution. Second, I looked at those trends that elicited the most vocal responses from the literary community, which made it possible to analyze what particular elements (premises, ideas, techniques, promises, etc.) of scientific experiments attracted the closest attention of the lay public. Third, I focused on those components of scientific experiments that provided an important nexus of scientists' relations with their Bolshevik patrons but went largely unnoticed by writers and the public at large. As such, an examination of these long histories put the similarities and differences, convergences and divergences, continuities and ruptures in their developments in Russia and elsewhere into sharp relief, and enabled me to discern the specific impact of the Bolshevik Revolution on both scientific and literary experiments, as well as their interrelations and interactions.

Many observers (myself included) have approached the history of 1920s Russia with the hindsight of the tragic events that were to come—famine, Gulag, and terror—looking for the early signs of Stalinism and often creating a linear, almost teleological account of events leading seamlessly from the Bolshevik Revolution to Stalin's "revolution from above" and beyond.[12] This book's focus is different: it looks at the 1920s as a distinct period in its own right and on its own terms. I have deliberately constructed this book as a series of essays, each following its own chronology and narrative structure, rather than attempting to impose a singular chronological or thematic narrative on the bewildering complexities and fantastic riches of 1920s science, fiction, and their interactions. Indeed, one of my goals in writing this book was to highlight the multitude of actors involved, the polyphony of voices raised and silenced, the diversity of pathways opened and closed, and the multiplicity and multidimensionality of stories generated at the junction of scientific and literary experiments.

None of the essays offers a definitive or comprehensive analysis of its subject: each of the cases I examine deserves a monographic book-length treatment in its own right. In fact, an essay about the work of Alexander Bogdanov—a prominent Bolshevik, physician, philosopher, science fiction writer, and the first director of the Moscow Institute of Blood Transfusion—who advanced a vision of "physiological collectivism" and rejuvenation through blood exchanges, which I had originally conceived as part of this study, refused to submit to my wishes and burst out of this project to become a separate volume.[13]

Many protagonists of my essays, as well as the majority of literary texts I examine, are unknown to Western audiences, and little known even to Russian ones. The accounts of their dramatic life stories included in the book illustrate the multiple contexts of their research and writing, while brief summaries of amazing and amusing literary experiments provide textual evidence illuminating the various ways writers addressed and assessed scientific experiments. To make the text accessible to a wide audience, I have sought to avoid professional jargon and polemics with the specialized literature in the five fields with which this book intersects.

Although each of the essays could stand on its own, I hope that this book exceeds and transcends the mere arithmetical sum of its constituent chapters, presenting a dynamic, polychromatic, and multidimensional picture of 1920s Russian science, fiction, and their interactions. Taken together, the essays depict the period's hopes and fears, promises and expectations, desires and anxieties, and dreams and realities embodied in scientific and literary experiments. Through the lens of published and archival materials, they examine why and how biomedical science came to occupy such a prominent place in the stories of numerous litterateurs and in the culture and society of post-revolutionary Russia more generally, documenting the rapid rise of science's funding, public resonance, and cultural authority in the aftermath of the Revolution. Using popular-science and science fiction writings of various authors, the essays show which particular elements of scientific activities (ideas, techniques, institutions, individuals, promises, and premises) captured the public's attention and which went unnoticed. They illuminate the complex relations of science to other constituent components of Russian culture, such as religion and literature, by comparing and contrasting the actual scientific experiments and their fictional incarnations. Tracing the similarities and differences of these relations in Russia and the West, the chapters assess the role of the Bolshevik Revolution in continuities and ruptures in the developments of certain scientific ideas, institutions, and ideologies, as well as certain literary traditions and cultural practices. They chronicle the transformation of knowledge generated by experimental biology and medicine into a powerful cultural resource, a resource that a panoply of actors used to their own ends, but which also fed public expectations that science was about to bring to humanity its ultimate gift—immortality.

What follows is a result of my parallel reading of—rather than cross-gazing at—certain scientific and literary experiments aimed at the abolition of death and the pursuit of eternal life. With Bulgakov's novella "Fateful Eggs" as its entry point, the first chapter identifies the common (political, institutional, cultural, economic, ideological, demographic, etc.) contexts that shaped the interactions of various actors, ranging from anarchists-biocosmists to physiologists and from popular novelists to Bolshevik Commissars, involved with this particular cluster of experiments in the aftermath of the Revolution. These multiple contexts were generated by the historic coincidence of the Bolshevik Revolution with three major scientific revolutions[14] unfolding during the first decades of the twentieth century: an experimental revolution in the life sciences, which dramatically changed understandings of life and death, health and disease, human nature and human future; a revolution of scale, which marked the transition from "small" to "big" science and made science a mass profession; and a revolution in science's public visibility and cultural authority, aptly manifested in the explosion of "popular science" in the media and the spectacular rise of a new literary genre, science fiction.

Chapter 2 presents literary and scientific experiments with isolated organs. Focusing on Alexander Beliaev's 1925 novella "Professor Dowell's Head" and Sergei Briukhonenko's extensive research on extracorporeal blood circulation conducted on isolated dog heads, it compares the goals and outcomes, techniques and values, and traditions and innovations in the fictional and actual investigations of "severed heads" whose life was sustained by elaborate scientific machinery. It follows the transformation of a severed head from a frightening symbol of certain death into both an exciting emblem of "science's victory over death" and an epitome of scientists' "ethical deafness," examining the attitudes of scientists, science government patrons, newspaper reporters, writers, literary critics, and the lay public toward the possibility of life without a body.

Starting with Alexander Beliaev's story "Neither life, nor death," the next essay explores actual and fictional investigations of anabiosis. It follows Porfirii Bakhmet'ev's discovery of the "border between life and death" by freezing and thawing living organisms: insects, bats, and fishes. It documents Bakhmet'ev's efforts to further this research and build a special laboratory for his studies, halted by his untimely death in 1913. It chronicles the period of "suspended animation" that research on anabiosis entered following the Revolution, despite the enthusiastic popularization of anabiosis by the media. It analyzes numerous fictional experiments conducted by scores of writers, including Vladimir Mayakovsky, Aleksei Tolstoy, and Boris Pil'niak, with "freezing and thawing" of humans inspired by Bakhmet'ev's discovery.

Chapter 4 focuses on actual and fictional experiments with internal secretions—hormones—which both scientists and writers saw as a powerful tool for manipulating the shape and functioning of the human body. Opening with Porfirii Bakhmet'ev's vision of future science presented in his fictional story "The Billionaire's Last Will," this chapter shows how, for scientists, hormones also became a convenient instrument for influencing their Bolshevik patrons and for insinuating their own dreams of "big science" into their patrons' visions of the country's future. It documents the joint efforts—and different goals—of scientists and their patrons in conducting a unique experiment: building large institutional structures for a new biomedical discipline, endocrinology. It also reveals that although the transformative power of hormones over human body and personality did fire up writers' imagination, the institutional embodiment of that power completely eluded their attention.

Through the lens of Mikhail Bulgakov's novella "The Dog's Heart," the next essay describes the "rejuvenation craze" that swept Russia during the 1920s, following the widely publicized experiments on sex glands (testicles and ovaries) by the Austrian biologist Eugen Steinach and the French surgeon Serge Voronoff. The essay portrays numerous Russian scientists testing and debating various techniques of manipulating sex gonads and equally numerous writers mocking, glorifying, or decrying their results and quickly extending the promise of rejuvenation into a certainty of immortality. In addition, it documents the efforts to rejuvenate old revolutionaries and to build a "monkey farm" to ensure a domestic supply of monkey glands for rejuvenation. Finally, it examines the attitudes of some members of the lay public toward the intertwining of science and sex that raised hopes of science fulfilling their wildest dreams of changing human (sexual) identities and getting control over human nature.

Chapter 6 brings the separate case studies together to illuminate the place of biomedical sciences in the culture and society of 1920s Russia. It investigates the role that the new knowledge of human biology played in answering the "big" questions of human nature and human destiny, which had haunted humanity for millennia and assumed a particular urgency and saliency in the aftermath of the Revolution. It also demonstrates the particular cultural affinity of the Bolsheviks' visions of the country's future with the promises of eternal life and hopes for conquering death generated by concurrent biomedical research. Furthermore, it examines collective, though not necessarily coordinated, efforts of scientists, their Bolshevik patrons, and their literary fans/critics, which transformed esoteric biomedical knowledge into an influential cultural resource, a resource that facilitated the establishment of large specialized institutions, inspired numerous science fiction stories, displaced religious beliefs, and gave the millennia-old dream of immortality new forms and new meanings in Bolshevik Russia.

A brief Epilogue sketches the profound impact of the "revolution from above" on the biomedical research, science fiction writings, and revolutionary visions of the 1920s, as well as the continuing quest for immortality in post-Soviet Russia.

St. Petersburg—Toronto, January 2012.

To align the name of the great artist with the revolution—a revolution, which he clearly misunderstood and from which he tacitly removed himself—might at first glance seem strange and artificial. Could we call a mirror something that obviously does not reflect phenomena correctly? But our revolution is an extremely complex phenomenon. [. . .] And if what we have before us is indeed a great artist, then, he must reflect at least some of the essential features of the revolution in his works.
VLADIMIR LENIN, "Leo Tolstoy as a Mirror of the Russian Revolution" (1908)

In our days the very air smells of death,
To open windows, as to open veins.
BORIS PASTERNAK, "Break" (1919)

1 "The Ray of Life": Science in Revolutions

IN 1925, MIKHAIL Bulgakov, a thirty-three-year-old journalist and writer, was at the peak of his literary career (see Figure 1.1).[1] In the first months of that year, *Russia*, a well-known "literary-political journal," began publishing the opening chapters of his first novel, *The White Guard*.[2] Almost simultaneously, *Subsoil*, a popular literary almanac, printed Bulgakov's "fantastic" novella entitled "Fateful Eggs."[3] The novel attracted little notice beyond a narrow circle of litterateurs and its publication was not even finished, because the journal *Russia* was closed down and its editor-in-chief Isai Lezhnev expelled from the country (for his allegedly anti-Bolshevik views). But the novella stirred quite a commotion among publishers, critics, and the reading public. In May to June, *Red Panorama*, a popular Leningrad illustrated weekly, serialized a shortened version under the title "The ray of life."[4] During that summer, the novella was republished as the centerpiece in a collection of Bulgakov's short stories, entitled *Diaboliad*, so popular that it was reissued just a few months later.[5]

Set a few years into the future, in 1928, the novella tells the story of a professor at the Moscow University Zoology Institute, named Vladimir Ipat'evich Persikov. By pure accident, Persikov discovers that the irradiation of amoebas with light rays of a particular *red* spectrum tremendously speeds up their reproduction and increases their size. Intrigued, the professor begins to investigate the rays and quickly finds out that they are present only in the light of an electric lamp—and absent in sunlight. With the help of his assistant (who bears the generic name Petr Stepanovich Ivanov), the professor constructs a special apparatus to isolate the rays and study their properties. An expert in amphibians, Persikov begins his experiments by irradiating fertilized frog's eggs. In just two days, thousands of tadpoles hatch from the irradiated eggs. The next day, the tadpoles turn into huge frogs "so vicious and voracious that half of them immediately devour the other half." Right away, the survivors begin to lay eggs and in two days produce the next generation of innumerable frogs, which overflow Persikov's laboratory and have to be killed by poison.

The media gets wind of Persikov's work, and reporters besiege his institute. Persikov reluctantly grants an interview, and the media—newspapers, magazines, radio, and moving picture documentaries—popularizes and sensationalizes his discovery of "the rays of life." Although annoyed by the media frenzy, Persikov receives tangible benefits from this publicity: his work comes to the attention of the country's highest agencies. Various commissariats and "the Kremlin" call him, while the secret police provides a security detail for his institute to protect his discovery from foreign spies. Fascinated by the effects of the rays on "the egg's deiteroplasm," Persikov plans to expand his experiments to other egg-laying animals and uses his newly acquired access to state agencies to get the necessary equipment and materials from abroad. With the help of his assistant, he constructs three much larger incubators for his experiments with the red rays and orders eggs of various reptiles and birds from Germany and the United States.

At this very moment, a "chicken plague" hits the Soviet republic, killing chickens all over the country with horrifying speed. A supreme commission comprised of the heads of various government agencies, including Narkomzem (the People's Commissariat of Agriculture) and Narkomzdrav (the People's Commissariat of Health Protection), is set up to deal with the problem. Due to his newly found notoriety, Persikov becomes a member of the commission, even though he is not an ornithologist. He reluctantly participates in the commission's meetings and conducts research on the physiological and anatomical changes the plague produces in chickens, trying to isolate a bacillus that causes the disease. In a few weeks, the plague kills off all the chickens in the republic and the urgency of Persikov's work on finding its agent subsides. He returns to his studies of the mysterious red rays and gives a report on his discovery to a large gathering of Moscow scientists. The presentation turns into a triumph: his fellow scientists give Persikov a standing ovation.

But Persikov's report also causes a major disruption in his research. An old Bolshevik, with a telling name, Rokk (*rok* means "fate" in Russian), who attended the presentation, comes up with an idea of using Persikov's ray of life to alleviate the chicken crisis. The idea finds immediate support among the powers that be and on the government's orders, Rokk requisitions Persikov's equipment. He then sets it up at a state farm outside of Moscow, with the goal of producing giant chickens by irradiating chicken eggs that he orders from abroad with Persikov's rays.

As fate would have it, Persikov's and Rokk's egg orders get mixed up. Rokk receives the eggs Persikov has ordered for his experiments and, being completely ignorant of zoology, doesn't realize it. He places the eggs into Persikov's incubators, causing a herd of monsters—giant snakes, crocodiles, and ostriches—to rapidly multiply, kill everyone, and destroy everything in sight (including Rokk's farm with Persikov's equipment). Meanwhile, all the artillery, airplanes, and poisonous gases deployed by the Red Army fail to stop the monsters' march on Moscow. In the ensuing riots and panic, the terrified mob kills Persikov and sets his institute on fire. Just on the verge of ultimate disaster, however, unusual "August frosts" descend upon the country, killing the monsters and their eggs. The Soviet republic is saved, but Persikov's discovery is obliterated. All attempts at recapturing the red rays undertaken by Persikov's former assistant and his successor as the director of the restored Zoology Institute fail. End of story.

Ever since its appearance, first contemporary observers, and then literary scholars, have hailed—or condemned–"Fateful Eggs" as a brilliant satire of the Bolshevik Revolution, its leaders, and Soviet society as a whole.[6] Regardless of their response, practically all contemporary Soviet literary critics noticed "Fateful Eggs."[7] As one of them characterized the

FIGURE 1.1 Mikhail Bulgakov, ca. 1926. According to numerous memoirs, Bulgakov carefully choose his attire for this photograph, intentionally adopting the image of a "bourgeois" litterateur (including a monocle, a bowtie, a dinner jacket, and a white, starched shirt). Perhaps it was a deliberate "challenge to the proletariat" on the part of Bulgakov, reminiscent of the famous "challenge to the bourgeoisie" issued by the Russian futurists and epitomized by Vladimir Mayakovsky's "yellow jacket." *Courtesy of NIOR RGB.*

novella, "The political point of the story is clear: the Revolution gave birth to 'monsters' and only such a miracle as a minus eighteen degree frost could save us from them."[8] Even *New York Times* Moscow correspondent Walter Duranty enthusiastically greeted the appearance of the novella, though in his article he misspelled its author's name and somewhat misrepresented its contents.[9] Furthermore, according to Vladimir Mayakovsky, the country's most acclaimed "futurist" poet who visited the United States during the summer and fall of 1925, some US newspapers mistook Bulgakov's novella for a description of actual events that had happened in Soviet Russia.[10] Later commentators have analyzed in detail the story's social, political, literary, and even zoological contexts, as well as its moral and philosophical dimensions.[11] They have examined its structure and meaning and have searched for its literary predecessors, ranging from Leo Tolstoy to H. G. Wells.[12] They have also attempted to identify the prototypes of its major characters: most popular being Vladimir Ipat'evich Persikov as a stand-in for Vladimir Il'ich Lenin, the leader of the Bolsheviks,[13] and Alexander Semenovich Rokk for Lev Davydovich Trotsky, the second in command among the Bolshevik leadership during and immediately after the Revolution.

Yet illuminating and insightful as they are, these analyses seem to overlook a key feature of "Fateful Eggs": it is not a story of the Bolshevik Revolution, its leaders, or Soviet society, *as such*. Unlike Bulgakov's other stories that appeared alongside the novella in his collection *Diaboliad*, "Fateful Eggs" focused specifically and deliberately on one particular segment of 1920s Soviet reality—namely, biological science and scientists in their relations to the new Soviet order.

Science and the Bolsheviks

In fact, Bulgakov's "fantastic" novella depicts very realistically and quite accurately—up to the smallest detail—the evolution of these relations from the Bolsheviks' seizure of power in October 1917 to "the present" (even though Bulgakov stretches that evolution for four more years into the future, beyond the autumn of 1924 when he actually wrote the story). He identifies its major actors: scientists, their state patrons, the media, and the public, as well as the main trends and modes of their interactions, pinpointing a critical nexus of these interactions—the popular media—and painting a succinct portrait of its workings.[14] The novella vividly protrays the contemporary fascination with research into the "mysteries of life and death": their subjects, goals, means, promises, and uses, revealing the Bolsheviks' utilitarian attitude toward, and the general public's awe of, science and its practitioners. "Fateful Eggs" captures the rapidly rising public visibility and cultural authority of science and scientists in Soviet Russia. Indeed, Bulgakov's choice of the subject for his novella is one of the best illustrations of this growing importance of science in Russian life of the 1920s. No matter how fantastic Bulgakov's story might appear to its readers, his depiction of the early 1920s scientific scene in Soviet Russia is much closer to reality than later commentators could imagine.

Seen from this particular perspective, "Fateful Eggs" presents a virtual guidebook to the history of early twentieth-century Russian (biological) science. Like the fictional Persikov, who reads in four foreign languages and speaks French and German fluently, university professors in tsarist Russia belonged to the country's highly educated elite. Like Persikov, whose knowledge of his subject is rivaled only by that of two other scientists in the entire world (one in Rome and another in Cambridge), many Russian scientists enjoyed solid international reputations. Also like Persikov, who lives in a five-room apartment in a prestigious neighbourhood and has a housekeeper who takes care of his everyday needs, Russian professors of that era knew no material difficulties. Indeed, Moscow University professors during the Imperial era, such as Persikov, had a civil servant rank equal to that of a military general, with all the attendant perks and privileges.

The Bolshevik Revolution stripped all of these privileges from Russian scientists. The victorious "proletarians" pegged scientists as part of the "bourgeoisie" to be harassed and eventually liquidated. Under the policy of War Communism introduced by the Bolsheviks during the Civil War, many Russian scientists suffered considerable material hardship. In the novella's flashbacks, the new regime requisitions three out of the five rooms in Persikov's apartment in 1919 and, (though Bulgakov doesn't mention it in "Fateful Eggs," he does so in several other short stories) most likely, gives the requisitioned rooms to some members of the "previously oppressed classes." During the civil war years, many Russian scientists could not continue their research because they lacked supplies, equipment, and even utilities. Similarly, Persikov's specimens die from starvation and cold (the temperature in his institute stays permanently at −5° Celsius, independent of the outside temperature), and the same fate (from the same causes) befalls his institute's animal keeper. Many Russian scientists greeted the new regime with distrust and open hostility. So does Persikov blame the head of Narkompros (the People's Commissariat of Enlightenment, headed at the time by Anatolii Lunacharskii), who is responsible for science and education in the Bolshevik government, for the death of his animals and their keeper. Many Russian scientists considered emigration a viable option, and some did leave the country, but the majority stayed in Russia. So, too, distraught by the circumstances of his life, Persikov threatens to emigrate, since his reputation "could have secured his appointment to the zoology department at any university in the world," but decides to stay. Despite all of the hardships of War Communism, many Russian scientists strove to continue their work the best they could. Likewise, twice a week, bundled up in his fur coat and fur hat, Persikov lectures at his institute on the subject of "the reptiles of hot climates" to just eight (instead of his customary hundreds) students and spends the rest of his time holed up in his room in front of a small stove in which his housekeeper burns his gilded chairs.

But, as Bulgakov put it in his novella, "everything in the world eventually comes to an end. Ended the years of 1920 and 1921, and in 1922, there began something of a reverse movement." Indeed, by the end of 1921, the Bolsheviks had won the Civil War: the Red

Army had driven out both the "White guards" and the allied expeditionary forces that supported them.[15] The new rulers paid a steep price for the victory: the economy was shattered and cities depopulated, factories stood still and fields empty, epidemics were rampant, and famine reigned supreme. Faced with severe economic crisis, in late 1921 the Bolsheviks abandoned War Communism and adopted NEP—a New Economic Policy, which proved highly successful in reviving the country's economy, repopulating its cities, and restoring its agricultural production. Although under NEP the Bolsheviks preserved state control over banking and key industries, they allowed private ownership and initiative in trade, services, and the small-scale production of consumer goods.[16] They launched an enormous campaign to combat illiteracy and popularize science.[17] They greatly expanded the entire system of education—from primary schools to universities—and promoted the rapid growth of all branches of science, medicine, and technology.

The country's new rulers did not forget about scientists. Indeed, one of the first acts of the Bolshevik government at the end of the Civil War was a special decree "On Improvement of Scientists' [Living] Conditions" issued in December 1921 by the highest government agency, the Council of People's Commissars (SNK), presided over by Lenin. A few months later, in June 1922, the Central Commission to Improve Scientists' Living Conditions (TsKUBU) opened (in a magnificent mansion requisitioned from a wealthy Moscow capitalist) a House of Scientists, overseen by the head of Narkomzdrav Nikolai Semashko and located on that very street where Bulgakov placed Persikov's apartment in his novella (see Figure 1.2). The House of Scientists was the major center for distributing special government rations, as well as food packages sent by the American Relief Administration, to Russian scientists. It also became a kind of scientists' cultural club, which Bulgakov featured in "Fateful Eggs" as the "TsKUBU huge auditorium" where Persikov gives a report on his discovery of the red rays to fellow-scientists.[18]

Persikov's life from 1922 onward mirrors that of many scientists in Soviet Russia. The introduction of NEP inaugurates a new era in the professor's collaboration with the regime: the very same Narkompros that Persikov blamed for the death of his animals and their keeper in 1919, now strives to provide the conditions necessary for the resumption of his teaching and research. The institute's heating is restored and a new keeper is appointed. With the keeper's help, Persikov catches some toads and fills up the institute's terrariums (that had remained empty for three long years) with specimens for his studies. He begins lecturing more and more, up to thirteen times a week, at the institute, at the university, and at the workers' courses (*rabfaki*), with the number of students growing accordingly: in 1925, he famously fails seventy-six (!) of his "Marxist" students during the spring examinations, because they could not name anatomical differences between amphibians and reptiles. In 1926, Narkompros completely renovates Persikov's institute and supplies it with all of the necessary equipment and materials imported from Germany. Persikov regains his high social status, aptly manifested in getting back all five rooms in his apartment, which he fills "with 2500 books, stuffed birds and animals, diagrams, and preparations." In December 1926 he publishes his first post-revolutionary work: a

FIGURE 1.2 The façade of the Moscow House of Scientists that looks onto *Prechistenka*, the very street where Mikhail Bulgakov placed the apartment of Professor Persikov in "Fateful Eggs." Before the Revolution, the building belonged to the textile magnate Ivan Konshin. In 1908–1910, Konshin's widow financed its complete reconstruction by Anatolii Gunst, a well-known Moscow architect, who turned the building into one of the city's most luxurious mansions. In early 1916, Aleksei Putilov, a prominent industrial and financial tycoon and one of the richest men in the country, bought the mansion from Konshin's heirs for nearly half a million rubles—an enormous sum in those days. But he did not have a chance to really enjoy his house. After the Bolshevik coup, Putilov fled the country and the new rulers expropriated the mansion. *Photo by the author, December 2011.*

lengthy study of a particular genus of amphibians. The next year, he astonishes the scientific world with a monumental volume on the embryology of amphibians. Issued by the country's main publishing house "Gosizdat," the volume is immediately translated into six languages, including Japanese.

Nearly all natural scientists in Russia followed a similar trajectory—from initial rejection and animosity to subsequent collaboration and close personal contacts—in their relations with Bolshevik agents and agencies, while the regime moved from neglect to almost unlimited support of science and scientists, showering them with numerous privileges.[19] During the early 1920s, scientific life in Russia quickly revived. Old research laboratories and institutes were renovated and supplied with heat, electricity, equipment, and personnel. A large number of new research establishments in every field of knowledge proliferated throughout the country. International scientific relations (particularly with another pariah of Versailles—Germany) resumed. New professional periodicals and books were published. Numerous conferences and meetings were convened in practically all branches of science. And, as Bulgakov correctly depicted in "Fateful Eggs," the Bolshevik government generously funded all of these endeavors.

Bulgakov pinpointed quite accurately a major factor in Russian scientists' eventual compromise with what they had at first perceived as an illegitimate and abhorrent regime: patriotism. It is Persikov's love and pride for his country that underlies his decision to stay in Russia and motivates his call to the secret police when a foreign spy comes to his institute with an offer to buy his discovery. It is also his feeling of obligation to his country and its people that to a considerable degree guides the professor's involvement in teaching zoology to the new cohorts of proletarian students, his reluctant participation in the government commission to alleviate the chicken crisis, and his research on the causes of the chicken plague. Indeed, these very sentiments inspired the post-revolutionary work of many Russian scientists.

At the same time, Bulgakov captures unmistakably the Bolshevik leadership's pragmatic and paternalistic attitudes toward Russian scientists. The Bolsheviks treat Persikov as a big child who does not really understand anything beyond his scientific specialty. Bulgakov uses the same phrase—"gentle voice"—several times to describe the way the Kremlin bosses are patronizing toward Persikov. But the same Kremlin bosses are always ready to exploit the professor's special expertise—as well as his discoveries—for their own purposes. In fact, it was the Bolsheviks' desperate need for scientists' specialized knowledge to revive the country's industrial and agricultural production and to rebuild its education and public health systems that proved the decisive factor in the Bolsheviks' attitude toward the "bourgeois" scientific community they inherited from the Imperial era.

What is more, in his portrayal of the early 1920s science scene, Bulgakov goes beyond the science-state (and the scientists-Bolsheviks) relationship, identifying a very important nexus of science's connections with the new Soviet society as a whole: the media. It is the media—daily newspapers, weekly magazines, radio, newsreels, and documentary films—that to a considerable degree carry out the regime's extensive campaign for the popularization of science.[20] Science becomes not merely a newsworthy subject, but a focus of concerted efforts at bringing "science to the masses" and "knowledge to everyone" on the part of a variety of government agencies. During the 1920s, the number of popular science periodicals, brochures, pamphlets, and books grew exponentially, taking a large share (more than a fifth, by some estimates) of the publishing market in Russia.[21] Biological sciences figured prominently in this campaign. According to contemporary—far from complete—data, from November 1917 to October 1925, the number of popular science books and brochures dealing with biological questions grew thirtyfold, up to 240 titles per annum. By October 1925 the total number of popular books and brochures on various biological subjects issued since the Revolution had reached 642, with an average print run of five thousand copies each.[22] The number of articles on the same subjects in daily newspapers, weekly magazines, and popular science journals is difficult to estimate, but practically all Soviet periodicals carried such articles on a regular basis (at least once a week).

Bulgakov portrays with great precision and sarcasm the role of journalists in bringing science to the masses. As a staff writer for *Whistle*, a popular newspaper of the railroad-workers' trade union, and a contributor to various other dailies and weeklies not only in Moscow and Leningrad but also in the provinces, he was intimately familiar with the workings of the Soviet press. Sensationalism, exaggeration, ignorance, trivialization, total lack of understanding of science and its workings, oversimplification, failure to check reported facts, arrogance, and complete disregard for scientists' own views are all characteristic features of newspapers' "scientific" reportages in "Fateful Eggs." In the media reports on Persikov's discovery, names are misspelled, "facts" invented, and absurd conjectures abound. In the newspapers (and in public perception), Persikov's work transmutes into "nightmarish rays of life," reflecting journalists' propensity for sensationalist headlines such as "nightmarish murder," "nightmarish findings in the catacombs," and a "nightmarish chicken disease."

Unsurprisingly, the media reports generate wild rumors and speculation among the public, fostering not so much an understanding as a deep fear of science and scientists, which ultimately results in Persikov's murder and the torching of his institute by a mob. Prior to this, however, such publicity also brings Persikov's work to the attention of high-level government officials and facilitates his access to state resources. Bulgakov thus pinpointed the role of the media as a critical mediator of relations between scientists and their state patrons, as well as between science and the general public.

In fact, it is quite likely that Bulgakov himself "lifted" the major component of his plot—the ray of life—straight from the pages of *Izvestiia*, the country's major newspaper frequently mentioned in "Fateful Eggs." In late May 1924, the daily carried a lengthy article on "Biological rays" subtitled "the brilliant discovery of Russian scientists."[23] The reporter, a certain F. Davydov, recounted investigations of two Russian scientists, Alexander Gurvich, professor of biology at the Crimean University in Simferopol, and Nikolai Kravkov, chair of pharmacology department at the Military-Medical Academy in Leningrad. Both Gurvich and Kravkov had published their results in German scientific journals during the previous year, and Davydov apparently felt the urge to acquaint the Russian public with their "brilliant discoveries."

Gurvich's article appeared in the first specialized periodical for experimental embryology (founded by Wilhelm Roux in 1894 and known popularly as the *Roux Archive*) under the title "On the nature of a specific inductor of cell division."[24] The article described Gurvich's experiments with onion roots. Gurvich noticed that if a root was "injured" (for instance, by a cut made with a scalpel), the cells in the damaged area began to divide rapidly. When Gurvich placed two roots close (2 to 4 millimeters) to each other and injured one of them, a strange phenomenon occurred—the cells of the intact root began to divide, too. Cell division (called mitosis), however, took place only on one side of the intact root—the one closest to the injured root. This observation led Gurvich to the idea that dividing cells of the injured root emit particular "rays," which stimulate the nearby cells of the intact root to start dividing as well. Gurvich named these rays "mitogenetic" and launched extensive investigations of their nature and properties.

Kravkov's article came out in one of the first specialized periodicals for experimental medicine and was entitled "On the sensitivity limits of the living protoplasm."[25] The article reported the results of Kravkov's latest experiments with his favorite research tool: a severed rabbit ear. Kravkov had invented this tool primarily for his studies on the effects of various poisons and pharmaceutical preparations: since blood vessels in rabbit ears are readily visible, it is quite easy to observe their changes under the action of various substances introduced into the blood stream. In the early 1920s Kravkov expanded his studies from chemical to physical agents. He placed a piece of copper or silver close (within 0.5–1.0 centimeter) to the ear and observed that the ear's blood vessels contracted; when he removed the piece the vessels expanded. He obtained similar effects by placing a magnet in close proximity to the ear. These observations led him to suggest that the protoplasm is quite sensitive to certain electromagnetic waves supposedly emitted by metals and magnets.

Both Kravkov's and Gurvich's articles were highly technical, describing in detail (and specialized language) the intricate techniques and sophisticated apparatuses the scientists used in their research. Both were quite careful in presenting their work, wording their findings in tentative terms and calling for further investigation of the phenomena they described. Davydov's report of Gurvich's and Kravkov's "sensational research" omitted all the technicalities and presented their findings in much more definitive and superlative terms: "This is a remarkable experimental proof of [Gurvich's] audacious hypothesis"; and "This bold suggestion found complete confirmation in the following outstanding experiments [conducted] by Kravkov!"

It seems quite likely that Davydov's article provided Bulgakov with that "rubbish" out of which his novella "shamelessly grew." But why did he decide to write the novella in the first place? In his previous works he had depicted either the Civil War or the contemporary life in NEP Moscow but never dealt with either science or scientists. Other writers, however, did feature science and scientists prominently in a variety of literary works in 1920s Russia. Viktor Goncharov, a popular writer, portrayed a geologist in his novel, *The Valley of Death* (1925). Andrei Belyi, one of the most popular novelists of the time, chose a mathematician for his novel, *Moscow* (1926). Others described astronomers, physicists, chemists, and engineers, working on creating wonderful new technologies of space flight or terrifying "rays of death," as did Aleksei Tolstoy in his extraordinary popular "fantastic" novels, *Aelita* (1923) and *The Hyperboloid of Engineer Garin* (1925–26). Indeed, the notion of the rays of death—certain mysterious rays that exploded or killed everything in their path—was very popular in contemporary fiction.[26] In 1924, exactly at the time when Bulgakov was working on "Fateful Eggs," several novels entitled *The Rays of Death* appeared in book format and were serialized in various periodicals.[27] The next year, a full-length motion picture, titled *The Ray of Death*, directed by Lev Kuleshov, a leader of Soviet cinematography, and scripted by Vsevolod Pudovkin, a rising film director (who

FIGURE 1.3 A poster for the movie *The Ray of Death*, 1925. It presents the movie as "a cinematic novel in eight parts" and contains easily identifiable portraits of its main protagonists: a "Soviet engineer"—the inventor of a death-ray apparatus; a "fascist," who steals the apparatus to suppress a workers' uprising; and a "woman-shooter," one of the leaders of the uprising. It also lists the names of producers and actors involved. *Courtesy of the Russian National Library.*

also played one of the lead characters), hit the screens of movie theaters around the country (see Figure 1.3).[28]

Why, then, did Bulgakov decide not to follow this fad, but picked a *biologist* as the main protagonist of his novella and rays of *life* as the major element of its plot? Bulgakov's choice was certainly not accidental. His first profession, like that of his famous predecessor Anton Chekhov, was medicine. The son of a well-known theologian, he had graduated (with distinction) from Kiev University's Medical School and had practiced medicine for several years prior to embarking on a career as a professional writer. Almost simultaneous to the publication of "Fateful Eggs," the first stories from his autobiographical work "The Notes of a Young Doctor" appeared on the pages of *Red Panorama* and *Medical Worker*—a magazine published by Narkomzdrav. Furthermore, the moment he had finished "Fateful Eggs" in the late fall of 1924, Bulgakov started working on a "sequel," eventually entitled "The Dog's Heart," which told the story of a prominent surgeon conducting daring experiments on rejuvenation in NEP Moscow, thus expanding on and elaborating the major themes and subjects of the first novella.[29] Obviously, even though he gave up his medical profession for literature, Bulgakov was acutely attuned to, well informed about, and attentively followed—at least through popular science publications—the latest developments in contemporary research in experimental biology and medicine.

Bulgakov's medical training and great talent enabled him to sense and capture in his novellas the particular salience and importance of biomedical sciences in post-revolutionary Russia. In October 1925, just a few months after the publication of "Fateful Eggs," *Herald of Knowledge*, a popular-science magazine, carried an article entitled "Biological rays." Written by Leonid Vasil'ev, a Leningrad physiologist working at the Institute of the Brain, the article recounted in detail Gurvich's latest research.[30] Although Vasil'ev's article itself was quite accurate and careful in reporting the newest developments in investigating mitogenetic rays, the cover of the magazine bore a bold announcement: "Revolution in biology: The rays of life!" (see Figure 1.4). Preoccupied with uncovering social, ideological, and political connotations of Bulgakov's novellas, most critics and commentators have failed to notice that "Fateful Eggs" and "The Dog's Heart" documented and

FIGURE 1.4 The cover page of the October 1925 issue of *Herald of Knowledge*, a biweekly popular-science magazine, which carried an article on mitogenetic rays discovered by Alexander Gurvich. The headline reads: "Revolution in biology: The rays of life!" *From* Vestnik znaniia, *1925, 19–20.*

satirized not merely the Bolshevik Revolution, but yet another revolution as well, one that unfolded alongside, and in close interaction with, the Bolshevik Revolution: the experimental revolution in biomedical sciences.

"Visionary biology" and "revolutionary dreams"

A historical coincidence dramatically influenced the development of Soviet biomedical sciences. Just as Russia was going through its brutal political revolutions in the first decades of the twentieth century, a subset of natural sciences—one that nowadays is called the life sciences but at the time was represented by the two interconnected and overlapping fields of experimental medicine and experimental biology—was undergoing its own revolution.[31] The very names of these fields allude to the essence of this "mini" revolution: the introduction of experimental methods, largely borrowed from physics and chemistry, into the study of life and death, health and disease. Begun in the last decades of the nineteenth century, this revolution reached its apex in the 1910s and 1920s.[32] Armed with the new experimental methods, numerous researchers around the world enthusiastically tackled the mysteries of basic life processes and their pathological changes, including metabolism, reproduction, nervous and endocrine regulation, cell division, psychological and behavioral patterns, variability, aging, immunity, growth, evolution, and heredity.

The advances of experimental research were quickly taken up in various branches of medicine, leading to new therapeutic treatments, surgical procedures, preventive measures, and diagnostic techniques. In the first decades of the twentieth century, successes in tissue and organ transplantation, immunizations, blood transfusions, cell and tissue cultivation, serodiagnostics, hormone-, vitamin-, x-ray-, chemo-, and sera-therapies, and in deciphering the basic mechanisms of heredity and embryonic development generated a euphoric vision: science could *control* life, death, and disease.[33] Captivated by this "visionary biology," as Mark B. Adams has aptly termed it,[34] many scientists around the world came to believe that experimental techniques could provide them with the means to improve dramatically the wellbeing of humanity, control the spread of diseases, extend the lifespan, manipulate human reproduction, and create new life forms. Indeed, to many scientists, it seemed that the new discoveries would enable them to control not just human life, disease, and death, but human destiny, and perhaps even to fulfill the millennia-old dream of immortality.[35] Russian scientists were no exception.

As elsewhere around the world, in Russia the development of experimental biology and medicine had begun in the last decades of the nineteenth century and was epitomized by the establishment in 1890 of the Imperial Institute of Experimental Medicine in the country's capital, St. Petersburg.[36] In subsequent years, a number of Russian scientists became engaged in research on new, exciting biomedical subjects. Nobel prizes awarded to Ivan Pavlov (1904) and Elie Metchnikoff (1908) marked the international recognition of Russian contributions. But despite the considerable efforts of many scientists, the institutional development of Russian experimental biology and medicine lagged behind their intellectual advances. Thus, Nikolai Kol'tsov, one of the leading proponents of "visionary biology," had spent nearly a decade trying to secure funding to establish an Institute of Experimental Biology in Moscow until he finally succeeded at the end of 1916.

The situation changed rapidly after the Bolshevik Revolution, which coincided not only with the experimental revolution in the life sciences, but with yet another revolution that affected all natural sciences: the revolution of scale.[37] At the start of the twentieth century, science began its transformation from a small-size endeavor of individual researchers, who made their own simple instruments and often financed their own pursuits, into a huge industry-like enterprise that involved large specialized institutions, hundreds of workers, complex machinery, and ever more resources.[38] This revolution made science a *mass profession* with its own hierarchies, division of labors, recruitment patterns, managerial styles, training programs, practices, ideologies, and cultures. Scientists all over the world desperately sought patrons and partners to provide the support and funding necessary for this emerging "big science."[39]

In Russia, they found such a partner—the Bolshevik state. No patron was more willing or more enthusiastic in its support of science than the Bolshevik regime. In the space of just two decades after the Revolution, the combined efforts of scientists and the new Bolshevik government transformed Russia from a modest province of world science into one of its great centers, with a simultaneous tenfold increase in the number of people engaged in scientific research and teaching. One area of natural sciences benefited particularly from this newly emerging alliance of Russian scientists and Bolshevik rulers—experimental biology and medicine.

The Revolution fostered a virtual explosion of research in numerous subfields of "visionary biology" and resulted in the creation of large specialized institutions characteristic of "big science." Within just one decade, a number of disciplines and research directions previously underdeveloped or nonexistent in Russia became quickly institutionalized. Journals, societies, research institutes and laboratories, conferences, and teaching departments in a variety of new biomedical disciplines, from biochemistry and biophysics to endocrinology and hematology, proliferated throughout the country.[40]

Of course, in certain ways these intellectual and institutional developments mirrored similar processes unfolding during the same time elsewhere in Europe and the Americas. During the 1920s, investigators around the globe conducted similar research and strove to build specialized institutions to advance their studies of new exciting biomedical subjects.[41] What made the Russian case special was that after 1917 science in Russia became an exclusively state-sponsored enterprise. If elsewhere in the world the drive to advance and institutionalize the new directions in biomedical research was promoted and supported largely by wealthy individuals, universities, foundations, and corporations, in Russia the sole patron was the Bolshevik state.[42]

How did Russian scientists win the ear of state agencies and agents who fostered the unprecedented expansion of research in experimental biology and medicine in new

Soviet Russia? And why was the Bolshevik leadership willing to commit resources of the new state (scarce as they were) to building large research institutions in these new fields and to popularizing the new discoveries of "visionary biology"?

The ideal of controlling and manipulating life processes that underpinned early twentieth-century biomedical research fit in well with the Bolsheviks' general technocratic and utilitarian attitudes toward science and its place in the new social order they were creating.[43] It was exactly these pragmatic attitudes that Bulgakov parodied in "Fateful Eggs" by appointing professor Persikov to the "supreme commission" set up to fight the chicken plague and by turning over his equipment to Rokk to alleviate the chicken crisis in the republic. The application of various advances in experimental biology and medicine did indeed promise to help the new regime in addressing urgent problems it confronted, such as creating artificial foodstuffs to combat pervasive famine and producing vaccines to check the spread of epidemics. Not surprisingly, among the first research establishments founded by Narkomzdrav at the very beginning of the Bolshevik regime during the Civil War of 1918–1921 were an Institute of the Physiology of Nutrition and an Institute of Vaccines and Sera.

The Bolshevik commitment to the protection of health of the country's population was heightened by anxieties over the health of the ruling elite. Although in 1922 Nikolai Tikhonov, a popular young poet, surmised that "nails made out of Bolsheviks" would be "the world's strongest," the country's new rulers—Joseph Stalin's pseudonym notwithstanding—were not made of steel.[44] They were human beings with the usual human frailties, weaknesses, and health problems. Concerns about the health of the "party old guard" played a significant part in the Bolsheviks' support for new research and institutions in a variety of biomedical fields, from endocrinology to hematology.[45] As Bulgakov vividly portrayed it in "The Dog's Heart," enterprising scientists cleverly exploited this "human" side of the Bolsheviks to satisfy their own material needs, advance research interests, or promote institutional agendas.

Yet the immediate utility (either real or imaginary) of biomedical investigations was certainly not the only reason the Bolshevik government generously supported research in various branches of visionary biology. To a large extent the growing union between the newborn Soviet state and "visionary biologists" stemmed from a particular cultural atmosphere of ever-present death and high hopes for the future, which permeated the country.

The decade of 1914–1923 in Russia was a reign of death, which plunged the country into an unprecedented demographic catastrophe.[46] Russian losses on the battlefields of World War I were higher than those of any other combatant country. The war decimated Russia's western provinces and resulted in the death of some two to three million soldiers, mostly young, able-bodied men (nobody counted civilian casualties). The Civil War that came on the tails of the Revolution engulfed the entire country: from the Crimea in the south to Arkhangelsk in the north and from Vladivostok in the east to Petrograd in the west. In addition to more than three years of fierce fighting, all belligerents in the Civil

War—be they "Reds," "Whites," or "Greens"—exercised indiscriminate terror, executing tens of thousands of "enemies." The Civil War compounded the effects of the collapse of the old civic and economic order prompted by the Revolution and forced large-scale migration of the population,[47] which was highly conducive to the spread of epidemic diseases: typhus, typhoid, cholera, and Spanish flu. The exact figures will never be known, but available estimates indicate that some eight to ten million people perished in the Civil War: killed in its battles, executed as "enemies," and succumbed to disease, cold, and starvation. As if these horrendous losses were not enough, just as the Civil War ended in 1921, the worst drought in half a century hit the country's southern regions, resulting in a catastrophic famine that killed millions of people. In sum, in just one decade, some fifteen to twenty million people out of a population of 140 million in Russia died of "unnatural" causes.[48] The number of injured, maimed, and disabled both physically and psychologically remains unknown and is difficult even to imagine.

Furthermore, combined with a sharp decline in fertility and increased infant mortality, the "decade of death" heavily distorted age and sex distributions in the Russian population. Women outnumbered men by 5 to 4, and in some regions the ratio was even higher. In certain age groups (20–30), women outnumbered men by 2 to 1. The age group from 1 to 30 years old, which in 1897 had constituted more than 60 percent of the population, dropped to well under 50 percent in 1920, while the cohort aged 20–29 shrunk by 150 percent. By the end of "the decade of death," Russia had "aged" considerably: the cohort of 40–60-year-olds comprised almost one third of the entire population in 1920.[49]

Horrific as they are, these numbers do not convey the full impact of the "decade of death" on those who survived it. In February 1921 Vladislav Khodasevich, an influential poet and literary critic, provided a succinct characterization of this particular generation:

> Not renegades, nor degenerates, they are simply new people. Many among them, as yet-non-shaving youth, virtually boys, had been sent into the trenches, had seen mountains of dead bodies, themselves had ripped open countless guts, had burned down towns, had destroyed roads, had trampled down fields, and just yesterday [they] came back, carrying along their psychical contagion (*psikhicheskuiu zarazu*).[50]

The omnipresence of death evoked a wide range of responses, shaping attitudes, interests, beliefs, emotions, behaviors, and imaginations. It turned some people into habitual killers, others into perpetual victims. It led some to hedonism, others to asceticism. In some, it ingrained all-embracing hatred, in others it fired up all-forgiving love. It converted some into devout mystics, others into militant materialists. But it did not leave a single soul untouched, becoming a formative experience for several generations of Russia's population, and, especially, for the country's biomedical community.[51]

Yet although death was all-pervasive throughout the 1914–1923 period, so were high hopes and great expectations for a bright and wonderful future ahead. Of course, separate segments of Russian society nurtured different hopes and different expectations, but nearly all shared one vision—of renewal, revival, remaking, and rebuilding of the country. The Revolution turned this vision into what Richard Stites has perceptively termed "revolutionary dreams" of creating a "new world."[52] Inspired by these "dreams," the country experimented on an unprecedented scale with new ideas and new practices in every facet of life, be it the arts, industry, state administration, education, laws, technology, social policies, health care, literary works, or community services.[53] The dreams of a "new world" being brought about by the Revolution included visions of a new state, a new society, a new family, a new art, and even a new life . . . without death! (See Figure 1.6.)

The belief in the possibility of "abolishing death" and "achieving eternal life" became a prominent component of the revolutionary dreams circulating in Russian society at the time. Perhaps the most forceful and succinct articulations of this particular dream came from a small, short-lived, but very active group of anarchist writers, poets, and artists, who called themselves Biocosmists-Immortalists. In January 1922 the group published a manifesto on the pages of *Izvestiia*.[54] The manifesto advanced a "new ideology" based on "the notion of personality that grows in its strength and creativity until it grounds itself in immortality and cosmos" (that is, outer space), which its authors named *biocosmism*.[55] The biocosmists considered "the right to being (immortality, resurrection, rejuvenation) and to the freedom of movement in cosmos" as constituting a "fundamental and inalienable right of personality." In November 1922 the group's Petrograd branch even founded a special journal, tellingly called *Immortality*.[56] According to biocosmists' "commandments," death was "logically absurd, ethically impermissible, and aesthetically ugly."[57]

Certainly the dream of abolishing death and achieving immortality predated the Revolution and the decade of death, yet its formulations in 1920s Russia differed drastically from the previous renderings. Thus, many later commentators have linked biocosmists' programs to the mystical cosmism of Nikolai Fedorov (1829–1903)[58] and his "common cause" of abolishing death and resurrecting the dead.[59] The biocosmists themselves, however, denied any connection to Fedorov's ideas, stating forcefully that Fedorov's "philosophy is completely alien" to biocosmism.[60] For Fedorov and many of his followers, the theological, the mystical, and the occult constituted the main foundations of their "common cause." For biocosmists, however, God was "rubbish," in the words of one of their leaders Alexander Iaroslavskii. Indeed the founder of the group's Moscow branch Alexander Agienko (who wrote under the pen name of Alexander Sviatogor) eventually become one of the principals of the Bolshevik "League of Militant Godless." As Sviatogor, put it: "We are not getting caught in the quagmire of religion or mysticism. We are too rational [for that] and declare war on religion and mysticism."[61]

For biocosmists, as for many others in post-revolutionary Russia, immortality was more than a mystical fancy or a literary *épatage* (from the French verb *épater*). It was a realistic goal and a program of action and, as their numerous pronouncements demonstrated, it was contemporary *experimental biomedical research* that made such a goal attainable and such a *bio*cosmic program doable. In contrast, in his *Philosophy of the*

Common Cause written during the 1890s, Fedorov appeared to be highly skeptical about experimental biology and medicine, even though these fields were quickly developing during that exact time:

> As natural science in general, so too its application, medicine, should move from conducting *laboratory, pharmaceutical-therapeutic experiments* in hospitals and clinics to the utilization of the telluric-solar and psycho-physiological force controlled by knowledge; [they] should move to the elimination of sickness in general, and not limit themselves to the treatment of individuals.[62]

The biocosmists held entirely different views on how their goals could be achieved. The first and only issue of the journal *Immortality* carried—on its second page—an article by the *pharmacologist* Nikolai Kravkov entitled "Data on and perspectives for the revival of tissues of the dead," which summarized his numerous experiments with organs and tissues taken from cadavers.[63] A short poem written by Alexander Iaroslavskii, which followed the article, hailed "the laboratory of Kravkov, wherein the flowers of immortality grow" and declared that biocosmists were "overthrowing death with the hand of immortal Kravkov."[64] Biocosmists also regularly referred to the "rejuvenation" experiments by the Austrian biologist Eugen Steinach, which in the early 1920s attained enormous popularity around the world, as proof that science was indeed capable of fulfilling the dream of eternal life. They even invited Steinach to publish his latest research on the pages of their oracle, *Immortality*.[65] Biocosmists also followed attentively the current fascination with anabiosis—the state of "neither life, nor death"–enticed by the active popularization of the biophysicist Porfirii Bakhmet'ev's experiments with freezing and thawing of living organisms: insects, fishes, and bats. Iaroslavskii even wrote a long "Poem of Anabiosis," presenting "freezing and thawing" of humans as the main tool of the "biocosmic revolution" that would soon recreate the entire planet.[66] Like many of their contemporaries, the Russian biocosmists firmly believed that humanity was, in the words of one of them, "on the eve of science's victory over death."[67]

The dream of immortality, appealing as it might have been for many individuals, in and of itself probably was not a major factor in the Bolsheviks' support for visionary biology. But, associated with that dream, the ideology of materialism and atheism certainly was. Atheism and materialism were two cornerstones of the Bolsheviks' doctrine, Marxism, and they spared no effort to combat its perceived "enemies": religion and idealism. Just three months after they had seized power, in January 1918, the SNK issued a special decree, signed by Lenin, on the separation of the state and the church. In line with Lenin's dictum (formulated in his 1905 article "Socialism and Religion") that "religion is opium for the people," the decree called for the expropriation of all properties belonging to religious institutions and for the complete abolition of the teaching of religious doctrines in all schools of the newborn Soviet republic. Although the decree allowed (at least temporarily and on paper) the freedom of conscience and worship, the Bolsheviks dismissed religious beliefs as ignorant superstition and waged a militant propaganda campaign against religion. In 1922, they began to publish a special newspaper, *The Godless*, that fostered the establishment of the All-Union League of Militant Godless, with branches throughout the entire country.[68]

The Bolshevik attitude toward idealism was no less aggressive, even though major actions against it had to wait until the end of the Civil War. In early 1922 the Communist Academy, the center of Bolshevik scholarship established in 1918 as a counterweight to the "bourgeois" Russian Academy of Sciences, began publishing a new monthly tellingly entitled *Under the Banner of Marxism*.[69] The editorial opening the journal's first issue called upon Marxists in all fields of knowledge to "unfold the banner of militant materialism," and to launch a broad attack on idealism.[70] To drive the point home, the journal's third issue carried an article entitled "On the Significance of Militant Materialism," written by none other than Lenin himself. Lenin charged scientists with the task of "the struggle against the pressure of bourgeois ideas and the bourgeois worldview." He insisted that every "scientist must be an up-to-date materialist, a deliberate follower of the materialism presented by Marx, that is, he must be a dialectical materialist."[71] Lenin's article became the manifesto of Communist scholars, and its imperative tone—must be—set the scene for the expulsion from the country, just a few months later, of some two hundred prominent intellectuals—philosophers, sociologists, historians, and economists—accused of idealism.[72]

In contrast to the "suspect" humanities, experimental biology and medicine provided a solid foundation for the regime's militant materialists and militant godless, thoroughly undermining a theistic worldview that had dominated human thought for millennia. Contemporary biomedical research radically challenged the previous views on such issues as body and soul, life and death, health and illness, human origins and human evolution.

In particular, the advances of experimental biology and medicine profoundly transformed understanding of life and death. A theistic worldview portrayed life as the result of a divine creation that animated (literally) dead matter by infusing it with an immortal soul/spirit. Accordingly, this worldview presented death as a *momentary event* of an immortal soul leaving a mortal body, which could be reversed only by a divine/demonic intervention resulting in resurrection and life after death.[73] The experimental view of life based on manipulating life processes at all levels of organization—from a whole organism to its constituent organs, tissues, cells, and molecules—left literally no place in the body for an immortal soul.

In 1902 Porfirii Bakhmet'ev, a leading proponent of the nascent field of biophysics, published an article under the provocative title "The seat of the soul."[74] Contrary to what one might have expected in an article with such a title, it described research that did not involve humans, animals, or indeed any living objects at all. Instead, Bakhmet'ev studied the "behavior" of droplets of a particular chemical (paranitrotoluol). His experiments demonstrated that the droplets "behaved"—that is, reacted to various chemical and physical changes in their environment—in a way that closely resembled the purposeful behavior of live unicellular organisms. These experiments led Bakhmet'ev to propose that "the seat of the soul" should now be placed not in the brain, the heart, or even the cell, but in molecules, and that scientific investigations of "the soul" should eventually become just a subfield in the general studies of energy and its different forms.

As this example suggests, experimental biology replaced "the soul" with the intricate physical and chemical interactions of body parts and divine/demonic intervention with science and its practitioners who could "create life" in their laboratories. "The Body is the Master of the Soul"—forcefully asserted the title of a popular-science brochure published in 1925 by one of the Communist scholars.[75] Accordingly, the experimental worldview presented death as a *gradual process*, which unfolded in stages (with different organs, tissues, and cells of a body dying at different speeds and different times) and which at its various points could be slowed down, stopped, reversed, and otherwise manipulated by scientific techniques. The new worldview negated the theistic understanding of resurrection and life after death. Instead, it provided vivid examples of scientific "resurrection" and "life after death": mummified human fingers that grew nails and severed heads of dogs kept alive by elaborate scientific machinery.

As one might expect, the dream of abolishing death inspired a variety of experiments and was, in turn, reinforced by the successes of contemporary experimental biology and medicine. During the early 1920s, Soviet Russia was flooded by books, brochures, and pamphlets (to say nothing of countless articles in professional journals and especially popular periodicals) bearing such titles as "Immortality as a subject of natural science," "Life and death," "What is death?," "Aging and death," "The problems of death and immortality," "Death and revival," "Death from the viewpoint of modern science," and "The enigma of death."[76] In 1925, Georgii Shor, a renowned pathologist, even suggested that a special discipline dealing with death be instituted. Following in the footsteps of Elie Metchnikoff, he named this discipline "thanatology" for the Greek god of death *Thanatos*.[77]

Not only medical specialists, ranging from pathologists and bacteriologists to endocrinologists and psychiatrists, but also biologists of all stripes, engaged in studies of death. Virtually every Russian biologist of note published a work on death. Death figured prominently in the research and publications of biochemists, histologists, physiologists, cytologists, protistologists, biophysicists, morphologists, physical anthropologists, and even ichthyologists.[78] In 1925, *Book-carrier*, a bibliographic weekly that provided advice on new books for "worker's libraries," published a lengthy survey of the "New popular-science literature on life and death."[79] Needless to say, these popular-science publications in turn generated a storm of newspaper reportages and magazine articles issued under such enticing headlines as "Miracles of science," "Death and resurrection on the operation table," "Science above god," "The dead are revived," "Life after death," and "Science that conquers death."[80]

Certainly publications of these works by state-run publishing houses fit in well with the Bolsheviks' militant antireligious campaign, bringing to the populace "the view of modern science" on such theologically charged issues as death, relations between body and soul, immortality, life after death, and resurrection.[81] As the Bolshevik propagandists openly proclaimed, "The most important task of the popularization of natural sciences, in our worker's state, at this particular moment, is *the struggle against the old ideology* [read—*religion*], whose remnants are still firmly embedded in the masses."[82] To give but one example, in October 1924 (exactly at the time when Bulgakov finished "Fateful Eggs" and began thinking about "The Dog's Heart"), the major venue for the popularization of science—the Moscow Polytechnic Museum—organized a "public dispute" on the question "What is 'the soul'?" Reported in the country's major dailies, the event featured Nikolai Semashko, the head of Narkomzdrav, as a keynote speaker. Semashko referred extensively to the advances in contemporary experimental biology and medicine as proof of his main theses: "the soul" does not exist; "the immortal soul" is an invention of priests; material—biological and social—factors explain adequately the complexities of the human psyche and human behavior.[83]

Четверг, 20 сентября
Вечерняя Москва
Наука, побеждающая смерть
Опыты С. С. Брюханенко с отрезанной собачьей головой, продолжающей жить, благодаря искус-
ному восстановлению кровообращения.
Справа — проф. Кулябко, присутствующий при опытах.

FIGURE 1.5 A front page of the popular daily *Evening Moscow*, featuring original photographs taken during experiments with the severed head of a dog that accompanied an article entitled "Science that conquers death." The caption reads: "Experiments of S. S. Briukhanenko (*sic*!) with a severed dog head that continues to live thanks to the artificial blood circulation. On the left—Professor Kuliabko attending the experiments." Following almost exactly Bulgakov's depiction of scientific reportages in "Fateful Eggs," the author of this article misspelled the name of its main protagonist, Sergei Briukhonenko. *From* Vecherniaia Moskva, *September 20, 1928, 1.*

Yet despite the obvious vested interest on the part of the Bolshevik state in promoting the scientific understanding of life and death, one can infer wide popular interest in these subjects, for a large portion of the countless publications dealing with "issues of life and death" came from private publishers operating on a for-profit basis and playing to the public's demands. Furthermore, under NEP, even state-owned enterprises, including book and newspaper publishing, were expected to turn a profit, suggesting that they, too, were attentive to the public's interests (see Figure 1.5).

Science/Fiction Revolution

This public interest in "science that conquers death," along with the broader social and cultural implications of the budding union between visionary biologists and revolutionary dreamers, found vivid expression in numerous works of fiction, of which Bulgakov's novellas presented perhaps the best, but by no means the only or even the first, example. Indeed, Bulgakov's novellas both documented and embodied the coming of yet another "scientific revolution": the dramatic upsurge in the public visibility and cultural authority of science in general, and biomedical sciences in particular, in post-revolutionary Russia as well as in other counties of Europe and the Americas.[84] Clearly detectable in the enormously increased media coverage of scientific advances vividly portrayed in "Fateful Eggs," this revolution was also manifest in the rapid development of a new literary genre, *science fiction*, and especially a subset of this genre that focused on the "issues of life and death," as did Bulgakov's novellas. Of course, these issues had always attracted the close attention of the literary community.[85] But the 1920s fiction that examined them differed profoundly from similar literature of previous decades. And to a considerable degree this difference derived from the new understanding of life and death offered by experimental biology and medicine.

A lengthy novella with the telling title "Professor of Immortality," published in 1891 in the popular "thick" journal *Russian Herald*[86] by Konstantin Sluchevskii, a well-known novelist, illustrates this difference perfectly.[87] Although any of the 1920s writers could have used its title,[88] the novella differs radically from any of the works published in the 1920s. Its main protagonist is actually not a "professor" but a doctor of medicine, named Petr Ivanovich Abatulov, who had earned his degree some twenty years earlier and had since practiced medicine in a faraway village in the lower Volga region in the south of Russia. The majority of his patients are local peasants and members of a nomadic tribe (Kalmyks) who live nearby. Abatulov deals daily with horrible wounds, fatal diseases, and death, which prompt his wife to call his office "the butcher shop." Busy with his practice, the doctor nevertheless continues his "scientific studies." He reads scientific books and journals and writes a long treatise on . . . immortality. And it is because of this work that his acquaintances confer on him the (ironic) title of "professor of immortality."

Abatulov claims to have proven by "natural-science evidence" that "the human soul cannot be mortal." His "evidence," however, is purely theoretical. For Abatulov, Charles Darwin's theory of evolution leads to the "logical necessity to acknowledge the immortality of an individual soul" (336). He bases his own theory on two "scientific premises." First is that "all the organisms on Earth, since the earliest times, are constantly improving (*sovershenstvuiutsia*)"; and second, that "any once-achieved improvement is preserved" in subsequent generations (311). As he explains to a visitor, "It is Darwin, but not exactly Darwin." Indeed, Abatulov's first premise rests on the idea of biological progress, which in the late nineteenth century became closely intertwined with Darwin's concept of evolution.[89] His second premise simply restates the most popular understanding of heredity of his day—that characteristics acquired by an organism in its lifetime are inherited. In articulating his views, Abatulov refers specifically to Herbert Spencer and Ernst Haeckel, two of the most influential evolutionists of the time who had ardently supported both the notion of biological progress and the concept of the inheritance of acquired characteristics.[90]

According to Abatulov, Darwin had demonstrated that biological evolution is an inexorable law of nature, much like the law of the conservation of energy. Since biological evolution had "produced" a progression from microbes to man, it is logical to suppose that this progression must extend further: from a mortal (man's) body to an immortal soul—as the next step in biological organisms' continuous "improvement." On the other hand, since the human "psychical (*psikhicheskaia*) activity"[91]—scientific theories, philosophical doctrines, symphonies, and works of arts, which Abatulov summarily names "psychical organisms"—represents the "highest level [of evolution] achieved by Nature," it is inconceivable that the soul that produces all of these "psychical organisms" perishes with the death of the body. It would contradict Abatulov's second premise that improvements, once gained, are preserved. Referring to multidimensional geometries developed by Nikolai Lobachevskii and Bernhard Riemann (both of whom are directly mentioned in the text), Abatulov suggests that after the body dies, the soul continues its existence in "another dimension."[92]

Although Abatulov is convinced that the human soul is immortal and he does believe in his own way in the existence of God, he has no use for organized religion and its rituals. He does not pray or go to church, and instead of customary icons, a copy of the famous painting "Crucifixion" by the prominent Russian artist Karl Briulov hangs in his office. His wife's accidental death by drowning, however, brings him back into the embrace of the church (literally of a local priest, who comforts Abatulov after he hears the terrible news). Faced with the death of his beloved, Abatulov begins to doubt his "scientific" theory and finds consolation in his faith.

Much like other fictional (and theological) writings of the late nineteenth century on the same subject, Sluchevskii's novella represented one of the numerous attempts at the reconciliation of newly available biological knowledge—Darwin's theory of evolution, Haeckel's monism, Spencer's progressivism, and contemporary notions of heredity—with a theistic worldview. It did not address the issue of bodily immortality at all, focusing instead on the immortality of the soul. Quite tellingly, Abatulov's theory is purely speculative and has no grounding in any experiments. What is even more important, the novella is not futuristic in any sense. It deals with the "here and now" and does not attempt to look beyond and over its protagonist's life.

The Russian fiction of the 1920s on "issues of life and death" was entirely different, as Bulgakov's novellas clearly indicate. The contemporary advances of experimental biology and medicine often supplied crucial elements to plots, settings, and the characterization of protagonists. The majority of writings focused first and foremost on the immortality of the physical body. And to a considerable degree, this literature was openly futuristic, depicting a future in which bodily immortality had become a fact of life.

These writings formed a subset of a new literary genre—scientific fantasy (*nauchnaia fantastika*), as it was called in Russia, or scientific romance, scientific fiction, scientifiction, and science fiction (SF) in the English-speaking countries—which brought together issues of science, future, and human destiny.[93] Without going into details of the origins and permutations of this genre (and the unending debates surrounding them[94]), one can safely state that the appearance of the words *science* and *scientific* as modifiers of such nouns as *romance*, *fiction*, and *fantasy* signaled unambiguously the arrival of yet another "scientific revolution": the dramatic rise of science's public visibility and cultural authority in contemporary societies. Not unexpectedly, this appearance occurred exactly at the time when science (and technology) began to occupy an ever-larger place in the life of people in industrialized nations.

In the decades around the turn of the twentieth century, science expanded tremendously both the space and time dimensions of the human universe. Instead of a few millennia since its creation by God(s), science made the universe at least several billion years old. Instead of visible "heavens" and a "small" solar system, science exploded the Cosmos to distances of unimaginable magnitude: thousands upon thousands of light years. At the same time, new technology "shrank" our home planet: a voyage around the world, which in previous centuries had taken years, could now be made in mere days, while communications between continents, which previously had taken months, now could be virtually instantaneous. The wonders of modern technology (telegraph, telephone, radio, phonograph, electric lights, x-rays, cinematography, airplanes and automobiles, to name but a few, highly symbolic advances of that time) rapidly penetrated the very fabric of daily life. Furthermore, scientists actually managed to convince the public that all these technological wonders were nothing but results of the *application of science*, in the process greatly elevating the authority of their own profession in contemporary societies.[95] And it was exactly the recognition (first of all, by writers) of this newly found *cultural* importance and authority of science that gave birth to the new genre of *science* fiction.[96]

As elsewhere in the world, this new genre flourished in 1920s Russia, filling the pages of daily newspapers, weekly magazines, popular-science and adventure journals, and books.[97] In his 1922 essay about an acknowledged founder of the genre, H. G. Wells, one

FIGURE 1.6 This 1921 painting by Konstantin Iuon (1875–1958), an eminent Russian artist, is entitled "The New Planet." It presents a visual embodiment of the contemporary "revolutionary dreams": a new world—Soviet Russia—emerging out of people's anguish and sufferings and illuminating the universe. It pictures the Bolshevik Revolution as an event of cosmic proportions. *Courtesy of the State Tretyakov Gallery.*

of the pioneers of Soviet SF, Evgenii Zamiatin, foresaw that "Russia, which during the last few years has become the most fantastic country of modern Europe, will undoubtedly reflect this period in its fantastic literature."[98] As Bulgakov's novellas demonstrate, Zamiatin's prediction quickly came true.

Science fiction took Soviet Russia by storm, and visionary biology inspired much of this popular genre, documenting scientists' intentions to seize control not only over such physical characteristics of the universe as energy, space, and time, but also over its "organic" parameters, such as life and death, health and illness, emergence and evolution. Dozens of novels, poems, and innumerable short stories projected future applications of contemporary advances in various fields of experimental biology and medicine (with their promise of abolishing death or at least prolonging life), exploring their possible impact on individuals, societies, and the world at large. Some of these works, such as the gynecologist Fedor Il'in's novel *The Valley of New Life* (1928) or the zoologist Boris Fortunatov's *The Island of Gorilloids* (1929), were written by biologists and physicians who themselves pursued exciting new lines of research in experimental biomedical sciences. Others were created by professional poets, journalists, and litterateurs, including Alexander Beliaev, Valerii Briusov, Mikhail Bulgakov, Vladimir Mayakovsky, Boris Pil'niak, Andrei Platonov, Aleksei Tolstoy, and Evgenii Zamiatin, to mention only a few well-known names. Scores of much less-remembered and sometimes completely forgotten authors, such as Sergei Beliaev (no relation to Alexander Beliaev), Fedor Bogdanov, Nikolai Dolin, Evgeniia Fortunato, Mikhail Gireli, Viktor Goncharov, Alexander Iaroslavskii, Andrei Irkutov, Mark Krinitskii, Vadim Nikol'skii, Iakov Okunev, Vsevolod Valiusinskii, and Mark Volosov, tried their hand at fictional experiments with various projects and techniques of visionary biology, ranging from "biological rays" to rejuvenation, blood transfusions, tissue extracts, organ transplants, and artificial insemination.[99]

Works of fiction took as a given that science could and would gain control over life and death. In contrast to scientific experiments that investigated the concrete mechanisms of life in search of the means to gain control over various life processes and to apply this control to fight diseases, aging, and ultimately death, fictional experiments had entirely different goals. They investigated the role of science and scientists, state (and private) agents and agencies, technologies and personalities in bringing about a "new world" without death, trying to envision what that world would be like. Some of them were overly enthusiastic, hailing the coming era of eternal life. Others were deeply pessimistic, suggesting that instead of immortality, science would only bring forth herds of horrible monsters and unthinkable catastrophes, as it "did" in "Fateful Eggs." Some writers made fun of science, its practitioners, and its pursuits. Others were frightened by the powers afforded to science, scientists, and their patrons by new discoveries and experimental techniques. Yet whether laughing or crying, singing praises or predicting doom, many writers shared with the scientists they portrayed or imagined an unshakable belief in science and its ability to deliver to humanity—in just a decade or two, or in a century at most—the ultimate gift: eternal life.

Of course, some writers—for instance, Bulgakov—did not share in this belief and warned readers of the dangers inherent to scientists "playing god" and striving for control over human life, human nature, and human destiny. But practically all of the fictional explorations of the prospects open to humanity by experimental biology and medicine (whether their authors themselves believed in them or not) enjoyed enormous popularity with readers, testifying to the tremendous cultural authority that science in general, and biomedical sciences in particular, enjoyed in contemporary Soviet society. It was this cultural authority that fueled both public fears of and hopes for a science that would conquer death.

We are born to make a fairy-tale into reality.
PAVEL GERMAN, "The Aviators' March" (ca. 1924)

You had always been a fiery preacher of a theory that after a man's head is severed, his life ends, he turns into ash and seizes to exist. I'm glad to tell you, in the presence of my guests, even though they serve as proof for an entirely different theory, that yours is both solid and witty. However, all theories deserve one another.
MIKHAIL BULGAKOV, *Master and Margarita* (1940)

2 "Professor's Head": Isolated Organs

Beheaded for science

FOR MILLENNIA, BEHEADING—SEPARATING head from its body—was a sure way to inflict death. The very phrase "capital punishment" derives from the Latin word *capita* meaning "head" and, in many cultures, decapitation (deriving from the same word) was a preferred method of putting to death kings and commoners, criminals and ideological opponents.[1] Immortalized in numerous paintings by Botticelli, Caravaggio, Durer, Klimt, Moreau, Titian, and a host of other artists, the image of John the Baptist's head delivered to Salome on a dinner plate became a recognizable symbol of violent death.[2] In 1920s Russia, the image of a severed head on a dinner plate came to symbolize something entirely different, indeed the opposite: the victory of science over death.

Just as the *Red Panorama* was finishing the serialization of Mikhail Bulgakov's "Ray of Life," the Russian reading public welcomed another literary depiction of an exciting project pursued by visionary biologists. From June 16 to July 6, 1925, the readers of a Moscow daily, *The Worker's Gazette*, eagerly awaited each issue of the newspaper in anticipation of the next installment of a "scientific-fantastic story," entitled "Professor Dowell's Head."[3] Written by a former jurist, amateur musician and actor, and now debutant littérateur, Alexander Beliaev (1884–1942) (see Figure 2.1),[4] the novella became an instant hit: in just a few weeks, it was serialized by *World Tracker*, a newly established popular-science and adventure magazine.[5] A few months later, at the beginning of 1926, a slightly expanded and revised version of the novella came out in book format, with the first print-run of 4,000 copies.[6] A decade later, Beliaev reworked the story into a full-blown novel by the same name.[7]

FIGURE 2.1 Alexander Beliaev, ca. 1906. Born in Smolensk to a family of an Orthodox priest, Alexander began his education at a local theological school (seminary). But he did not take his vows, instead entering a local law school, the Demidov Lyceum of Jurisprudence, in 1902. The photo captured the newly minted attorney at law taking notes while questioning a witness during one of his assignments. *Courtesy of TsGALI SPb.*

The story unfolds in an unidentified city in the United States. It begins with a meeting of a certain Professor Kern and a recent medical graduate, Miss Adams, who is desperately looking for a job to support her aged mother and a younger sister. Kern offers Miss Adams a position as his assistant. He puts forward one condition, however—she must be "mute as a fish" and must not talk to anyone about anything she sees or hears in Kern's laboratory. Miss Adams is mystified. On a tour through the laboratory, she discovers the reason for Kern's insistence on secrecy:

> Miss Adams turned her head and suddenly saw something that made her shudder as if from an electric shock.
>
> A human head stared at her—a head alone, without a body.
>
> It was mounted on a square platform, made of glass and supported by four long, shiny metal legs. From its severed neck, from the aortas [sic!], through the holes in

> the glass, two tubes ran to tanks. A third, thicker tube came out of the head's throat and ran to a large cylinder. The tanks and the cylinder had valves, manometers, thermometers, and some other unknown [to Miss Adams] instruments.
>
> The head looked at Miss Adams attentively and mournfully, blinking its eyes. There could be no doubt: the head was alive; severed from its body, it led its own separate and conscious existence. [. . .] Live human thought shone in its eyes.[8]

Despite her profound shock (see Figure 2.2), Miss Adams thinks she recognizes the head. It looks very much like the head of Professor Dowell, a prominent scientist and surgeon "renowned for his experiments on the revival of separate organs excised from fresh cadavers," whose public lectures Miss Adams had frequented during her studies in medical school. Kern confirms her impression—the head indeed belongs to Kern's mentor and colleague, Professor Dowell, who had succumbed to an, "as yet, incurable disease." In his will, Professor Dowell had donated his body for "scientific experiments," and, Kern followed his wishes. "Unfortunately, I was able to revive only the head," Kern exclaims. "Not all at once!" he confidently asserts. "As it is, we, scientists, intrude upon 'unshakable laws of nature,' challenge death itself, and take away the livelihood from miracle-makers and from God himself." Kern explains that Miss Adams's duties in his lab will be to take care of Dowell's head, and she accepts the job.

Kern shows Miss Adams how to operate the machinery supporting the head, but stringently forbids her touching a valve on the big cylinder, warning that it will certainly kill the head. After a few weeks in the lab, Miss Adams and the head establish a kind of simple communication, and the head repeatedly indicates that it wants Miss Adams to open the valve on the big cylinder. Thinking that the head wants her to end its dismal existence, Miss Adams refuses. In time, deeply moved by what she perceives as horrible suffering, she gives in to the head's wishes and opens the valve. As it turns out, her action does not kill the head, but instead gives it a voice! The cylinder contains compressed air, which blown through the head's throat makes it able to speak. And speak it does!

The head tells Miss Adams the story of Dowell's research on the revival of human heads and of Kern's role as his assistant. It tells her of Dowell's last bout of asthma that led to his death, and she begins to suspect that, instead of helping his professor, Kern actually murdered Dowell with a lethal injection of morphine. It tells her how—by torture—Kern forced the revived head into continuing Dowell's research, which Kern now presents as his own, exploiting ideas and insights the head provides. Miss Adams is outraged and wants to expose Kern. But the head begs her not to do so. The exposure would not bring Dowell back to his normal state. The head does not care about Kern stealing Dowell's fame or the money that might be generated by his work. It does not need either. All it wants now is to continue the research, which would be impossible if Kern is exposed.

In preparation for a public presentation of "his" research, Kern revives the heads of two cadavers: one male (the victim of a traffic accident), another female (a murder victim).

FIGURE 2.2 Miss Adams and Dowell's head. The illustration depicts very realistically the profound shock Miss Adams is experiencing at the first encounter with the severed head of Professor Dowell. Compare this with Figure 2.7. *From Aleksandr Beliaev,* Sobranie sochinenii v vos'mi tomakh *(Moscow: Molodaia gvardiia, 1963), vol. 1.*

The male head soon dies and Kern demonstrates the female head at a highly publicized meeting attended by the lights of the scientific community and members of the high society. His presentation causes much excitement. "One of the leading American scientists" praises Kern's genius as the manifestation of "almighty science," of "victory over death," and of "America, which gives birth to such minds and which brings to the world the greatest scientific achievements."

At this moment, "a whirlwind of long-stored anger and hatred" brings Miss Adams to the podium. She shouts that Kern had killed Dowell, and tortured Dowell's revived head into continuing his research. "Don't believe him," cried Miss Adams, pointing at Kern. "He is a thief and a murderer. He stole the work of Professor Dowell." Amidst general confusion, Kern calmly announces that due to her daily encounters with the severed heads, Miss Adams suffers from delusions and "has lost her mind." Miss Adams is taken away and placed in a mental institution. A few months later, the son of Professor Dowell, Arthur, who lived in England, comes back to America. His return is prompted by press reports on the incident with Miss Adams's outburst during Kern's demonstration. Arthur finds Miss Adams and arranges for her release from the mental institution. Together, they convince the authorities to search Kern's laboratory. They find Dowell's head in very bad shape. But before the head finally expires, it confirms to the police investigators Miss Adams's accusations. Kern is arrested. End of story.

The novella's first readers most likely perceived it as pure fantasy. But less than three months after Beliaev's story appeared on the pages of *The Worker's Gazette*, its major "scientific-fantastic" idea—the possibility of keeping a severed head alive in a laboratory—came true. On September 18, 1925, Sergei Briukhonenko[9] (1890–1960) (see Figure 2.3), a young doctor from the Moscow Chemical-Pharmaceutical Institute, demonstrated to the Second All-Union Congress of Pathologists in Moscow a special apparatus—the inventor called it an *autojector*—he had constructed to keep the severed heads of laboratory animals alive. The apparatus included two electric pumps, which supplied the severed head of a dog with citrated blood through a system of rubber tubes. The outflowing blood was oxygenated in a special vessel, warmed to 37–40° Celsius, and pumped back into the head. The autojector thus imitated in certain ways the functions of both the heart and the lungs and kept the severed head "alive" (it exhibited various reflexes) for one hour and forty minutes.[10] Brief accounts of Briukhonenko's demonstration appeared in reports on the congress in medical journals but the popular press somehow missed the exciting news.[11]

In May 1926, together with a colleague at the Chemical-Pharmaceutical Institute, physiologist Sergei Chechulin, Briukhonenko presented the results of further investigations

FIGURE 2.3 The young doctor Sergei Briukhonenko in the trenches of the First World War, ca. 1915. Born in a small town in the Volga region to a family of an engineer, Sergei graduated from a classical gymnasium and enrolled in the Moscow University Medical School. He received his doctor diploma in the spring of 1914 and with the beginning of the war during that summer was mobilized to the army. This photograph provides a vivid illustration to Vladislav Khodasevich's characterization of the cohort of "yet-non-shaving youth, virtually boys, [who] had been sent into the trenches [and] had seen mountains of dead bodies." *Courtesy of ARAN.*

to the Second All-Union Congress of Physiologists in Leningrad.[12] Originally their report had not been on the congress's program. Apparently, Chechulin, who served as the congress's executive secretary, put the report on the agenda at the last moment, when the congress was already in session. On the morning of May 28, just before the closure of the congress, Briukhonenko demonstrated his autojector in action, while Chechulin summarized their experiments with dogs' severed heads. A brief description of the presentation appeared in the congress' proceedings, but for the next six months, Briukhonenko's research vanished from the public eye.

In November 1926, however, *Evening Moscow*, a popular daily, carried an article entitled "Experiments with the revived head of a dog," devoted to research conducted at the Chemical-Pharmaceutical Institute.[13] The reporter (identified only by his initials "Vl. O.") did not mention Briukhonenko's name, but described his autojector in detail and praised the invention as an "absolutely new and never before used apparatus that allows to reproduce mechanically the basic functions of an organism." The article briefly summarized the experiments with the severed head of a dog demonstrated to the congress of physiologists just a few months earlier. The reporter also described new experiments conducted with Briukhonenko's apparatus. According to "Vl. O.," the researchers succeeded in keeping a dog alive after its heart and lungs had been "taken out" and replaced by the machinery: "The dog, without the heart and the lungs, continued to live under chloroform for two hours." The dog reacted to pain stimuli, opened its mouth, and moved its legs. Its eyes reacted to light. The article prophesied great future for this line of research. The next day a similar article, though unsigned, appeared in the *Red Gazette*, a popular Leningrad newspaper.[14]

Some six months later, on May 22, 1927, *Komsomol Truth*, another daily with a country-wide circulation, published an article by its correspondent, Grigorii Grebnev, with the sensational title "Head, severed from the body, lives."[15] The article explicitly referred to "Professor Dowell's Head," stating that Beliaev's "fairy-tale" that "could only entice a smile" became a reality in Briukhonenko's experiments: "The body lays dead. Only its severed head is alive. It does not bark, the dog's lungs are dead, but the mouth opens and the tongue pushes out a piece of cotton soaked in iodine solution. If someone moves a piece of paper in front of the head, its eyelids close in fright." The reporter hailed Briukhonenko's invention as an "artificial heart" that enabled the "resurrection" of the "dead head" and kept it alive for nearly twenty-four hours. He forecasted that Briukhonenko's work would interest "the entire scientific world," for it "opens incredible perspectives for all of medicine, and particularly, surgery." In the reporter's view, it would make possible "the most complicated surgical operations on practically all internal organs of the human body, including the heart." Grebnev predicted that it would be possible "to take out a diseased heart, cure it, and put it back; all the while the life of the organism is sustained by the artificial heart." He excitedly proclaimed that Briukhonenko's experiments showed the possibility "of fighting against death even after it had occurred." Grebnev lamented that, despite the "enormous importance" of his research, Briukhonenko worked under the most pitiful conditions: one small room that could barely accommodate an operating table and two researchers, with almost no equipment, chemicals, or experimental animals. He concluded his report with a call to "the Soviet public to help the talented scientist in his research."

The "Soviet public" answered the call instantly. The newspaper's next issue (which came out two days later) announced that Dmitrii Fursikov, an eminent physiologist and director of the Communist Academy Institute for Studying the Brain, was ready to provide Briukhonenko with a laboratory and all the necessary materials at his institute.[16] But as the subsequent events indicate, Briukhonenko did not need to take up Fursikov's offer. Evidently, the newspaper publications induced the director of the Chemical-Pharmaceutical Institute Oscar Steppun to expand the institute's own support for Briukhonenko's research.

Exactly one year after the publications in *Komsomol Truth*, Briukhonenko's work reappeared on the public scene. This time, it produced a frenzy in academic circles and especially in the popular press. At the end of May 1928, at the next congress of Soviet physiologists held in Moscow, Briukhonenko presented five (!) reports: an event unprecedented for a scientific conference. Three reports dealt with issues of crucial importance for his experiments—the coagulation and stabilization of blood.[17] The fourth one, supplemented by a demonstration, described the use of his autojector for establishing the system of artificial blood circulation in a dog.[18] And the last one, presented jointly with Chechulin, detailed various experiments with a severed dog head kept alive by the autojector.[19]

The scientific community enthusiastically greeted Briukhonenko's newest research. Just before the opening of the congress, the Chemical-Pharmaceutical Institute released a special issue of its *Proceedings* devoted in its entirety to Briukhonenko's investigations.[20] Entitled "The study of new methods for artificial blood circulation and blood transfusion," the volume consisted of four articles written by Briukhonenko (one coauthored with Chechulin and one with Steppun), which detailed the investigations he presented to the congress.[21] After the congress, *Journal of Experimental Biology and Medicine*, a leading academic periodical in the field established in 1925 under the aegis of the State Institute for People's Health Protection (GINZ), invited Briukhonenko to submit his findings for publication. In the next few months, the oracle of Russian experimental biomedicine carried three articles by Briukhonenko (one coauthored with Chechulin).[22] Soon thereafter, the Moscow Institute of Blood Transfusion invited Briukhonenko to join its staff.[23] The inventor also entered into collaboration with Nikolai Terebinskii, a well-known Moscow surgeon, in developing surgical operations on a stopped heart. Using Briukhonenko's autojector in his experiments on dogs, Terebinskii worked out techniques for complex operations on heart valves, which paved the way for many subsequent developments in open-heart surgery.[24]

The media frenzy over Briukhonenko's studies far outstripped the enthusiasm of the academic community.[25] Already during the congress, Briukhonenko's work had captured journalists' attention. Press coverage of the congress in daily newspapers and weekly magazines highlighted his reports as one of the greatest accomplishments presented at the congress.[26] After the congress, accounts of Briukhonenko's experiments appeared in such popular-science magazines as *Hygiene and Health*, *Sparks of Science*, *Want to Know Everything*, *Scientific Word*, *Searchlight*, *Man and Nature*, *Spark*, and *Herald of Knowledge*.[27] Even *Women's Magazine* and *Screen* (a popular magazine about the cinema) carried articles on "the life of organs outside the body" and on "revived heads" respectively, lavishly illustrated with photographs of the "revived head," the autojector, and its inventor.[28] The front covers of two weeklies, *Hygiene and Health* and *Sparks of Science*, featured a picture of the severed head of a dog on a dinner plate attached to Briukhonenko's apparatus.[29] (See Figure 2.5).

Popular dailies had a field day with Briukhonenko's work, publishing articles with such enticing titles as "A page from the book of life and death," "On the border of life and death," "The future's smile," "Severed head forced to live," and "Science that conquers death."[30] In sensationalizing Briukhonenko's studies, many of these publications followed almost to the letter the media representation of Persikov's "nightmarish rays of life" depicted by Mikhail Bulgakov in "Fateful Eggs." The majority of press reports omitted all of the technicalities of Briukhonenko's research. Very few of them paid any attention to the complicated and time-consuming surgical operation Briukhonenko had to perform to sever the head from its body and to connect it to the autojector or to the variety of problems related to the actual construction of the "artificial heart," as Briukhonenko's autojector came to be labeled in the press. Furthermore, three out of Briukhonenko's five talks at the congress had dealt with issues critical for a proper functioning of his autojector—the coagulation, stabilization, and oxidation of blood. Briukhonenko's studies of blood coagulation, and especially of various chemicals that could prevent the clotting of blood in the rubber tubes of his apparatus, had attracted particular attention from the scientific community. Indeed, it was exactly because of the importance of his work for perfecting the technique of blood transfusions, which at this very time was quickly developing in Russia, that after the congress, the Moscow Institute of Blood Transfusion invited Briukhonenko to join its staff.[31] But these questions received little or no coverage in the media.

As the titles (not to mention the contents) of the newspaper publications clearly indicate, the media focused predominantly on one feature of Briukhonenko's work: its "fight against death." As one of the reports predicted, "very soon, we will witness the greatest achievement of science—its victory over death."[32] Obviously, the journalists perceived (and heavily exploited) the image of the "severed head" first and foremost as a potent symbol of certain death: the head that was supposed to die after being severed from its body continued to "live." They were thrilled and enthralled by the fact that in Briukhonenko's experiments a severed head was "forced to live," as one reporter put it. Supported by the autojector, it exhibited various "signs of life": movements and reaction to light, touch, and chemicals. Yet disconnected from the autojector, the head "died again." The reporters drew obvious conclusions from Briukhonenko's research—it was now possible to repair the body the same way as "a car is repaired by replacing worn parts by new ones" (see Figure 2.4). In "the bared teeth of the severed dog head" they saw "a gentle smile of the future."[33]

Nearly all of the media reports reiterated that Briukhonenko's experiments proved unambiguously that science had the power to effectively control the life and death not just of cells, tissues, and "simple" organs such as legs, ears, or even hearts, but of the most important "organ" of all—the head. In the popular imagination, a head (and even more so

FIGURE 2.4 This 1924 cartoon, entitled "A human-repair workshop," appeared in the popular satirical magazine *Crocodile*.
A caption on the bottom reads:
"**The Man Operated-Upon** (shaking the ashes of his cigarette into the sawed-off top part of his scull): I'm here for the third time, and you still didn't fix my brain. I need [it] to run my department.
The Operator: Well, you can do that even without a brain."
Another caption placed in the top right corner and printed in a much smaller font-type with the reference—"from newspapers"—says: "At a meeting of surgeons, Dr. [Iustin] Dzhanelidze has demonstrated a patient who eleven years ago had undergone an operation of stitching his heart and two wounds in his lungs." Although this caption provides a justification for the "medical" subject of the cartoon, as well as some actual information on the possibility of repairing various human organs (e.g., hearts and intestines hanging on the clothesline in the picture), the thrust of the cartoon is not the medical feat in itself, but a satirical comment on the bureaucratization of Soviet life. The second caption merely makes sure that its prospective readership will take the "medical" message as a given.
From Krokodil, *1924, 7: 13.*

a human head) represented the brain: the "organ" of cognition, memory, thinking, and, ultimately, personality–"the seat of the soul."[34] In his famous canon "I think therefore I am," the originator of the concept of reflex, Rene Descartes, even equated thinking with human existence itself. And Briukhonenko "revived" not just the head of a fish, frog, or even rabbit—the usual subjects of physiological experiments—but the head of a dog. This had a profound impact, especially on the Russian public, since dogs customarily served as a substitute for humans in Russian physiological research and theorizing. Ivan Pavlov's famous studies of conditional reflexes, which laid the foundation for his theory of "higher nervous activity," widely publicized in Soviet Russia as the materialist explanation of human psyche and its basic characteristics (thinking, cognition, memory, etc.), were all conducted on dogs. Of course, in Briukhonenko's experiments, there was no way to ascertain that the "revived" head still possessed the higher nervous activity (he did not even record the electrical activity—EEG—of the brain),[35] but the head certainly exhibited simple reflexes, and this was good enough for the journalists. "The experiments of Drs. Briukhonenko and Chechulin demonstrated for the first time the possibility of the survival and preservation of functions of the nervous system under artificial conditions," raved one reporter, proclaiming boldly: "These experiments force us to reconsider the established views on the essence of the process of life and death."[36]

Yet no matter how sensationalist and/or erroneous it was, the publicity brought to Briukhonenko tangible benefits in almost exactly the same way as it did to Professor Persikov in Bulgakov's "Fateful Eggs." Amidst the media hype, on September 12, some three months after Briukhonenko's presentations at the physiology congress, *Evening Moscow* reported that the highest agency in charge of medical research—Narkomzdrav—had provided a large sum (30,000 rubles) for the expansion of Briukhonenko's work.[37] The next day, *Izvestiia* brought this piece of news to all corners of the country.[38] The announcement in turn generated a new wave of articles in the popular press, celebrating "the brilliant discovery" and "the great achievement of Soviet science."[39]

Moreover, reports on Briukhonenko's research found their way into the Western media. In early October, Briukhonenko submitted three articles (two of them in collaboration with Chechulin) to an influential French periodical, *Journal de Physiologie et de Pathologie Générale*.[40] On October 13, 1928, two days after the articles had arrived at the journal's editorial office, one of them—accompanied by the photographs of a dog's head attached to the autojector—appeared on the pages of *Monde*, a pro-Soviet weekly edited by Henri Barbusse, under the headline "Sensational experiment."[41] In January 1929, Briukhonenko's article entitled "Alive bodiless dog head," appeared in a popular German magazine, *Die Umschau*.[42] Two weeks later, the French journal *La Presse Médicale* published an eyewitness account of Briukhonenko's experiments written by J. Fernan-Perez, a Spanish physician who had visited the Soviet Union the previous year.[43] The February 1929 issue of *Science and Invention*, a popular-science magazine published by the "godfather" of US science fiction, Hugo Gernsback, carried an account of Briukhonenko's "living head."[44] By September, the news

FIGURE 2.5 The cover page of the May 1928 issue of the popular-science monthly *Sparks of Science*, featuring a very faithful drawing made from an original photograph that accompanied several of Sergei Briukhonenko's articles published in specialized periodicals. The only element introduced by the artist in the drawing (and absent on the original photo) is the lamp that illuminates the severed head and the autojector and that was perhaps meant to signify the light science sheds on the mysteries of life and death. *From* Iskry nauki, *May 1928, 5.*

had reached South America: *El Día Medico*, a popular medical magazine in Buenos Aires, published Briukhonenko's article on the "Continuation of the life of the central nervous system under artificial conditions" supplemented by an extensive editorial commentary.[45]

Publications on Briukhonenko's experiments in the Western media did not go unnoticed. In December 1928, Briukhonenko received an indignant letter from Konrad Mass, a German writer and the mayor of the small town of Schwerin, who condemned his research as "abominable" and "diabolic."[46] In mid-March 1929 a German daily, *Berliner Tageblatt*, and a US daily, *The New York Times*, simultaneously published a letter written by the prominent Irish dramatist and ardent anti-vivisectionist George Bernard Shaw.[47] The letter responded to an anonymous correspondent, who had asked Shaw's opinion about Briukhonenko's experiments. Shaw found these experiments "frightfully interesting."[48] Unknowingly, Shaw envisaged a situation not dissimilar to the one depicted in Beliaev's novella:

> The experiment should be tried on a scientist whose life is endangered by an incurable organic disease, say cancer of the stomach, whereby humanity is threatened with the loss of services of his brain. What is easier than to save such genius from the death bed by cutting off the head, thereby freeing the brain from disease, and keeping up artificial circulation in the arteries and veins so that the great man may continue to lecture and advise us without being impeded by body infirmities.

With his characteristic wit, Shaw dwelled on the wondrous possibilities Briukhonenko's work seemed to offer: "A university in which all chairs were occupied by a row of finest brains in the country with nothing but pumps attached to them—briefly, where the whole system of teaching was purely cerebral—would be [an] enormous improvement on the present state of things." Shaw was even "ready" to try the procedure on himself: "I am greatly tempted to have my head cut off so that I may continue to dictate plays and books independently of any illness, without having to dress and undress or eat or do anything at all but to produce masterpieces of dramatic art and literature" (see Figure 2.6).

FIGURE 2.6 This drawing illustrated a Russian translation of George Bernard Shaw's letter to the *New York Times* and *Berliner Tageblatt*, published by *Evening Moscow* under the rubric "A literary panopticum." The headline reads: "About the Head of Bernard Shaw." The caption on the lectern announces: "Today is a lecture by Bernard Shaw." *From* Vecherniaia Moskva, *May 25, 1929*.

If the writer Shaw found Briukhonenko's experiments "frightful" and somewhat humorous, the British scientist John Desmond Bernal found them inspiring. Indeed, it seems likely that Briukhonenko's work provided Bernal with a major idea—later incarnated in the popular image of "the brain in a vat"[49]—that Bernal developed in his 1929 futuristic essay, provocatively entitled *The World, the Flesh and the Devil.* In the essay's section on "the flesh," Bernal envisioned the possibility of preserving the human brain—and thus life—even after the entire body had failed. The essay introduced a situation similar to the one imagined by both Beliaev and Shaw, but the conclusions Bernal drew from it differed considerably:

> Sooner or later some eminent physiologist will have his neck broken in a super-civilized accident or find his body cells worn beyond capacity for repair. He will then be forced to decide whether to abandon his body or his life. After all it is brain that counts, and to have a brain suffused by fresh and correctly prescribed blood is to be alive—to think. The experiment is not impossible; it has already been performed on a dog and that is three-quarters of the way towards achieving it with a human subject.[50]

Bernal likely learned of Briukhonenko's experiments from his Cambridge colleague, prominent experimental biologist J. B. S. Haldane.[51] Haldane had visited the USSR in the spring of 1928 and had attended the Congress of Physiologists in Moscow. He had personally observed Briukhonenko's demonstration of the autojector and had been impressed by what he had seen.[52] Clearly Bernal was impressed, too. He expanded on the obvious implications of Briukhonenko's experiments, particularly on the possibility of replacing the body with intricate machinery that could accomplish all of the functions—but would not be impeded by the frailties—of the flesh. He imagined not just the head, but the brain removed from its body (with all its sensory organs—eyes, ears, taste buds, and so on—also replaced by mechanical parts delivering electric signals directly to the brain). He foresaw the brain supported by "an elaborate, automatic contrivance"—"an artificial heart-lung-digestive system"—that would supply it "with fresh oxygenated blood and drained of de-oxygenated blood." He prophesied that such "mechanized humanity" would be a first step toward human immortality!

Art and Life

Some students of Beliaev's writings have claimed that his literary experiment somehow precipitated or even prompted Briukhonenko's scientific one.[53] On the other hand, one of Briukhonenko's biographers has claimed that the autojector directly inspired Beliaev's story.[54] The nearly simultaneous appearance of the first version of the "Head of Professor Dowell" (in June 1925) and the first public demonstration of the autojector (in September 1925) clearly undermines both of these claims. Briukhonenko had been working on his apparatus for at least two years prior to Beliaev's literary debut, but he had made no public reports on his work until after Beliaev's publication of his novella. The scientific and literary explorations of the idea that a severed head could be "revived" and kept alive by special machinery in a laboratory obviously emerged and evolved in the minds of their respective progenitors quite independently.

Furthermore, neither the scientist nor the writer could claim priority in originating this idea: severed, but "resurrected by scientific means," heads had figured prominently both in science and in fiction for quite some time. Briukhonenko's autojector continued a long line of similar apparatuses constructed for the perfusion of isolated organs and artificial (extra-corporeal, as it would be called later) blood circulation.[55] Indeed, the idea that a severed head could be kept alive by supplying it with blood dates back to at least the early nineteenth century and is commonly associated with the eminent French physiologist Julien Jean César Legallois (1770–1814).[56] The growth of physiology as a distinct scientific discipline in the course of the nineteenth century was accompanied by the proliferation of special techniques for keeping alive various organs (legs, hearts, lungs, kidneys, etc.) that had been separated from the body of an animal. For instance, in the 1850s, another French physiologist, Charles Brown-Séquard, conducted experiments on an isolated head of a dog, injecting blood into its severed arteries, and reported that the head had exhibited various reflexes.[57] In the last quarter of the nineteenth century, the perfection of perfusion techniques by famous German physiologist Carl Ludwig spurred a virtual explosion of research on isolated organs and fostered the development of special machinery and saline solutions for their perfusion.[58] At the turn of the twentieth century, isolated organs became a major experimental tool in studying the physiological functions of separate organs, as well as a preferred tool for pharmacological research on the effects of various preparations introduced into perfusion solutions.

Not unexpectedly, the issues of the survival and revival of organs and tissues taken from dead bodies—particularly human bodies—was a focus of special attention for scientists all over the world, including Russia. Circa 1900, Aleksei Kuliabko (1866–1930), a researcher at the Physiological Laboratory of the Imperial Academy of Sciences in St. Petersburg, spent several years studying physiology and pharmacology on isolated hearts and intestines of chickens and rabbits.[59] In February 1902, Kuliabko reported that he had successfully revived a rabbit heart forty-four hours after it had stopped beating.[60] In September, after several failures, he succeeded in reviving a human heart more than twenty hours after its owner's death.[61] Five years later, in 1907, he developed a technique of artificial circulation in a severed fish head.[62] Around the same time, Nikolai Kravkov (1865–1924), the chair of the pharmacology department at the Military-Medical Academy in St. Petersburg, began his studies on the effects of various poisons on isolated rabbit hearts and ears.[63] From 1910 to 1913, a young Moscow surgeon, Fedor Andreev (1879–1952), in his research on the effects of electricity on the heart, succeeded in reviving a dog killed by an electric shock. He injected a mixture of adrenalin and saline solution in the arteries of the dead dog and observed the complete restoration of the heartbeat, breathing, and nervous functions.[64]

Briukhonenko's experiments with an isolated head thus continued a long tradition of research conducted by numerous Russian and foreign scientists prior to World War I. Indeed, in 1926, in the first published account of his experiments with the autojector, Briukhonenko referred by name to five researchers who had constructed similar machines, emphasizing the improvements and differences of his own apparatus.[65] Two years later, in the first detailed publication on the autojector, Briukhonenko provided a list of his predecessors, which included no fewer than nineteen names.[66]

Similarly, Beliaev's story continued a long tradition of constructing fictional plots using the revival and transplantation of separate organs by "scientific means." This tradition also goes back to the early nineteenth century and was famously inaugurated (just a few years after Legallois had published his ideas) in Mary Shelley's 1818 classic *Frankenstein, or, The Modern Prometheus*, portraying a doctor who creates a living creature out of separate organs and tissues taken from cadavers. At the end of the nineteenth century, H. G. Wells further developed this tradition in his depiction of an ambitious surgeon transforming animals into humans in *The Island of Dr. Moreau* (1896). Wells's work directly inspired Maurice Renard, a pioneer of French SF, to explore a similar idea—transplanting a human brain into another (animal) body—in his 1908 novel *Le docteur Lerne, sous-Dieu*.[67] All three novels were translated into Russian and were regularly reissued by various publishers both before and after the Revolution.[68] Beliaev was undoubtedly familiar with these classics.

Furthermore, if the aforementioned novels explored just the general ideas of revival and transplantation, in 1908, Carl Grunert, a pioneer of German SF, published a short story entitled "Mr. Vivacious Style," whose plot revolved around a severed human head living in a laboratory.[69] In this story, also set in the United States, a prominent journalist named Vivacious Style is decapitated in a railroad accident.[70] A certain "Dr. Magnus Magician," who happened to be at the scene of the accident, picks up Mr. Style's severed head and brings it to his laboratory. He attaches the head to an apparatus for blood transfusion filled with "sanguinum," a chemical substitute for human blood that Dr. Magician had created, and the head is revived. The good doctor constructs a special machine that blows compressed air through the head's throat and enables it to speak. Mr. Style's head resumes his journalistic crusade against racial segregation in the United States, dictating his works to a phonograph and publishing them in newspapers. In the end, the head is killed by Mr. Style's political enemies. Before World War I, Grunert's stories regularly appeared in Russian translations in popular magazines, as well as in separate editions, and Beliaev might well have read "Mr. Vivacious Style."[71]

Despite numerous parallels between "Professor Dowell's Head" and its literary predecessors—particularly Grunert's story—Beliaev himself denied that he had simply "borrowed" the idea from some other writer.[72] In 1939, recounting the sources for his novella, Beliaev claimed that its plot was inspired by his own personal experiences:

> "Professor Dowell's Head" is to a large degree an autobiographical story. Once, a disease kept me in a "plaster bed" for three and a half years. This disease was accompanied by the paralysis of the lower half of my body. And although I could use my hands, my life during those years was reduced to a life of a "head without a body," I could not feel my body at all. [. . .] It was at this time that I lived through and felt everything, which a head "without a body" could experience.[73]

Beliaev undeniably used his personal experiences in the literary ornamentation of the story, for instance, in vividly describing Dowell's terrible feeling of helplessness when a large beetle flew into the lab and crawled onto his head. But it seems unlikely that the story's major "scientific-fantastic" idea came from these experiences.

Indeed, an examination of the story's first publications points to other sources. The publication in *The Worker's Gazette* was prefaced by the following statement: "Scientists have been able to revive a heart taken from a fresh corpse. Other parts of animal and human bodies have also been revived. It was these experiments that lay the foundation for the printed [below] scientific-fantastic story."[74] The editorial afterword was even more specific: "Beliaev's story 'Professor Dowell's Head' is of course a fantasy and a fiction; surely nobody has been able to revive a human head. The story, however, has a solid scientific basis. It is founded on a whole series of experiments by Russian and foreign scientists—physiologists and surgeons."[75] The editorial referred to the research on tissue cultures by Franco-American surgeon Alexis Carrel, who in the 1910s had captured the world's attention with his experiments on growing the cells of an organism outside the body. Carrel claimed that the tissue cultures of chicken heart he had cultivated in his laboratory were potentially "immortal."[76] The editorial also mentioned experiments with isolated rabbit ears and even wombs, as well as human fingers, by Russian pharmacologist Nikolai Kravkov. In addition, it quoted the experiences of Vladimir Oppel', a prominent Russian surgeon, with the revival of a human heart that had stopped during a surgical operation: "What was before us was [an instance of] true resurrection. The man had died—the heart had stopped, the breathing had seized, the pupils had dilated. A half an hour later, the man was breathing and an hour later he was speaking."

An editorial foreword to "Professor Dowell's Head" in *World Tracker* was even more detailed in its references to scientific research that served as the underpinnings of Beliaev's story. In addition to recounting Carrel's and Kravkov's experiments, it also informed its readers that the Russian physiologist Aleksei Kuliabko had succeeded in reviving a severed head—"albeit a fish head, so far, but a head, nonetheless."[77] The editorial emphasized that "the idea [of reviving a human head] is already floating in the air." "So far, it is only a scientific fantasy [*nauchnaia fantastika*]," and "a brave jump into 'tomorrow,'"—the editors admitted,—"yet this scientific fantasy stands on the lines of scientific discoveries of today; it is simply a *logical* end of that path to which the experiments conducted in this field *lead*."[78]

Furthermore, in 1928, in an essay expressively titled "Fantasy and Science" and published in a volume of his collected stories, Beliaev himself emphasized that the entire genre of scientific fantasy is always grounded in contemporary scientific research and insisted that to balance science and fantasy is the most important task of any writer of scientific fantasy.[79] What is more, the volume was appended with a list of "recommended popular-science writings" relevant to the plots of the stories compiled in the volume, which Beliaev himself had apparently used in constructing those plots. The same year, in the afterword to his new novel *The Amphibian Man*, Beliaev explicitly referred to certain scientific research that led him to the idea of an "amphibious man" created by the implantation of shark gills into a human child.[80]

It is conceivable that the nearly simultaneous appearance of Beliaev's story and Briukhonenko's experiments in 1925 was simply a coincidence. Each, in his own field, was developing further ideas that had been explored by numerous predecessors prior to World War I. Furthermore, just a few months after the publication of Beliaev's story and Briukhonenko's demonstration of his apparatus, yet another story (allegedly translated from French and entitled "Professor's secret") about a severed head living in a laboratory appeared on the pages of *Red Panorama*, a popular illustrated weekly that only a few months earlier had published Bulgakov's "The Ray of Life."[81] One might suggest thus that the timing of Beliaev's and Briukhonenko's works was more than coincidental. Both literary and physiological experimentations with severed heads stemmed from a particular cultural milieu of mid-1920s Russia, which was at one and the same time permeated by omnipresent death and by high hopes for the future, instigated by the decade of death and the subsequent "resurrection" of the country.

In no other place in the country was this revival more rapid, more noticeable, or more exciting than in the capital of the new, Soviet Russia—Moscow.[82] Mikhail Bulgakov, who came to Moscow in September 1921, captured this resurrection in a ten-page essay, published in April 1923.[83] Set as a series of diary entries, the essay proffered snapshots of Moscow life, chronicling the dramatic change the city and its inhabitants had undergone in less than eighteen months: from desolate, dark, cold, desperate, empty, dead, hungry, frightened, and frightening "naked times" of the fall of 1921 to a glittering, illuminated, crowded, bursting with life, plentiful, and cheerful "Mother-Moscow," which "even at night sleeps without turning off her blazing eyes," in the spring of 1923.

Yet although in the dazzling carnival of NEP it seemed that the decade of death was now all but forgotten, it left an indelible impression on Russia's biomedical community, making "issues of life and death" a subject of intense research. This research filtered rapidly from scientific circles to the popular scene, filling the pages of numerous daily newspapers and weekly magazines and becoming the subject of popular lectures and exhibits. For instance, in 1922, the scientific community attentively followed the experiments of Russia's foremost pharmacologist Nikolai Kravkov on the revival of animal and human organs, published nearly simultaneously in Russian and German specialized periodicals.[84] Two years later, newspapers and magazines widely publicized Kravkov's latest research on isolated organs, presented during a lecture tour in Moscow in March 1924.[85] The reporters were particularly fascinated by the fact that long dead and mummified human fingers could be "brought back to life" and even grow nails. Perhaps the recent (on January 21, 1924) death of Vladimir Lenin, the leader of the Bolsheviks, which had sparked a country-wide campaign for his commemoration and "immortalization," amplified public fascination with Kravkov's work.[86] The uproar over these experiments in the public media was further fueled by Kravkov's own sudden death in mid-April, just a few weeks after his visit to Moscow.[87]

In the spring of 1925, at the time when both Beliaev and Briukhonenko were working on their respective experiments, a special event devoted to "The problem of life and death" was held at the Moscow Polytechnic Museum—a leading venue for public education and the popularization of science in the city. In its auditorium that seated nearly two thousand people, a prominent Leningrad ichthyologist and popular-science writer, Petr Shmidt, and a pioneer of "resurrection" techniques, surgeon Fedor Andreev, delivered lectures on the essence of life and death from the viewpoint of modern science, providing examples of "apparent death," "suspended life," "revival," and "resurrection of isolated organs taken from cadavers." Reports on this event appeared on the pages of Russia's major newspapers with country-wide circulation.[88] The editors of *World Tracker* were, thus, not exaggerating in their foreword to "Professor Dowell's Head": in the mid-1920s, the idea of reviving a severed head was indeed "floating in the air."[89]

The authors of the first experiments with severed heads, Beliaev and Briukhonenko, breathed deeply of that air. Both had lived through the "death and revival" of their country and its capital, Moscow. Furthermore, each had personally undergone a similar process of "death and revival," and the decade of death had left a strong personal imprint on both of them. In 1915, Beliaev became very sick with tuberculosis of the spine, and in search of relief he moved to Yalta, a popular TB resort in the Crimea. He spent more than three years in bed, paralyzed, on the brink of death, struggling with the disease. His wife left him, and his mother, who had accompanied him to the Crimea, died during the civil war years. Miraculously, Beliaev recovered, but the disease stayed with him for the rest of his life. He came to Moscow in late 1922 and began working as a jurist in various government agencies. After the success of his first literary experiment, Beliaev abandoned his judicial work and became a freelance journalist and SF writer.

Briukhonenko's encounters with death were as personal as Beliaev's and no less powerful. Almost immediately after graduating from Moscow University's Medical School in the spring of 1914, Briukhonenko was drafted into the army and served as a doctor with an infantry regiment on the front lines, experiencing firsthand the horrendous killing machine of World War I. After the Russian army collapsed following the Bolshevik Revolution, he returned to Moscow and spent the civil war years working in a large hospital engaged in fighting a horrific epidemic of typhus, which took his wife and infant daughter.[90] According to his unpublished memoirs, it was during this hospital work that he first came to the idea of using a severed head in investigating the high fevers that accompany typhus.[91] It also seems quite likely that the hospital's director—one of the pioneers in the development of "resurrection" techniques, Fedor Andreev—actively encouraged Briukhonenko's interest in severed heads. In 1922, Briukhonenko left medical practice for good and became a full-time researcher at the recently created Chemical-Pharmaceutical Institute. For both Beliaev and Briukhonenko, early experiments with severed heads launched new careers, as a writer and as a researcher, respectively.

Experiments and Values

Commentators have often reduced analysis of the interrelations between science and fiction to two simple formulae: "life imitates art" or "art imitates life."[92] Yet as we saw, the relation between "art" and "life" is much more complex.[93] The appearance of Beliaev's fictional experiment a few months ahead of the scientific ones may well have prepared the Soviet public for the favorable reception of Briukhonenko's real-life studies: as the journalist Grebnev noted in May 1927, Beliaev's "fairy-tale" that "could only entice a smile" became "a reality" in Briukhonenko's experiments.[94] But, as we have seen, Beliaev's fiction did not imitate Briukhonenko's science, nor did Briukhonenko's science emulate Beliaev's fiction. Rather, both of them used the idea that a severed head could be kept alive to explore issues of life, death, and revival.

A major source of inspiration for both the fictional and the scientific experiments lay in research on isolated organs, tissue cultures, and transplantations that had been pursued by numerous Russian and foreign scientists for several decades. What made this research a focus of attention on the part of both the scientific and the literary communities in the 1920s, however, was a general atmosphere of revolutionary fervor and wide-scale experimentation, which enveloped Russian society in the aftermath of the Revolution and the decade of death. In this atmosphere, issues of life, death, and revival captivated the imagination of both scientists and the general public. A revolution-inspired belief—that anything was now possible—encouraged literary and physiological explorations of previously unthinkable possibilities that research on isolated organs and transplantations seemed to be opening up. But of course, the goals of Briukhonenko's and Beliaev's explorations were as different as were their respective vocations.

Briukhonenko's first experiments were aimed at examining the possibility of keeping a severed animal head alive in order to investigate certain physiological mechanisms: initially, the regulation of body temperature. His 1925 report indicating that this was indeed technically possible was barely noticed by the scientific community and completely ignored by the media and, hence, the public at large. At the next stage, after he had perfected his technique (including the surgical operation of severing the head from the body, the construction of the autojector, and blood stabilization and oxygenation), he began to explore numerous new possibilities that this technique opened up: investigations of the brain's reactions to various preparations introduced into perfused blood and the effects of "death and revival" on the brain. It was exactly these new possibilities—most importantly, the performance of surgical operations on the stopped heart by temporarily replacing its functions with the autojector and the control of blood clotting during blood transfusions—that so excited the scientific community and ignited wide scholarly interest in his work. The ultimate purpose of Briukhonenko's experiments was to investigate life, death, and revival as *physiological processes* with an eye on applying whatever knowledge he could gain from these experiments to medical practice. He certainly succeeded in reaching his goals: his research paved the way for creation of the first Soviet technologies for the industrial production of several anticoagulants, including heparin, for development of new techniques of open-heart surgery, and, eventually, for construction of a heart-lung machine in Russia.[95]

Beliaev's literary experiment assumed that keeping a severed human head alive was already an accomplished feat of science, and the writer set out to investigate what effects this feat might have on human life, human relations, and human society writ large. The ultimate purpose of Beliaev's experiment was to explore life, death, and revival as *parts of human existence*. Through the eyes of a head without a body (and in the subsequent, 1938 version of the story, also a head transplanted onto another body), Beliaev examined the role of science and the scientist in "modifying" this inevitable part of human existence—death—while questioning the value of such "modifications." It is quite telling that, unlike in Grunert's story, the head belongs to the scientist who had invented the "resurrection" technique. Characteristically, in his ironic commentary on Briukhonenko's "frightfully interesting" experiments, George Bernard Shaw suggested exactly that: "I would of course, expect one or two vivisectionists to submit themselves to the experiment to prove to my satisfaction that it is practicable and not dangerous, but I assume that would not mean any serious difficulty."[96] Professor Dowell gets a taste of his own medicine, as it were, and his head is not at all happy with the application of his science, lamenting to Miss Adams: "I'm deprived of everything available to live people. And only mind (*soznanie*), as damnation, is left to me. [...] When I lost my body, I lost the world." Nor did other revived heads in Beliaev's story enjoy their "resurrection." Miss Adams puts it bluntly: "I'd prefer death to such resurrection."

One might speculate that Russian scientists' pointed explorations of the issues of life and death and their enthusiastic pursuit of various scientific means of "revival," "rejuvenation," and "resurrection," which were extensively covered and sensationalized by the popular media in the early 1920s, prodded Beliaev (much as they did Bulgakov) into questioning the value—particularly, the ethical value—of these pursuits. Beliaev's story could be seen as one of the earliest attempts to deal with that question, and the answer it provides is certainly in the negative: Dowell's head dies and his disciple Kern is arrested as a criminal (in the later, 1938 version of the story, Kern shoots himself).

Beliaev was not the only one troubled by this question. The first newspaper reports on Briukhonenko's actual experiments in November 1926 provoked a lengthy comment from Ivanov-Razumnik (Razumnik Vasil'evich Ivanov, 1878–1946), an eminent Russian anthroposophist and writer.[97] In a letter to a fellow anthroposophist and writer, Andrei Belyi (Boris Nikolaevich Bugaev, 1880–1934), Ivanov-Razumnik described these reports as "terrifying":

> Today [it's] a dog, tomorrow—[they] will separate the head of a man sentenced to the "highest measure of punishment" (a vile phariseism in the very turn of the phrase) and will "attach it" to the apparatus. The day after tomorrow—instead of the apparatus, there will be a body of another man, also sentenced to the "highest measure" and beheaded: the head of one [transplanted] onto the body of the other. One more step and the head of Beethoven will be attached to the body of a snake. [. . .] Even though "tomorrow" is a century away, and the "step" is a million miles long, the matter is in the principle and in the *first step*.[98]

The main issues that deeply troubled Ivanov-Razumnik had to do with what should be the limits of scientific experimentation, who should set the limits, and what should be allowed. He complained about the "ethical deafness" on the part of "people of science," who do not want even to think about the ethical implications of their work. On the other hand, he considered the Bolshevik government utterly immoral and incapable of providing science and scientists with any ethical guidance. Indeed, he obviously feared that Briukhonenko's research could give the new rulers yet even more arbitrary power over the life and death of their subjects. He admitted that, in theory, Briukhonenko's experiments could pave the way to saving human lives. Presaging the subsequent media portrayal of these experiments, he even envisaged a situation in which the head could temporarily be separated from the body in order to perform a complicated surgical procedure that would have been impossible on an intact organism. He conceded that in time this could become merely a "surgical operation, though unusual for now, but so was the case for tracheotomy and trepanation in their times."[99] What for Ivanov-Razumnik remained an open issue was whether the promise of saving human lives could be used to justify *any* scientific experimentation. In the end, he subscribed to the utilitarian view of ethics with a particular anthroposophist twist: "The goal justifies the means—is an awful formula, but only if the means are other *human beings*."[100]

Of course, despite their apprehension, Russian writers (Beliaev, Bulgakov, or Ivanov-Razumnik) could not provide definitive answers to the ethical issues inherent in the rapid development of experimental biology and experimental medicine in the 1920s. Nor were the issues they addressed an exclusively Russian/Soviet concern. As witnessed, for example, by the well-known polemical writings of J. B. S. Haldane in *Daedalus* (1923) and Bertrand Russell in *Icarus* (1924), which in turn inspired such famous scientific and literary expositions as J. D. Bernal's *The World, the Flesh and the Devil* (1929) and Aldous Huxley's *Brave New World* (1932), scientists, philosophers, and writers in Britain were haunted by similar concerns.[101]

In 1926, a leader of British experimental biologists, Julian Huxley (older brother of Aldous Huxley), conducted his own, far less known, literary experiment in a story evocatively entitled "The tissue-culture king." The story tells of a British scientist captured by natives in a remote area of Africa. Upon arriving at the tribe's "capital," the scientist meets another captive, a certain Dr. Hascombe. In exchange for his life, Hascombe had promised to make the king of his captors "immortal." The doctor had established a laboratory for his experiments with the king's tissue cultures and filled it with "monsters" and "circus freaks": two-headed toads, dwarfs, giants, and mustached girls. Huxley ended his story with the following sentiment:

> The question I want to raise is this: Dr. Hascombe attained to an unsurpassed power in a number of applications of science—but *to what end did all this power serve*? It is the merest cant and twaddle to go on asserting, as most of our press and people continue to do, that increase of scientific knowledge and power in itself be good. I commend to the great public the obvious moral of my story and ask them to think what they propose to do with the power which is gradually being accumulated for them by the labors of those who labor because they like power, or because they want to find the truth about how things work.[102]

As their own portrayals of the powers unleashed by experimental biology and medicine demonstrate, both Beliaev and Bulgakov conveyed to their readers analogous "obvious morals" and asked them to think of similar questions. Yet, unlike in Britain, where such eminent experimentalists as Julian Huxley, J. B. S. Haldane, and J. D. Bernal directly contributed to 1920s public debates over the ethical concerns spurred by the rapid advances of their science, such a debate never surfaced in the Soviet Union.[103]

The ethical implications of both Beliaev's and Bulgakov's stories seem to have been lost on their readers. The contemporary reviewers of "Fateful Eggs," enraged and blinded by the "satirical portrayal of the communist experiment," did not seem to notice the novella's "bioethical" dimension.[104] One review of "Professor Dowell's Head," which appeared in 1926 in the Bolshevik magazine *The Press and Revolution*, completely ignored the ethical issues embedded in the story. Although noting that "an artistic treatment of the head-spinning possibilities," including the "resurrection of the dead," which modern science opened up, "must have certain meaning," the reviewer did not find any such meaning in Beliaev's story, dismissing it as a piece of mildly amusing entertainment.[105]

Neither did Briukhonenko's actual research spark a public debate. Ivanov-Razumnik confided his concerns in a private letter and never voiced them publicly. The Soviet scientific community, as well as Soviet society at large, seemed totally unperturbed. None of the several dozens of articles devoted to Briukhonenko's experiments, which appeared in the popular and academic press from 1926 through 1929, even mentioned their possible ethical implications. The Soviet public became enthralled rather than dismayed by the severed heads of dogs supported by intricate apparatuses, which gazed at them from the

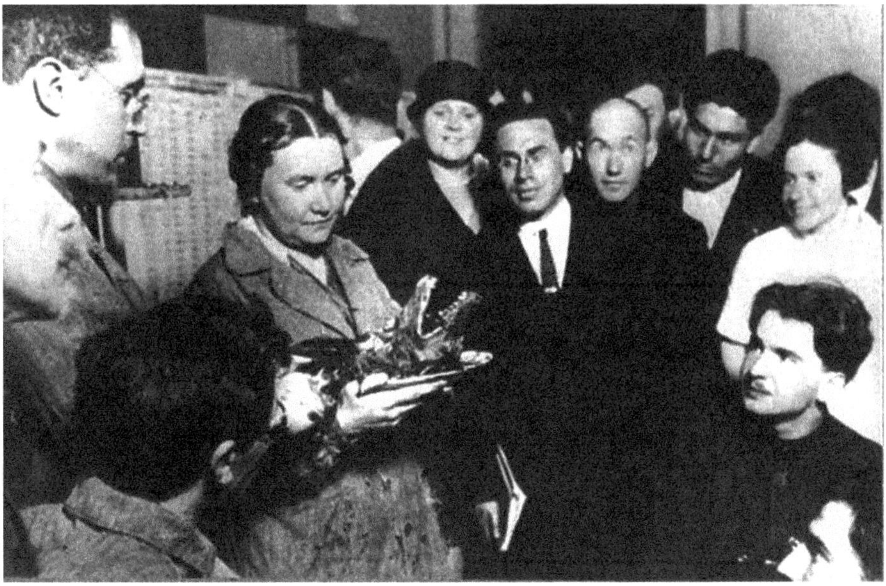

FIGURE 2.7 The public and the severed head, ca. 1928. This photograph of a public presentation of Briukhonenko's experiments captures the public members' amusement and excitement at viewing the severed head of a dog. Briukhonenko himself appears in this photo in the upper left corner. Compare this to Figure 2.2, which conveys Miss Adams's shock and fear at seeing the severed head of Professor Dowell. *Courtesy of ARAN.*

covers of popular magazines. The contemporary photographs of the public presentations of Briukhonenko's experiments captured the onlookers' amusement and excitement, not their disgust, revulsion, or fear (see Figure 2.7).

One might suggest that the decade of death desensitized the Soviet public to the ethical concerns of biomedical experimentation. In the public's eye, "science that conquers death"—as Briukhonenko's experiments were labeled in the popular press—was unquestionably moral.[106] To oppose such science on some ethical grounds seemed unthinkable. Indeed, Soviet commentators rebuffed attempts to criticize research with severed heads on "moral grounds." In the spring of 1929, for instance, *Evening Moscow* carried a small feuilleton entitled "Dogs and Humans."[107] The feuilleton cited a letter received by Narkompros from the president of the Frankfurt Society for Animal Protection, a certain Mr. von Artus, who expressed his outrage with the inhumane treatment of dogs in Briukhonenko's research: "The experiments of Dr. Briukhonenko are the work of devil, and those who conduct such experiments are the servants of hell." The author of the feuilleton sarcastically suggested that since "Mr. von Artus is so worried about the fate of the beheaded dog, he should offer his own head for the experiments on brain tissue, if, of course, he truly believes that this useful tissue is indeed present under his thick skull."[108]

The reactions to both actual and fictional experiments with severed heads seem to indicate that many among scientists, state officials, journalists, and the public in 1920s

Russia firmly believed in the ultimate "goodness" of science, which Huxley mocked in his exposé. They apparently subscribed to the view that science itself was value-free. It was a particular society that gave science and its applications a moral value and passed a value judgment on concrete scientific endeavors: in an immoral society, science would be immoral, while in a moral society, science would "automatically" become moral. The notion of value-free science explains why the Bolsheviks had no qualms about importing on a large scale "capitalist" science and, particularly, technology to their own "socialist" land, and it fits well with Lenin's remark: "To absorb everything valuable from European and American science is our first and foremost task."[109]

It seems likely that Beliaev quite deliberately set his story not in the Soviet Union, but in the United States—a country that, in the Soviet views of the time, represented the pinnacle of an "immoral" capitalist system, and where scientific and technological advances served to benefit not "the people," but a few capitalists in their pursuit of ever larger profits. The ultimate failure of Professor Dowell's science—symbolized by the death of his head and the arrest/suicide of his pupil Kern—was the logical ending for a story of science in an "immoral society."

Indeed, similar stories about the "failure" of science and scientists in capitalist settings were a staple of 1920s Soviet science fiction writings. For example, in 1924, another aspiring author, Andrei Irkutov, used this technique in a short story that appeared in *Struggle of the Worlds*, a popular adventure magazine, under the enticing title, "Immortality."[110] In this story a certain French professor, with the telling name Lois Mortel, develops a "special concoction" that makes people immortal, after thirty-year-long research and experiments on animals. At a highly publicized presentation in Paris, the professor announces that he is ready "to give his invention to all of cultured and civilized humanity," which "does not include savages and, of course, the Bolsheviks-communists." He publicly administers his concoction to a military officer, aristocrat, and nephew of the current president (who also carries a very telling name) viscount d'Hormone. Amidst a stormy ovation, Mortel tells the first immortal:

> You are guaranteed from any wearing down of your organs, from any changes induced by aging or disease. No poison is dangerous to you, no cold could affect your health. You can live as you like, without restricting yourself in anything. But remember, viscount, any violent, externally inflicted damage could end your life. You can die under the wheels of a streetcar or an automobile, from strangulation and a blow by a club, from a bullet or a knife. If medicine could not provide necessary help, all these things could kill you. Take care of yourself, viscount.

The "Committee of Immortality" is set up to decide how to use Mortel's invention. Rumors fly that only the rich would obtain immortality, but "workers would not." These rumors, however, have no impact on the workers, who appear totally uninterested in being immortal. As one of them puts it: "Do you think I want to paint fences forever?"

The Committee of Immortality, meanwhile, decides that injections of Mortel's concoction would only be given to those who volunteer for "a sacred crusade against the Soviet republic." The news reaches "Komintern" in Moscow that an "army of immortals" has been formed, but the news produces a very simple response from a man in charge (easily recognizable as Leon Trotsky): "An army of immortals. Excellent. We will take them with our bare hands." Indeed, frightened by the prospect of losing their precious "eternal lives," the immortal soldiers surrender at the first sight of armed opposition. Moreover, armed workers with no trouble take over power in France, since immortal police and gendarmerie also surrender at the first sign of violence. Thus, although intended by its bourgeois owners for "immoral" use against the first proletarian state, Mortel's scientific discovery of immortality concoction actually facilitates a workers' revolution. Tellingly, Irkutov's story did not mention any possible uses of Mortel's concoction *after* the victorious proletarian revolution.

If in Beliaev's fiction, set in the land of capitalism, Professor Dowell's head failed both literally and figuratively, in Soviet reality the scientific community, the media, the

FIGURE 2.8 The Commissar of Enlightenment Anatolii Lunacharskii (left) and the pioneer of the technique of studying severed heads Professor Aleksei Kuliabko (right) with the severed head of a dog attached to Briukhonenko's apparatus, ca. 1928. This carefully staged and often reproduced photograph is an epitome of both the scientific community's (represented by Kuliabko) and its patrons' (represented by Lunacharskii) approval of, and support for, experiments with severed heads conducted by Sergei Briukhonenko (left upper corner) and Sergei Chechulin (right upper corner). It could also be viewed as an illustration to Ivanov-Razumnik's notion of scientists' "ethical deafness." *Courtesy of ARAN.*

public, and the state patrons wholeheartedly embraced experiments with severed heads.[111] Briukhonenko's work was lauded on the pages of numerous professional and popular publications, while several state agencies provided funds for its continuation and expansion. Perhaps the best illustration of this unequivocal support came in the form of a photograph that accompanied several popular articles on Briukhonenko's research. In this staged photo, in the center, we see a laboratory table with the severed head of a dog placed on a large dinner plate and with the autojector in the background. Two distinguished gentlemen stare attentively and somberly at the head: on the right—the pioneer of research on severed heads Aleksei Kuliabko; on the left—the official patron of all "pure" science in Soviet Russia, the head of Narkompros Anatolii Lunacharskii (see Figure 2.8).

This photo embodied a popular sentiment that gained particular currency in Soviet Russia: in a moral society—and the first socialist state was by definition supposed to be moral, for its stated goal was to benefit the previously oppressed majority of the population—neither science nor scientists could be immoral, for they worked for the benefit of that society and that state. "Professor Dowell's Head" supported this assumption by picturing the failure of science in the "immoral" West, and launched Beliaev's literary career: during the next five years he published half a dozen novels and innumerable short stories, practically all of them unfolding somewhere in the "capitalist West." In contrast, Bulgakov's attempts to question and parody this assumption in "Fateful Eggs" and then "The Dog's Heart," set in the land of the victorious "proletarian" revolution, ruined his career as a writer. The secret police confiscated the manuscript of his second novella and Bulgakov never saw it published, while after its triumphal appearance in 1925, the first novella was kept out of print for more than half a century. From 1926 onward Bulgakov was unable to publish a single story and had to reinvent himself as a playwright.

As fate would have it, Briukhonenko's and Bulgakov's paths crossed a few years later. In fact, in 1933, the scientist Briukhonenko came to see the writer Bulgakov with a proposition that the latter write a play or a movie script based on his research. Briukhonenko even arranged for Bulgakov's visit to his laboratory and explained in detail his latest investigations, even though he could not demonstrate a severed dog head because of "the lack of a suitable animal."[112] In the end, Bulgakov declined the offer. But this encounter certainly left a visible footprint in Bulgakov's writings: the famous scene with the revival of Berlioz's severed head at Woland's ball in *Master and Margarita* was likely influenced by Briukhonenko's experiments.

'Tis not the whole of life to live,
Nor all of death to die.
JAMES MONTGOMERY, "The Issues of Life and Death" (1819)

No, the whole of me won't die –
In the sacred lyre, [my] soul
Will outlive my flesh
And will escape decay.
ALEXANDER PUSHKIN, "Exegi Monumentum" (1836)

3 "Neither Life, Nor Death": Anabiosis

LESS THAN A year after his debut with "Professor Dowell's Head," in May 1926, Alexander Beliaev published in *World Tracker* a new novella, entitled "Neither Life, Nor Death."[1] This time set in Britain, the novella opens with a conversation between Mr. Carlson, a "projecteur," as he defines himself, and Mr. Gilbert, a rich owner of coal mines. Gilbert is in deep trouble: "Exports are falling. Soviet oil is displacing competitors on Asian and even European markets. Banks don't extend credit. The government is no longer willing to subsidize coal mining. The workers make impossible demands and threaten to flood the mines." He simply does not know what to do. Carlson offers him a solution: "In a time of crisis, we will freeze the unemployed and put them in special storages. When the crisis is over and the demand for workers is on the rise again, we will warm them up, and . . . welcome back to the mines."

Gilbert is incredulous, considering Carlson's proposal to be "either craziness, or stupidity." But Carlson insists that his idea is nothing of the sort: it is based on "the greatest discovery that could make a man a millionaire." In fact, according to Carlson, it did make one man—the German fish-merchant Steinhaus—the richest man in Germany. Steinhaus used this very method—freezing caught fish, transporting it to a desired location, and then reviving it—to provide his customers with absolutely fresh, live fish. He made "tens, hundreds of millions." "But, that's only fish," says Gilbert after a pause, "and you are suggesting something totally impossible: to freeze people!" Yet Carlson insists that his idea is quite practicable and could bring Gilbert millions, for it is based on the discovery that living organisms can be frozen, kept for long periods of time at a low temperature in a state of "neither life, nor death," and then thawed and revived. This discovery, Carlson explains, had been made by a Russian scientist who froze and revived not just fish, but also such warm-blooded animals as marmots, hedgehogs, and bats. Alas, after the scientist's death his work was soon forgotten. According to Carlson, the name of this scientist was Bakhmet'ev.

Unlike in his previous story about the severed head of a fictitious "professor Dowell," the name of the "forgotten discoverer" in Beliaev's new novella was not invented. Bakhmet'ev *did* exist and indeed studied the phenomenon of "neither life, nor death"—anabiosis.[2] And, as Beliaev's novella demonstrated, at least in Soviet Russia, Bakhmet'ev's work was not forgotten at all.

The Science of Anabiosis

Porfirii Bakhmet'ev was born on February 26, 1860, into the family of a former serf-turned-successful-merchant in a small village, not far from Saratov, an administrative center of the Volga region.[3] The second of five brothers, Porfirii was initially schooled at home by a tutor hired by his father. Very early on, the boy demonstrated a true penchant for learning and in 1874 he was sent to a local "real school" (the Russian analogue of the German *Realschule*) in a nearby town. The boy was interested in everything, particularly physics and mathematics. As he recalled much later, "in the fifth grade of real school [. . .] I built an electrical machine and earned the moniker of 'antichrist'. [. . .] Based on information from the newspapers, I built a Bell-Edison telephone. This was the first telephone model in Russia."[4] The model aroused much interest among the villagers, who crowded into Bakhmet'ev's house to listen to "mysterious" voices seemingly coming from nowhere, and it won its inventor the title of "sorcerer." But the popularity the telephone brought to Porfirii also got him in trouble with the authorities: a local priest wrote a report to the police denouncing the young inventor's "anti-government opinions." The local police chief demanded that Porfirii be expelled from school. Fortunately, by the time this demand reached the school administration, Porfirii had graduated.

Porfirii was set on continuing his education and his father supported his determination with a generous allowance. In October 1880, Porfirii Bakhmet'ev enrolled in the department of mathematics and natural sciences at Zurich University. He immersed himself in his studies and his diligence quickly attracted the attention of his teachers. In the spring of 1881, Alfred Kleiner, a young *Ausserordentlicher* (associate) professor of experimental physics, offered Porfirii a position as his assistant. Porfirii helped the professor to prepare lecture demonstrations as well as the numerous experiments students were required to perform as part of their "Physics Practicum." Today remembered only as a doctoral adviser to Albert Einstein, Kleiner was an expert in measuring techniques, and his laboratory was well stocked with various apparatuses.[5] Impressed by his assistant's skills, Kleiner soon gave the young Russian permission to use the laboratory in his free time to pursue his own research interests.

Ever since he had built the model of the Bell-Edison telephone, Porfirii had been fascinated by electric and magnetic phenomena. Now he turned his boyhood interest into serious scientific investigations. Already in 1883, just two years into his university studies, Porfirii published three articles with the results of his experiments in the *Journal of the Russian Physico-Chemical Society.*[6] The next year, he published five more articles in the same journal, detailing his research on the interrelations of acoustic, magnetic, mechanical, electric, and thermodynamic properties in various metals.[7]

Yet even as his works were being published by Russia's leading physics journal, Bakhmet'ev was unable even to visit his Motherland. The available documents are unclear regarding the details of this situation. Apparently, someone had stolen Bakhmet'ev's passport in Zurich and had used it to enter Russia illegally in August 1882.[8] When Bakhmet'ev had discovered that his passport was missing and attempted to obtain a new one, the Russian mission in Switzerland rejected his application. Furthermore, in November 1883, the Russian Police Department issued a warrant for Bakhmet'ev's arrest if he ever tried to cross the borders of the Russian Empire. It seems likely that the police suspected Bakhmet'ev of connections to Russian revolutionaries who had their headquarters in Switzerland, particularly the notorious "People's Will" group responsible for the assassination of Emperor Alexander II in March 1881.[9]

In 1885, Bakhmet'ev finished his university studies, but with the road home closed, he stayed in Zurich. Kleiner allowed his former student to work in his laboratory, and Bakhmet'ev expanded his earlier experiments, focusing in particular on the studies of thermoelectricity and related phenomena.[10] In May 1889 he had to leave the laboratory, however, probably because of financial difficulties. Earlier that year, his father had died, leaving more than half of his fortune to Porfirii. Porfirii's brothers contested the will, initiating court proceedings to revoke it, and Porfirii decided to concede his part of the inheritance to his brothers. It was a noble decision, but it apparently forced Porfirii to start looking for a paying job. He tried to patent several of his inventions, including a safe automatic lighter for street gas lamps and what he called a "telephotograph," a device for the transmission of not just sound but pictures as well (an early version of television).[11] But this venture turned out to be a waste of time and money. Porfirii also began offering private lessons in physics and mathematics to those who wanted to enter Zurich's post-secondary schools. He became a popular and successful teacher: his classes attracted up to forty students and, reportedly, the majority of his pupils were able to pass entrance exams to Zurich University or Zurich Polytechnic.

Although at that time Bakhmet'ev still had not completed his doctoral dissertation, in early 1890 the Bulgarian Ministry of Education offered him a position as lecturer in the Department of Physics and Mathematics of the recently (in 1888) founded Higher School in Sofia.[12] The offer came with a price tag: Bakhmet'ev would have to become a Bulgarian citizen. In the late spring of 1890, he left Zurich for Sofia. He obviously liked what he saw and accepted the offer. In October 1890, Kleiner, who by that time had become the dean of the Philosophy Faculty, provided his former assistant with a complimentary letter of recommendation.[13]

Bulgaria turned out to be Bakhmet'ev's second Motherland: in 1892 he became a Bulgarian citizen (see Figure 3.1). He spent his most productive years as a teacher and researcher in Sofia.[14] He created a general course on experimental physics, which he taught for fifteen years, and in 1895 he became a full professor.[15] With a subsidy from the Ministry of Education, he organized a small Physics Institute where he continued his investigations of thermoelectricity and began new studies in geophysics, developing novel techniques for measuring the earth's electric currents.[16] In 1898, he was elected a corresponding member and, two years later, a full member of Bulgaria's premier learned society, the Bulgarian Association of Scholars, predecessor of the Bulgarian Academy of Sciences. He attracted a number of capable students who helped him with research,[17] made several good friends who shared his interests and labors, and eventually found happiness, marrying a beautiful young woman, Praskovia, who bore him three children.

FIGURE 3.1 A photograph of Porfirii Bakhmet'ev, perhaps taken for his Bulgarian citizenship application or travel documents, ca. 1890. *Courtesy of Ganka Kamisheva.*

In Bulgaria Bakhmet'ev also found a new scientific subject that would profoundly affect his subsequent research: entomology. As did some Russian children, during his years at the real school Porfirii had collected beetles and butterflies. But in Zurich he did not have much time for such diversions. A chance event rekindled his entomological interests: one day walking in the Sofia Botanical Garden, Bakhmet'ev chanced upon a large green caterpillar and became curious about what butterfly it would become. He brought the caterpillar home and placed it in a box, where it soon turned into a chrysalis. In a few months a beautiful Giant Peacock Moth (*Saturnia pyri* L.) emerged from the chrysalis.[18] This little incident turned a laboratory physicist into a passionate insect collector and field naturalist. In just a few years Bakhmet'ev became a leading expert on Bulgarian beetles and butterflies, publishing articles on his findings in Bulgarian, German, Russian, and Swiss entomological journals.[19] His passion culminated in a monograph printed in 1902 in the *Proceedings* of the Russian Entomological Society, which included descriptions of nearly one thousand species and subspecies of so-called "large" butterflies (*Macrolepidoptera*) in Bulgaria, along with a complete survey of all the relevant literature.[20]

It was this newfound passion for field entomology that also launched Bakhmet'ev in an entirely new direction in his laboratory research. It is hardly surprising that Bakhmet'ev applied his knowledge of mathematics, chemistry, and physics to his entomological studies, using, for instance, statistical analysis in his research on variations in butterflies.[21] But sometime in 1897 he moved from a classical naturalistic study of insects to what, in just a few years, he would call *biophysics*. He became interested in the question of how insects survive the unfavorable conditions of extremely cold winters or particularly hot summers. To answer this question Bakhmet'ev needed first to figure out the exact temperature of an insect's body and how it changed in relation to the temperature of environment. He was hoping to find the answers in entomological or physiological manuals, but his search through available literature failed to yield any definitive answers. He compiled numerous unconnected facts on experiments with the survival of different species of insects at low and high temperatures. But the data differed drastically from author to author and from species to species, with no apparent regularity. Furthermore, he could not find any reliable data on insects' body temperature per se. Thus, Bakhmet'ev embarked on his own investigation.

Measuring the temperature of an insect was no simple matter: employing the then standard mercury or alcohol thermometer obviously could not produce sufficiently accurate results. Bakhmet'ev's training as a physicist—and particularly his years of work in Alfred Kleiner's laboratory that specialized in various measuring techniques—came in handy. His longstanding interest in and numerous studies of thermoelectricity provided him with the necessary tool: an electric thermometer.[22] Using the well-known phenomenon of thermocouple—a connection of two different metals generates electric current that changes depending on the temperature of the connection—Bakhmet'ev devised a thin bimetal needle that could be inserted directly into the body of an insect. He used a sensitive string galvanometer (with an attached magnifying lens to observe the minuscule movements of its string) to measure the changes in the current generated by the needle. He then converted these measurements into temperature units. He also designed a simple thermostat—a glass jar turned upside down into either a vessel filled with the mixture of ice and salt or a water steam bath—to control the external temperature. In the fall of 1897, he began preliminary experiments.

Bakhmet'ev started with the same insect that had reawakened his interest in entomology, the Giant Peacock Moth. Europe's largest butterfly (up to eight centimeters long, with a wingspan of up to twelve centimeters), it has a very thick body and was an ideal subject to test his techniques and to calibrate his apparatuses. He quickly discovered that when the moth sat still, its body temperature was higher than that of the surrounding air only by a few tenths of a degree Centigrade.[23] But if the moth began moving its wings, its body temperature rapidly grew up to ten degrees higher than the air temperature: in one experiment, for example, with an air temperature of +15.5°, the moth's body temperature reached +25.6°. Bakhmet'ev then explored the upper and lower temperature limits of the moth's survival. His measurements indicated that the butterfly died if its body temperature reached +46° or −15.6°. If the moth was taken out of the jar at any temperature within these limits, it revived. If the limits were exceeded, the moth invariably died. In 1898, at the annual conference of German naturalists and physicians in Dusseldorf, Bakhmet'ev presented his methods of studying the temperature of insects and the first results of his experiments. He also published them in Russian and German scientific journals.

Investigating the lower temperature limits of insects' survival, Bakhmet'ev observed very unusual phenomena.[24] In his experiments with another large butterfly, Privet Hawkmoth (*Sphinx ligustri* L.), he placed the insect in the cooling jar with an air temperature of −20° and every 30 seconds recorded its body temperature. The temperature decreased steadily and in about an hour it dropped to 0°. Five hours and twenty-four minutes into the experiment, the moth's body temperature reached −9.3°. But four minutes later, instead of continuing its downward decline, it suddenly—in the course of thirty seconds–"jumped" up to −1.7°. Then the body temperature began to decrease again, but much slower than it had before. Bakhmet'ev took the moth out of the jar and placed it on his table at the room temperature. Almost immediately, its body temperature began to rise and in about an hour the moth started moving its wings. Bakhmet'ev returned the butterfly to the cooling jar, and in the course of three hours, its body temperature slowly decreased to −13.1°. But then, in just thirty seconds, it jumped up to −8.8°. About an hour later, the moth's body temperature reached −15.2°, and when Bakhmet'ev took it out of the jar the butterfly did not revive.

Bakhmet'ev at once understood the physics behind the unusual phenomena he observed: the "juices" in the moth's body slowly cooled down to a certain temperature at which they began to solidify, releasing in the process a considerable amount of energy and increasing the moth's body temperature. Yet although the general physics of the process seemed simple enough, the experiment raised numerous questions. Do all insect species exhibit such a jump in body temperature and, if yes, at what point? What temperature does the insect body reach after the jump in different species? What Bakhmet'ev found particularly puzzling was that the solidification of body juices did not immediately kill the butterfly: only when its body temperature again reached a certain point near or below the jump temperature, did the moth die. It was clear to Bakhmet'ev that the body juices stayed for a certain period of time in the state of supercooling—that is, the juices remained liquid even though their temperature was actually below their freezing point. Only after the jump did the juices begin to solidify, supposedly becoming completely solid with a further lowering of body temperature. What could explain the supercooling of the juices? And what factors determined the possibility of insect revival after its juices froze?

Intrigued, Bakhmet'ev launched a series of experiments to find answers to these questions. He discovered that all the insects he experimented with exhibited a jump, but the exact temperature at which the jump occurred (he named it the "critical point") differed from species to species and indeed was species specific. He learned that the temperature the insect body reached after the jump (he named it the "normal temperature of the solidification of body juices") was species specific, too. Bakhmet'ev plotted a graph depicting his observations and found out that its general shape was remarkably consistent in different species (see Figure 3.2). The body temperature decreased slowly to the "critical point" (K_1), then rose dramatically up to the "normal temperature of solidification" (N_1), then slowly decreased again to the same temperature as the "critical point." This time, however, it became "the point of death" (K_2), since after the insect body temperature had reached that point, it could not be revived. At a certain point between N_1 and K_2, all the juices apparently became frozen, but it was still possible to revive the insect at any point on the curve until the "point of death." Bakhmet'ev also discovered that starvation lowered both the "critical point" (K_1) and the "normal temperature of solidification" (N_1) considerably. He theorized that starvation leads to the concentrating of the juices, thus lowering the temperature at which they freeze. He actually squeezed juices from various species of butterflies and beetles and froze them in his cooling jar. It turned out, however, that such "free juices" froze at a temperature higher than N_1, which Bakhmet'ev explained by the fact that juices in the insect body were not "free," but contained in capillaries, which, as he well knew, lowered the freezing temperature of any liquid. Bakhmet'ev published the results of his extensive experiments in Swiss, German, and Russian journals and expanded further his investigations.

Indeed, for the next few years the research into the unusual phenomena he had discovered became his major preoccupation. By combining physics and biology, Bakhmet'ev developed a multi-pronged research program, conducting thousands of experiments on nearly one hundred species of insects and studying the physics of supercooling in various liquids under different conditions.[25] The detailed examination of the "critical point" (K_1) became a key focus of his research, with the investigation of its variations induced by the speed and duration of cooling and the amount of moisture in the surrounding air not only in adult insects but also in larvae and pupae, as well as in subspecies of the same insect species.[26] Bakhmet'ev also expanded his study on the upper temperature limits of

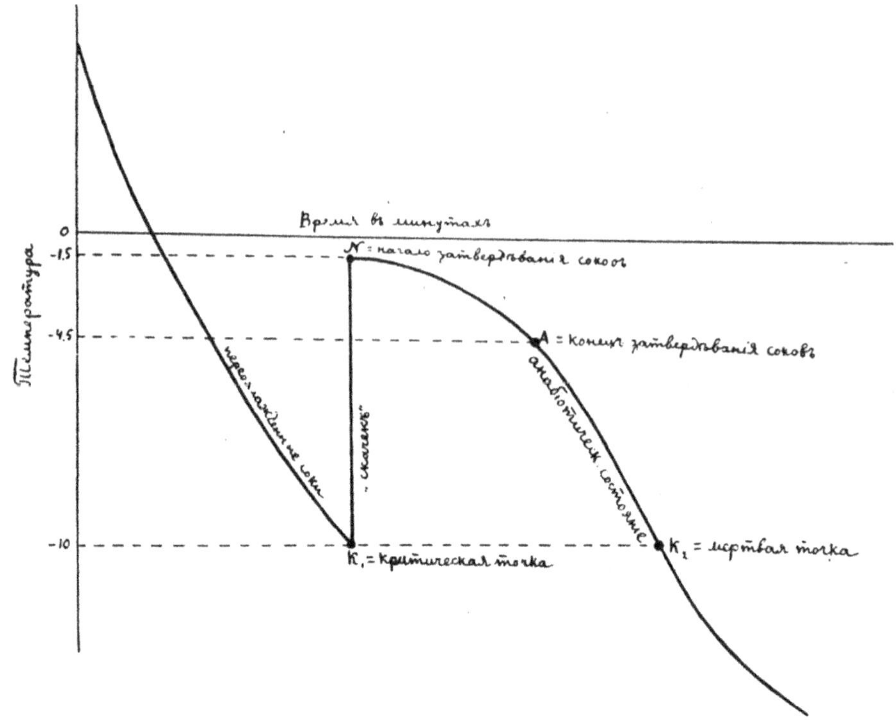

FIGURE 3.2 Porfirii Bakhmet'ev's original drawing of his "anabiosis curve" that appeared in one of his articles. The vertical coordinate line represents the body temperature, the horizontal, time elapsed from the beginning of the experiment. The first downward decline represents "the supercooling of the juices," from 0° until the "critical point" K_1 at the temperature of −10°. The upward line represents the "jump" of the body temperature from the "critical point" until the beginning of the "solidification of the juices" at the temperature of −1.5° (point N). The subsequent decline represents the "anabiotic state" between point A (the final solidification of juices at the temperature of −4.5°) to the "point of death" K_2 at the temperature of −10°. *From* Nauchnoe obozrenie, *1900, 1.*

insect survival.[27] He investigated the quantity of juices in various insects, introducing what he named the "juice coefficient" to measure the ratio between liquids and solids in the insect body, and explored the interrelation between the juice coefficient and the critical point in various insects.[28] In the process, he discovered that the juice coefficient differed among different species of insects, between adults and larva, and even between males and females of the same species, which, according to Bakhmet'ev, explained a large number of well-known but previously inexplicable facts in the geographical, ecological, temporal, and seasonal distribution of various insect species. Bakhmet'ev theorized that all these differences in the body temperatures of different insect species arose as a result of natural selection over the millennia of the struggle for existence as a means of protection against unfavorable environmental changes.

But what intrigued Bakhmet'ev most was the state of the insect between the "normal temperature of solidification" (N_1) and "the point of death" (K_2). His experiments suggested that at a particular point (A) within this range of temperatures (between N_1 and K_1), the insect's body juices became completely frozen, while the insect exhibited no signs of life at all. Yet until its body temperature reached the point of death (K_2), the insect could still be revived if returned to room temperature. Was it dead, then? Or "semi-dead"? Or "half-alive"?

In his search through the available literature on the temperature of insects and other animals, Bakhmet'ev found a description of phenomena closely resembling the ones he had observed in his experiments: "latent life" or "anabiosis." Introduced in 1891 by Wilhelm Preyer, a distinguished German physiologist, the term "anabiosis" referred to the state of a plant or animal organism in which it shows no visible signs of life yet is not dead, for it could "come back to life" under favorable external conditions.[29] The examples of anabiosis Bakhmet'ev found in the literature ranged from fish and frogs frozen into the ice of little ponds but revived once the ice melted to the dry branches and seeds of plants kept for years in herbariums yet able to grow again when placed in wet sand. Bakhmet'ev surmised that in his own experiments, insects with body temperatures between A and K_2 also were in an "anabiotic state," neither dead, nor alive, since, with all the body juices frozen, blood circulation was impossible, breathing stopped, and metabolism was absent, yet they could be revived.[30] Bakhmet'ev came to think that insects, fish, frogs, and other cold-blooded animals could apparently exist in such an anabiotic state for long periods of time—at least through the winter. He soon began wondering whether warm-blooded animals and, ultimately, humans, could be placed in a similar state.

In the spring of 1901, Bakhmet'ev submitted to the popular Russian magazine *Natural Sciences and Geography* a "biological sketch" with the enticing title "The recipe to live on to the twenty-first century."[31] He first surveyed current views on the prolongation of life, focusing specifically on the idea of a periodic slowing down of an organism's activities, as happens, for instance, during sleep or hibernation. Asserting that the best way to slow down the "life" (metabolism) of any organism was to lower its temperature, Bakhmet'ev recounted numerous studies that had appeared in the last ten years on the survival of various animals under low temperatures, giving exact bibliographic reference to every reported fact. He briefly summarized his own experiments and his discovery of the anabiosis curve with its points K_1, N_1, A, and K_2. He then introduced his plan for further research, but did so in a very unusual format—as a "scientific fantasy":

> A frozen man was found in an open field during a harsh winter. The knowledgeable people who had found him did not lift him by a hand or a leg but "as a whole," otherwise his limbs could have broken off, as easily happens in a frozen frog or a caterpillar. He was not brought into a warm room at once (as the famous [plant] physiologist [Julius von] Sachs had said, it is not freezing that kills the plant, but the way it is thawed), he was first rubbed thoroughly with the snow, and he awoke, [looking around] with bewildered eyes. A glass of cognac and further rubbing with alcohol made him completely conscious, and a few days later he fully recovered.[32]

"Of course," Bakhmet'ev admitted, "it is not always possible to revive a frozen man." But he thought it might be made possible.

Based on his own data as well as that reported in the literature, Bakhmet'ev designed a plan for investigating the ways of "freezing and reviving" warm-blooded animals. First, it was necessary to "transform" a warm-blooded animal into one whose body temperature would follow the changes in the temperature of its environment, which he thought was possible under the condition of narcosis induced by carbon dioxide.[33] The animal should then be placed in a "freezing bath" and slowly frozen to "point A," at which its blood would turn solid. The bath temperature should be kept steady at point A. An experiment with several animals would help determine the possible duration of keeping the animal in this frozen anabiotic state and the conditions of its thawing. Every six months, one animal should be taken out of the bath, so that by the end of five years, one could find out whether this method was practicable for longer periods of time. "I know that the present article might raise serious objections, particularly from physicians," Bakhmet'ev concluded, "but everyone should agree that the proposed experiments will produce many new facts illuminating the path that physiology must follow."[34]

In the spring of 1901, Bakhmet'ev summarized his research in a 170-page monograph, entitled *Experimental Entomological Studies*. Subtitled "Temperature interrelations in insects from the viewpoint of physical chemistry," it came out later that year in German.[35] In this book Bakhmet'ev critically reviewed previous investigations in determining the vital temperature of insects and provided a detailed account of his techniques and apparatuses. He recounted his numerous experiments with all of the data presented in tables and graphs. Bakhmet'ev concluded with an optimistic statement that his work "opens up a whole new area for researchers and, who knows, perhaps the time is near when one could create artificially the fabulous century-long sleep in humans too. Insects, at least, offer such a possibility."

Bakhmet'ev's book had great resonance and brought its author the recognition of his fellow naturalists. Germany's leading experimental biologist August Weismann wrote a complimentary preface to the monograph, emphasizing that by applying his "physicist's thinking" and the "newest measuring techniques [borrowed] from physics" to the "life phenomena of insects," Bakhmet'ev had obtained "exceptional results." Weismann noted that Bakhmet'ev's volume "gave for the first time a scientific explanation to the mysterious ability of insects to resist the cold" and provided "the foundation for a new direction in biological research."

Practically all major entomological and biological journals, from the *American Naturalist* to the *Russian Entomological Review*, from the French *Journal de physiologie et de pathologie général* and *Revue scientifique* to the Swiss *Societas entomologica*, and from the German *Die Umschau* to the British *Zoologist*, reviewed the book.[36] The reviewers—not only such prominent entomologists as Oskar Krancher, William M. Wheeler, and Reinhold von Hanstein, but also the leading physical chemist Wilhelm Ostwald, the paleontologist Hugo Roedel, and the zoologist Arnold Lang—were unanimous in their praise of the book. The tenor of these accolades is illustrated by the review written by the leading US entomologist, Wheeler, who characterized the book as "one of those rare publications which is full of interest not only to the specialist in entomology but to biologists in general."[37] He extolled the "startling, new, important, and far-reaching" results obtained "with the equipment of the trained physicist." Wheeler had no doubt that Bakhmet'ev's research "will ultimately form the basis for important work along theoretical lines in physiology, and for practical applications of great moment (in economic entomology, e.g.!), for they throw light on the geographical and climatic distribution of organisms, the resistance of animals and plants to cold and heat, and the problems of anabiosis."

The monograph brought Bakhmet'ev not only the praise of numerous reviewers around the world, but also honorific awards, particularly in Russia. In 1902, the Russian Imperial Society for the Acclimatization of Animals and Plants bestowed upon Bakhmet'ev its highest award, a large gold medal, while the Imperial Society of the Enthusiasts for Natural Sciences, Anthropology, and Ethnography elected him a permanent member.[38] A year later, the Imperial Academy of Sciences in St. Petersburg honored Bakhmet'ev with its most prestigious award, the Karl von Baer medal, while the Society of Naturalists in the city of Saratov elected him an honorary member.[39]

Bakhmet'ev was clearly excited by his discoveries and wanted to expand his research. On September 25, 1901, at the physiological section of the annual congress of German naturalists and physicians in Hamburg, he presented a long report on his studies of anabiosis in insects. Several prominent naturalists, including Arnold Lang, the rector of Zurich University, Justus Gaule, a renowned professor of physiology at the same university, and Ernst Haeckel, the most famous German biologist, approached Bakhmet'ev after his presentation to discuss in private his research and its future directions. Bakhmet'ev enthusiastically recounted his plan of studying anabiosis in warm-blooded animals. The three distinguished biologists approved of Bakhmet'ev's ideas and suggested that he begin his experiments with those animals that normally hibernate during the winter, such as hamsters, marmots, or bats, since these animals were "preadapted" to lowering their body temperatures and thus more suitable for preliminary experiments on anabiosis.[40]

The plan indeed looked very promising but there was one major obstacle to its realization: money. Bakhmet'ev conducted most of his experiments at his own house and paid all of the expenses out of his own pocket. However, in 1900 he managed to obtain a grant of $300 US dollars (not an insubstantial amount in those days) for his research on insects from the Elizabeth Thompson Science Fund in Boston.[41] But his projected studies of anabiosis in warm-blooded animals required considerable modifications of his techniques and apparatuses and thus would be much more expensive.

Bakhmet'ev began looking for subsidies. None of the German institutions he turned to could provide him with funds. He applied to the Imperial Academy of Sciences in St. Petersburg, but even though in 1902 the academy honored him with its Baer Medal and published his plan of research in its *Herald*, it didn't have any money to support his future studies. His home institution, the Sofia Higher School, refused to fund his "biological" investigations. Bakhmet'ev was deeply frustrated by the lack of funding for his research and vented this frustration by . . . writing fiction. In the fall of 1904, he published a hundred-page-long futuristic novella on how science should be organized, funded, and managed in *Natural Sciences and Geography*.[42] Only in 1905 did he manage to obtain a second small grant ($150 US) from the Elizabeth Thompson Science Fund.[43] But this time, another factor interrupted his work: politics.

Bakhmet'ev's life in Bulgaria was not all rosy: from the very beginning quite a few people had not been happy with the fact that a Russian was holding a professorial position in Sofia. As soon as news of his appointment became public in September 1890, a Bulgarian student in Zurich sent a long letter to the Bulgarian minister of education denouncing Bakhmet'ev as "a charlatan, a drunkard, and a liar" who would "sell his mother for a beer."[44] Its author warned the minister that Bakhmet'ev could "become a tool of foreigners against our homeland" and stressed that the position offered to Bakhmet'ev would be best filled by a deserving Bulgarian national (apparently meaning himself). This slanderous letter did not stop Bakhmet'ev's appointment, but the animosity toward the "foreigner" among some of his colleagues remained. Furthermore, it was not only certain Bulgarians who were offended by the policies of hiring Russian professors pursued by the Ministry of Education. Russian authorities were not pleased, either. In the spring of 1897, the Russian envoy to Bulgaria demanded that the ministry fire two Russian professors employed by the Sofia Higher School—the historian Pavel Miliukov (a future founder of Russia's Constitutional Democratic Party) and the physicist Porfirii Bakhmet'ev—as politically unreliable. The ministry did dismiss Miliukov, but not Bakhmet'ev.[45] To the contrary, when in 1904 the ministry initiated sweeping reforms of the system of higher education in Bulgaria, Bakhmet'ev was invited to advise the ministry officials and to take an active part in devising a plan for reform.[46] As a result of these reforms, the Sofia Higher School was expanded and transformed into Sofia University, with its faculty acquiring a considerable degree of autonomy in its internal affairs.

Ironically, Bakhmet'ev's foes among the faculty used their newly acquired academic autonomy to launch an attack on one of its architects.[47] In July 1905, the dean of the Faculty of Physics and Mathematics requested that the University Academic Board start disciplinary proceedings against Bakhmet'ev because of his "improper behavior" during the recent student unrest. At the same time, the Faculty filed criminal charges against Bakhmet'ev, accusing him of neglecting his teaching obligations and of defrauding the Faculty by receiving payments for lectures he had never actually delivered. The Board formed two "committees of inquiry," which during 1906 conducted lengthy investigations, collecting depositions from Bakhmet'ev's students, assistants, and colleagues. In March 1907, the Sofia District Court dismissed criminal charges as unsubstantiated, but the academic inquiry resulted in Bakhmet'ev's dismissal (though with a pension) from his professorship a few months later.[48]

Judging by Bakhmet'ev's scientific productivity during these years, 1905 through 1907, he did not pay much attention to the intrigue mounted by his colleagues. In 1906, he published a long treatise on "Biological analogies in crystals," demonstrating that some crystals exhibited properties similar to those of living organisms.[49] In February 1907, he finally submitted his dissertation on the "Current state of the question of earth electric currents" to the Philosophy Faculty of Zurich University.[50] His former professor, Alfred Kleiner, wrote a laudatory review of the dissertation, and, in June, the Faculty unanimously awarded Bakhmet'ev the degree of Doctor of Philosophy.[51]

Around the same time, Bakhmet'ev published at his own expense a second volume of his *Experimental Entomological Studies*, subtitled "The influence of external factors on insects."[52] This monumental, more-than-one-thousand-page-long tome presented a complete bibliographic survey of all known facts of insect life in relation to such external factors as climate, humidity, temperature, light, food, chemicals, electricity, magnetism, friction, gravity, and barometric pressure.[53] Bakhmet'ev compiled abstracts of more than one thousand publications (including thirty-two of his own) dealing with the influence of various external factors on the size, patterns of coloration, general appearance, and speed of development in various insect species. As Bakhmet'ev stressed in the preface, this tome was "intended primarily as a manual for those researchers who work in experimental entomology, because they will find in it, with little exception, everything known in this field by 1907."[54] Although he admitted that such phenomena as "seasonal dimorphism, protective coloration, mimicry, and parthenogenesis have only been touched upon in passing," he promised to prepare a special work with detailed treatment of these topics.

Bakhmet'ev also noted that he had finished a manuscript of the third volume of his *Studies*, which addressed the issues of individual variability in insects. Alas, that volume never appeared, though Bakhmet'ev did publish a series of articles on the subject, focusing in particular on the application of statistical methods to investigating insect variability.[55] For instance, he published an extensive paper on the butterflies of the Paleoarctic region. In this one-hundred-page-long article, based on his statistical studies, he proposed a "periodic system" of butterflies, which would allow entomologists to predict the existence of new species in that region, just as the periodic system of chemical elements allowed chemists to predict the existence of new elements.[56] None of these works even mentioned his ideas of investigating anabiosis in warm-blooded animals. Yet Bakhmet'ev did not abandon his attempts to find support for his plan.

On the Border of Life and Death

In late 1911, almost ten years after his last experiments with anabiosis in insects, Bakhmet'ev's persistence paid off. He obtained a substantial grant (3,500 French francs for two years) from the "Society for Promoting the Advancement of Experimental Sciences and their Practical Applications," named after its founder, the wealthy Russian merchant Khristofor Ledentsov.[57] Bakhmet'ev immediately began ordering the necessary equipment to set up a laboratory at his house. But an unexpected difficulty nearly derailed his plans: by that time, an electric streetcar line had been built on the street where Bakhmet'ev's house in Sofia was located and it interfered with the sensitive galvanometer he was using in his research. Bakhmet'ev was not going to give up, however. He decided to conduct his experiments at night when the tram was not running. Following the suggestion of colleagues Haeckel, Lang, and Gaule offered some ten years earlier, he began his research on animals that hibernate during the winter—bats—even though he personally was scared to death of these creatures.

With the assistance of two friends, at 1:56 am on February 8, 1912, Bakhmet'ev began his first experiment on a male specimen of a local bat (*Miniopterus scheibersii*), not much bigger than his favorite Giant Peacock Moth.[58] The bat's body temperature was 26,4° when Bakhmet'ev placed it in a cooling chamber at a temperature of −22°. In half an hour, at 2:26 am, the bat's body temperature reached 0°, but then the temperature decrease slowed considerably, exhibiting a small jump at −2.1° and reaching −4,0° by 2:56 am. At that point, Bakhmet'ev took the animal out of the chamber and placed it on a table (with the room temperature of 12°). To the touch, the bat was as hard as a rock and showed no signs of life. Five minutes later, however, it began breathing sporadically, at a frequency of 9 to 11 breaths per minute. In the next half hour, the frequency of breathing reached normal (about 21 breaths per minute). At 4:00 am the experiment was finished and the bat was placed in a holding box. The next day, the animal was breathing normally and periodically tried to escape from its box.[59] Bakhmet'ev immediately sent a short letter describing his experiment to the president of the Ledentsov Society that had funded his research, who forwarded it to the country's leading newspaper *Russian Heralds*, which in turn informed the public that it had been proven possible to put warm-blooded animals in a state of anabiosis.[60]

Bakhmet'ev continued his experiments with other specimens and discovered that bats invariably died when their body temperature reached −9.0° (after the jump, which usually occurred at the temperature between −1.9° and −2.1°), but could be revived if the body temperature was higher than −9°. Was the bat indeed in the state of "neither life, nor death," however? Was it truly frozen through and through? In one experiment, Bakhmet'ev took a bat with the body temperature of −7° out of the cooling chamber and opened up its chest cavity. It was completely frozen and the heart was hard to the touch. Within a few minutes, however, the heart began to contract spasmodically. It then started to beat regularly, with a frequency of 64 beats per minute. The observation continued for an hour, until the animal died from loss of blood. A peculiar feature of bats' biology also inspired Bakhmet'ev to investigate the possibility of placing sperm in the state of anabiosis. Although bats mate in the fall, the actual insemination occurs only in the spring. In between, while it hibernates during the winter, the female bat keeps the sperm in a special sack on its body. Bakhmet'ev was curious as to whether the sperm could survive the freezing and thawing of the bat. He conducted a special experiment freezing a female bat that carried the sperm and demonstrated that spermatozoids could indeed be frozen and then revived.

Encouraged by his first results, Bakhmet'ev applied for more funding from the Ledentsov Society to expand his experiments to other animals, such as rats, rabbits, and dogs. Since, according to the wishes of its founder Ledentsov, the Society was to support "primarily those discoveries and inventions that would bring maximum benefit to the majority of the population,"[61] Bakhmet'ev published several articles on "The theoretical and practical consequences of my investigations of anabiosis in animals."[62] He discussed the place of anabiosis in the "cycle of life and death," theorizing that freezing simply stopped life processes in the organism for a time, while subsequent thawing restarted them. He found an illuminating metaphor: a pendulum clock. If you stop the pendulum with your hand, the clock stops; yet if you move the pendulum with your hand again, the clock restarts. Thus, the clock with a stopped pendulum was "neither dead, nor alive," but was in the state resembling anabiosis. Obviously, the clock could stay in such an "anabiotic" state indefinitely (provided, of course, that it is protected from external influences that could damage its mechanism), until someone chose to move the pendulum again. Similarly, Bakhmet'ev hypothesized, a living organism could be placed in anabiosis for long periods of time and then revived.

Bakhmet'ev envisioned a whole series of "practical applications" for his discovery in science, agriculture, medicine, and trade. He thought that anabiosis could be used to control the number of pests damaging agricultural plants and forests by keeping parasites of the pests in an anabiotic state and then reviving and releasing them as necessary. He also suggested that peasants could increase profits by placing their cattle and beehives in anabiosis during the winter, thus saving on feed and heating. In addition, he proposed transporting cattle and horses in a state of anabiosis, thereby decreasing dramatically the usual losses associated with their transport by sea or rail. Most importantly, he opined that anabiosis could help fight infectious diseases, particularly tuberculosis, in cattle and even humans. Since TB bacteria die at a temperature of −6° in just a few days, keeping a frozen body at a lower temperature, say −8°, for a week will certainly kill all the bacteria and thus "cure" the organism of TB. Going a step further, Bakhmet'ev also proposed that it might be possible to find extinct animals and plants preserved in an anabiotic state in some remote northern territories. To illustrate this point, he referred his readers to the Imperial Academy of Sciences expedition sent to investigate a frozen mammoth found in Siberia. The expedition brought back samples of tissues from the mammoth's stomach, trunk, and ears. From these samples, researchers were able to extract and cultivate previously unknown bacteria, which, Bakhmet'ev surmised, were contemporaries of the mammoth and therefore had spent 100,000 years in the state of anabiosis!

Bakhmet'ev's efforts to publicize his latest work paid off: the Ledentsov Society gave him another two-year grant "for the continuation of research on the anabiotic state in warm-blooded animals."[63] Furthermore, at the end of the year, the Moscow City People's University—Russia's first private university named after the rich industrialist Alfons Shaniavskii, who had donated his fortune for its establishment in 1908—invited Bakhmet'ev to give a series of lectures on his studies. In early March 1913, Bakhmet'ev came back to his Motherland.

The timing of his return proved most propitious. Just a few months earlier, a group of distinguished scientists (many of them associated with Shaniavskii University) had created the Society of the Moscow Scientific Institute.[64] The society's goal was to raise funds for establishing four large research institutes in physics, biology, chemistry, and the humanities in Moscow. The society appealed to industrialists and merchants to support its efforts to advance science in Russia, and Bakhmet'ev's studies of anabiosis proved to be a very attractive project for prospective donors.

Indeed, the Moscow Refrigeration Committee—a group of scientists, engineers, and merchants involved in developing the Russian refrigeration industry—approached the society's governing council with the idea of establishing a special "laboratory of low temperatures" within the Biology Institute the society planned to found. A large part of the laboratory's mandate was to continue Bakhmet'ev's research on anabiosis. In late March, the society's council approved the plan of creating such a laboratory, and the Refrigeration Committee launched a fund-raising campaign.[65] The committee also invited Bakhmet'ev to deliver lectures on his research to its regular courses on refrigeration held at Shaniavskii University.[66]

The contacts with the Refrigeration Committee obviously fired up Bakhmet'ev's imagination. He came up with a new practical application for his research: the transportation over long distances of fish (especially, sturgeon) and caviar in a state of anabiosis. This could be of particular import for Russian merchants: the ability to deliver Russia's premier product—live sturgeon and caviar—to their customers not only throughout Russia but also abroad promised to increase profits considerably.[67] Indeed, Bakhmet'ev's suggestion could not have come at a more opportune moment: in 1913, under pressure from the US trade commission, the Russian Ministry of Trade and Industry explicitly forbade the use of two chemicals customarily used by Russian merchants for the preservation of caviar—hexamethylentetramin ($C_6H_{12}N_4$) and sodium tetraborate ($Na_2B_4O_7$)—due to their toxicity.[68] The United States had been one of the largest consumers of Russian caviar, importing up to four hundred tons annually. The loss of the US market was a heavy blow to the industry and Russian fish merchants were willing to pay generously for any chance to find new ways to preserve their premier product and to recapture the markets.

At the end of March, the Refrigeration Committee arranged an opportunity for Bakhmet'ev to conduct preliminary experiments with freezing and thawing fish at the refrigeration storage room of Moscow's Kazan' Rail Station.[69] Just a few days later, Bakhmet'ev went to Astrakhan—the center of sturgeon fishing and caviar production on the Caspian Sea—at the invitation of the well-known fish-merchant company "Brothers Shelekhov" to survey the company's refrigeration facilities. On April 8, he delivered a well-attended lecture on his research entitled "Neither life, nor death" at the local "people's university"[70] and promised to come back to Astrakhan later in the summer to experiment with anabiosis in fish and caviar.[71] In mid-April, Bakhmet'ev returned to Moscow to lecture for the courses on refrigeration.

The wide publicity accorded to Bakhmet'ev's research in newspapers, popular magazines, and professional periodicals quickly brought substantial benefits. In May, Moisei Gal'perin, a rich sugar merchant from Kiev, announced a gift of 12,500 rubles (an enormous sum in those days) earmarked for Bakhmet'ev's experiments on anabiosis in warm-blooded animals.[72] Bakhmet'ev asked Shaniavskii University's council to accept the gift on his behalf to build a special laboratory for his research under university auspices by the end of the year (while retaining 2,000 rubles for his experiments until the laboratory was ready). The university council accepted the gift and offered Bakhmet'ev a position on its staff as a professor. On May 30, Gal'perin's lawyer delivered the check to the university council and Bakhmet'ev began to order the necessary equipment.[73] He immediately contacted the Refrigeration Committee's secretary, engineer Mikhail Zarochentsev, requesting him to design special refrigeration chambers for his experiments. Zarochentsev enthusiastically attended to the task. In early June he even visited Gal'perin in Kiev and obtained a promise of additional funds for Bakhmet'ev's research when the initial gift ran out.[74]

Not to lose momentum in his fundraising campaign, during the summer Bakhmet'ev embarked on a lecture tour through southern Russia. From June through September, Bakhmet'ev visited Saratov, Tsaritsyn, Astrakhan, Khar'kov, Odessa, and Kiev. In every city he lectured on his research on anabiosis to overflowing auditoria.[75] He recounted his experiments with insects and bats and outlined possible applications of his discoveries, emphasizing in particular the possibility of extending human life and of curing tuberculosis by placing humans into a state of anabiosis. The local newspapers spread the message, publishing articles with such enticing titles as "On the border of life and death" and "The law of death." One journalist saluted Bakhmet'ev's "dreams of humanity's future, perhaps a very near future."[76] Another hailed his theory of anabiosis, which "had transferred the possibility of the artificial extension of life from the sphere of fantasy and legends into reality."[77] Practically all of the publications reiterated the "magic" number of −8°, at which TB bacteria would allegedly die in a body placed into an anabiotic state. Everywhere he went, Bakhmet'ev's lectures incited an enthusiastic response from the public: more than a dozen individuals sick with tuberculosis volunteered to be placed into anabiosis in the hopes of getting cured or simply to "sleep on" until medicine found a cure. Rumors flew that Bakhmet'ev had applied to the government for permission to conduct his "freezing and thawing" experiments on death-row inmates. Bakhmet'ev had to publish in the newspapers a special letter, explaining that Russian law forbade experiments on humans, even volunteers, which could lead to their death.[78]

Keeping his promise to local merchants, Bakhmet'ev spent nearly three weeks in Astrakhan, experimenting with freezing and reviving sturgeon and caviar. His visit clearly stirred up the locals: the governor of the region accompanied by an entourage of eminent fish merchants attended Bakhmet'ev's lecture on August 12.[79] The editor of a local newspaper excitedly recounted Bakhmet'ev's work and suggested that anabiosis could be used not only for transporting "live" fish and caviar, but could also become "a powerful tool for populating all suitable reservoirs with valuable species of fish," thus furthering "the rational organization of the fishing industry."[80]

The tour proved not only rewarding but also very tiring: in early September, Bakhmet'ev fell ill.[81] But even though he did not feel well, on September 13 Bakhmet'ev gave yet another public lecture on "the border between life and death" to a huge crowd gathered in one of Kiev's largest theaters.[82] Three days later he returned to Moscow. He had a high fever and was immediately put to bed. The attending physicians suspected that he had contracted malaria in Astrakhan, but their efforts to fight off the disease failed. On October 14, 1913, Porfirii Bakhmet'ev died at the age of fifty-three.[83]

Practically all Russian newspapers carried the sad news to their readers.[84] The Russian scientific community mourned the loss of its "prodigal son" in numerous eulogies that appeared in various scientific journals.[85] But Bakhmet'ev's death did little to curb the enthusiasm regarding anabiosis that his lectures and publications had generated among the Russian public.[86] In November 1913 the Imperial Academy of Sciences received a very unusual letter from Kiev. Apparently inspired by Bakhmet'ev's lecture, one of the city's inhabitants offered himself as a subject for experiments on anabiosis.[87] Fifty-eight years old, he was quite sick and, according to his doctors, had no more than two years to live. "Realizing that it is stupid to die without a purpose," he willingly placed himself "at the disposal of researchers of the academy," assuring his correspondents that he "did not suffer from alcoholism" and was of "sound mind." Members of the academy's Physics and Mathematics Division duly discussed the offer at their monthly meeting. They decided to send an inquiry to Shaniavskii University and to ask one of their fellow members, Russia's premier physiologist Ivan Pavlov, to write an assessment of the offer. In January 1914, the Nobelist Pavlov replied: "at the moment, the experiments on humans cannot even be contemplated. For a long time, experiments should be conducted with warm-blooded animals, particularly those that do not hibernate." A few weeks later, the academy also received a response from Shaniavskii University indicating that, although the university was still planning to establish a laboratory for research on anabiosis, no experiments on humans were being considered, "since the issues of anabiosis remain not fully investigated even in respect to animals."[88] The academy decided to reject the volunteer's offer.[89] The outbreak of the Great War in the summer of 1914 relegated research on anabiosis to the backburner.

Suspended Animation

The science of anabiosis got a new lease on life, as it were, in the new, Bolshevik Russia. Yet this time, it was not investigations into mechanisms of anabiosis and the building of special laboratories for such research, but the popularization and literary explorations of the biological phenomena of "neither life, nor death" that held the attention of Russian scientists and the Russian public. Indeed, during the 1920s, very few research articles on anabiosis appeared in Russia, particularly as compared to publications on the subject elsewhere in Europe and North America.[90] But during the same time period, a significant number of popular accounts and numerous literary renderings of "frozen and revived life" circulated in Russia. One man was largely responsible for making anabiosis a household name in Bolshevik Russia: Petr Shmidt (1872–1949).

Born in the country's capital on December 22, 1872, to a family of naturalized German immigrants (his father was a respected military topographer), Petr Shmidt graduated in 1895 from St. Petersburg University, with specialization in zoology (see Figure 3.3).[91] By that time he had published several articles in Russian and German zoological journals,[92] and the university's council chose him "to prepare for the title of professor"—a Russian analogue of what today we call graduate studies. The same year he began a teaching career, lecturing on zoology at the St. Petersburg Higher Women's Courses (*Bestuzhevskie kursy*) and on natural history at a private gymnasium. In a few years Shmidt also became a lecturer (Privatdocent) in zoology at his alma mater. During the same time, Shmidt mounted a two-year-long zoological expedition to the Pacific Rim of the Russian Empire, traveling through Korea, Japan, and Sakhalin and bringing back to St. Petersburg large collections of various specimens, particularly fishes, which provided material for his dissertation defended in 1904.[93] In 1906, he took an active part in establishing an Agricultural College in St. Petersburg and became a professor of zoology at the new institution, a position he would hold for nearly twenty-five years.[94] In the spring of 1908, he came to head the zoological section of a large expedition to Kamchatka funded by Fedor Riabushinskii, a scion of the prominent merchant dynasty. The expedition conducted detailed geographical, meteorological, geological, botanical, ethnographic, and zoological surveys of the large peninsula and neighboring islands, gathering mountains of data and producing several monumental accounts of its work. Shmidt prepared a huge tome with a general analysis of the expedition's zoological results.[95] His growing reputation as a leading expert in ichthyology resulted in his appointment as a senior zoologist to the Zoology Museum of the Imperial Academy of Sciences and as the curator of its ichthyology collections in 1914.

In addition to all his professional duties—teaching, research, expeditions, writing, and publishing—ever since his graduation, Shmidt also became engaged in what would be his lifelong passion—the popularization of biology. He lectured in "people's universities" and published articles on subjects ranging from zoopsychology and parasitology to cytology and gerontology in such popular-science magazines as *Knowledge for Everyone* and *Scientific Review*.[96] He also translated various popular treatises written by foreign colleagues[97] and compiled his own.[98] In addition, he became a member of the editorial staff of *Scientific Review*, a leading popular-science magazine, heading its bibliography section, wrote popular accounts of his numerous expeditions for a lay audience,[99] and published brochures in such popular-science series as "The library of a naturalist" and "Science and

FIGURE 3.3 A group photograph of students (easily identifiable by their student uniforms with metal buttons) and workers of the department of zoology at St. Petersburg University, ca. 1895. Standing left to right: G. F. Arnoldi, D. D. Pedashenko, M. N. Rimskii-Korsakov, A. O. Graftio, Petr Shmidt; sitting: Iu. N. Vagner and E. A. Shul'ts. In the next two decades, each of these individuals would become a recognized authority in his chosen subfield of zoological research, be it entomology, ornithology, or ichthyology. But only Shmidt would also become a prolific popular-science writer. *Courtesy of Sergei Fokin.*

arts."[100] In 1916, Shmidt produced a richly illustrated popular "introduction to biology," appropriately entitled *The Foundations of Life*.[101]

Most likely, it was through his work in popular science that Shmidt first learned of Bakhmet'ev's experiments on anabiosis—after all, Bakhmet'ev had regularly published accounts of his research in Russian popular-science magazines to which Shmidt also contributed. Indeed, Bakhmet'ev's first 1898 article on the "temperature of insects" appeared in *Scientific Review*, and the same issue included Shmidt's contribution on the Australian fauna.[102] Clearly, through the years, Shmidt had followed Bakhmet'ev's work attentively: in 1913, just one month after Bakhmet'ev's death, Shmidt published a lengthy article on anabiosis "dedicated to the memory of Bakhmet'ev" in one of the most popular "thick" magazines, *Russian Wealth*.[103]

Shmidt's professional studies were limited mostly to classical *faunistic* and *taxonomic* zoology, and especially ichthyology, and thus, they were far removed from the "issues of life and death," which held the attention of *experimental* biologists of the time. But in his numerous popular-science works, these issues were central. In early 1916, in the popular series "The library of a naturalist," Shmidt published a brochure, which—replicating the title of Bakhmet'ev's 1913 lectures—was entitled *On the Border of Life and Death*, and opened with a chapter dealing with these very issues.[104] In the subsequent chapters, Shmidt recounted in detail Bakhmet'ev's experiments on insects and his plan of studying anabiosis in warm-blooded animals. Shmidt lamented the fact that "unmerciful death had stolen from us the talented scientist" and expressed the hope that "his ideas will not be forgotten" and that "other researchers will follow the paths he opened."[105] Indeed, Shmidt himself offered a new path—"used by nature itself, but so far neglected by the investigators"—to achieving the state of anabiosis in animals: the combination of partial drying (desiccation) with the freezing of an organism. Later that year, Shmidt began his own studies following exactly this path.[106]

In the fall of 1916, assisted by one of his students, Shmidt began his research on anabiosis in earthworms at the Zoological Laboratory of the Agricultural College.[107] Shmidt's first experiments proved disappointing. He quickly found out that desiccation had no effect on earthworms' endurance of low temperature: both the normal and the partly exsiccated worms died between −1.6° and −2°C. But he also discovered that earthworms demonstrate significant tolerance to the loss of body water, easily surviving the loss of 20 to 30 percent of their body weight. He abandoned his idea of combining freezing and drying, focusing instead on investigating the possibility of inducing anabiosis by drying alone. The next summer, Shmidt began a new series of experiments with various methods of "exsiccation and revivification" to determine exactly how much water the worms could lose without losing their vitality.[108] He demonstrated that, if the exsiccation was conducted slowly and gradually, the worms could be revived after the loss of up to 61 percent of their body weight (that is, the loss of nearly 73 percent of the entire quantity of water in the organism!). It proved impossible, however, to keep the exsiccated worms for long periods of time at room temperature due to the rapidly developing microbial putrefaction. To inhibit the activity of the putrefying microorganisms, Shmidt placed the desiccated worms on ice and was able to revive them after forty-eight hours. There was a use for freezing after all. As he put it, "the combination of exsiccation and low temperature gives the most satisfactory results."[109] Shmidt believed that his experiments, however preliminary, had proven that "the phenomena manifested in the exsiccation of earthworms [...] with full right may be called 'anabiosis' ('over-life')."[110] Shmidt was convinced that "at suitable low temperatures one can preserve the vitality of the exsiccated worms during a very long time" and planned to "undertake experiments in this direction at the first opportunity." He submitted an article describing his research to the US *Journal of Experimental Zoology* and to the *Proceedings of the Petrograd Society of Naturalists*. But any opportunity to continue his experiments in the foreseeable future was shattered by the Bolshevik coup d'état in October 1917 and the Civil War that engulfed the country in its aftermath.

Although the Bolsheviks nationalized scientific and educational institutions, during the civil war years they did not have enough resources to fund any research that was not considered of vital importance for the new regime. At the same time, the Revolution liquidated all other sources of funding for science and education, including the personal wealth and incomes of individual researchers, which effectively put a stop on the scientific work of many Russian biologists and made their very survival extremely difficult. Indeed, the material conditions of life under War Communism were terrible: there were almost no goods, fuel, or electricity. The food rations distributed by the government were barely enough to sustain life.[111] With transport and distribution networks in shambles, cities and their inhabitants were hit the hardest: between 1917 and 1920, Petrograd lost more than half of its population to death and outmigration.[112]

Unable to continue his research, Shmidt focused his energies on its popularization, "recycling" many of his pre-revolutionary works and writing new ones. In addition to occupying his mind with something he loved to do, it supplemented his meager food rations with honoraria paid by the publishers, which he could use to buy food and fuel on the black market.[113] Shmidt published numerous magazine articles, brochures, and books on various biological subjects, ranging from the "psychical life of apes" to the "rejuvenation" of humans.[114] Much of his popularization efforts focused on "latent life." As early as 1919, while the Civil War was still raging throughout the country, Shmidt published an article on hibernation in the first issue of the just established popular magazine *In Nature's Workshop*.[115] A year later, in 1920, he issued a two-hundred-page volume on *The Riddles of Life*, which included texts of five public lectures on "the essence of life" he had delivered in various venues.[116] The third lecture presented a detailed overview of research on anabiosis, focusing in particular on Bakhmet'ev's discoveries. The same year Shmidt reissued his brochure on the "foundations of life," which also included a large section on anabiosis. The brochure went through three editions in as many years, eventually becoming a two-hundred-page volume.[117]

In late 1921, with the end of the Civil War and the introduction of NEP, life in Petrograd began to revive, and its population began to return to the city. The city's scientific community came back to life as well, resuming its research, publications, and conferences. In December 1922, the First Congress of Russian Zoologists, Anatomists, and Histologists convened in Petrograd, bringing together nearly two hundred scientists from all over the country (see Figure 3.4). Shmidt presented to the congress his research on the desiccation of microscopic animals (nematodes, rotifers, and tardigrades), which he had conducted during the summer at the Institute of Experimental Medicine.[118]

The following year, 1923, Shmidt expanded his old lecture on anabiosis into a separate volume of nearly two hundred pages, entitled *Anabiosis (the phenomenon of revival)*.[119] The monograph provided a detailed account of various forms of "latent life" and outlined research in this area carried out by numerous scientists, focusing in particular on Bakhmet'ev's contribution. Shmidt updated Bakhmet'ev's concept with his own research and analysis, concluding that anabiosis induced by freezing in Bakhmet'ev's experiments

FIGURE 3.4 A photograph of a plenary meeting of the First Congress of Russian Zoologists, Anatomists, and Histologists, held in Petrograd in December 1922, at which Shmidt presented his research on anabiosis. The second and third rows on the right side of the room are filled with students of Nikolai Kol'tsov, who himself is sitting at the head table in the presidium (second on the left). First on the right in the third row is Mikhail Zavadovskii. Note that most attendees are wearing their overcoats, which suggests that the auditorium was not properly heated. *Courtesy of Valeriia Kallinnikova.*

was in fact also a result of exsiccation, since the solidification of juices in the animal organism resulted in the loss of water by its tissues and organs. He recounted the possible practical applications of anabiosis envisioned by Bakhmet'ev, including freezing and thawing humans. "Of course, so far, this is only a dream that, though possibly realizable, would require long and consistent labors," Shmidt admitted. "Many such dreams, however, have been realized before our eyes by the miracle-making touch of the magic wand of science," he concluded optimistically.[120]

Thanks to Shmidt's untiring efforts, from 1923 onward anabiosis began to figure regularly on the pages of popular-science magazines and brochures.[121] Shmidt himself continued publishing and delivering public lectures on the subject. In 1924 he compiled a *Small Biological Encyclopedia* that included an entry on anabiosis.[122] In March 1925, he delivered a keynote address on anabiosis at a special event organized by the Moscow Polytechnic Museum on "The issues of life and death."[123] Two months later, he published this address in a popular literary journal, *New World*.[124] In subsequent years he published several more articles and brochures on the subject.[125]

But all of this active propaganda/popularization of anabiosis as a particular state "between life and death" was not supported by new research. Indeed, in the ten years since the Bolsheviks' seizure of power in the fall of 1917, fewer than a dozen articles on anabiosis appeared in Russian specialized scientific periodicals.[126] Only one of them described an attempt to induce anabiosis in warm-blood animals: a researcher at the Belorussian Agricultural Institute conducted a very preliminary (and unsuccessful) experiment with "freezing and reviving" a puppy.[127] Another article described the "influence of low temperatures on spermatozoa" in order to find out the necessary conditions to preserve the sperm for further use in artificial insemination.[128] The rest dealt with protozoa, insects, and earthworms.[129] Surprisingly, though one of Shmidt's students continued his studies of desiccation in earthworms, after 1922, the most active propagandist of anabiosis did not himself conduct any new research on the subject: he was far too busy with his ichthyological studies in the Russian Pacific.[130] None of the research published in the 1920s came even close to the technical sophistication of Bakhmet'ev's experiments. Moreover, the research published during the 1920s was not directed to the questions of the "border between life and death" per se. Rather, the majority of the publications examined such questions as the survival of various organisms during harsh winters and its influence on the geographical, seasonal, and ecological distribution of various species under study.

During the first decade after the Revolution, then, despite active popularization, actual investigations of anabiosis were in the state of "suspended animation." There was not a single attempt to establish a specialized laboratory to study anabiosis. An Institute of Experimental Biology created in late 1916 by Nikolai Kol'tsov with the support of the Society of the Moscow Scientific Institute and Shaniavskii University, which "inherited" the money donated by Moisei Gal'perin for a future laboratory of low temperature, did not conduct any research on anabiosis. Similarly, a special Institute of Biophysics, established in 1920 by Narkomzdrav, never tried to repeat and advance Bakhmet'ev's *biophysical* investigations.[131] The Soviet scientific community appeared largely uninterested in and incapable of undertaking detailed studies of "the border between life and death," to say nothing of building specialized research institutions for that purpose. Certainly, none of the researchers who did touch upon anabiosis in their studies had the training and expertise to continue Bakhmet'ev's investigations. Furthermore, the equipment necessary for such research (freezers, sensitive electric thermometers, special apparatuses for registering low electric currents, etc.), was quite expensive and difficult to obtain in post-revolutionary Russia.

A Poem of Anabiosis

Yet anabiosis did inspire extensive "experimentation" . . . in the pages of literary fiction. Of course, in their own experiments, writers did not pay much attention to the complexity of ideas, sophisticated techniques, intricacy of measuring and cooling apparatuses, and particularities of actual scientific experiments. Nor did they bother with protozoa, insects, worms, fishes, bats, or any other animals. Following Bakhmet'ev's "recipe," they were freezing and thawing humans.

Alexander Iaroslavskii (1891–1930), a poet, anarchist, and the founder of the northern group of biocosmists-immortalists, produced one of the first literary renderings of anabiosis.[132] In late 1922, he published a fourteen-page-long *Poem of Anabiosis*.[133] It was a poetic manifesto of "the new anarchism,"[134] which combined the belief in a continuing world revolution with the biocosmists' goals of achieving immortality, resurrecting the dead, and opening up the cosmos for human travel and habitation.[135] For Iaroslavskii, anabiosis became the main tool of a new "biocosmic" revolution. Answering those who perceived NEP as the "betrayal" and "death" of the goals of the socialist revolution, Iaroslavskii predicted optimistically that the Bolshevik October 1917 coup was but a first step to the "Biocosmic revolution—the only one that will triumph" on a planetary scale. As he put it, "[It is] coming, the revolution of biocosmists, [right] after October's Red Star." In accordance with its author's anarchistic views, the biocosmic revolution would destroy all states, including the newly born "proletarian" state in Russia. But it would do much more than that! It would resurrect the dead geniuses of the past (Iaroslavskii's curious list included Socrates, Euclid, Spinoza, Hiram Abiff, Plato, Aristotle, Shakespeare, Whitman, and Homer) to form a "Genius Council of the Wisest" that would "decide humanity's fates." Guided by the Council, an "army of biocosmists" would transform the planet by "cleaning up the entire Earth, with the wise broom of science." With the assistance of "chemist and physicist, engineer, architect, and poet," as well as "bacteriologist and biologist," it would control the weather, eliminate microbes and diseases, build a bridge over La Manche and tunnels through mountains, restore the "drowned continent of Atlantis," liquidate weeds and pests, "dress Sahara in flowers," make people immortal, and conquer the Universe by employing "our friend—the wondrous radium" and "[inter]planetary ships."

Iaroslavskii recognized that humanity—"not just the kings of wasted earth, but even you, the wise madman Lenin"—might not welcome the biocosmic revolution. "But what to do when a blind herd, like a mound of mud, impedes the world's uplifting?" asks the poet rhetorically. "To kill it and throw in the darkness [of death]?" This was not an acceptable course of action, since "every living [being] is sacred, even if it's infinitely stupid." Anabiosis solves the poet's dilemma of what to do with reluctant humanity:

> And so as not to murder anybody,
> Will freeze the world in states' bourgeois barracks,
> It is simpler and easier than to turn into a bloody lump
> Dozens and hundreds of living human armies.

Iaroslavskii does not dwell on the technicalities of freezing the entire planet and placing it in an anabiotic state. Rather, following Bakhmet'ev's metaphor, he presents it as merely "a five minute stop in life's roundabout."

Between Life and Death
Will ram a heavy wedge,
A third door opens,
For the world, anabiosis.

While entire humanity is "asleep," frozen in anabiosis, the army of biocosmists— "workers and young engineers"—recreates Earth.

And when all the work is done
And when, as a perfect toy, Earth—
Wake up, the living, again
On cities' squares and fields.

Anabiosis will allow biocosmists not only to recreate Earth, but also to abolish death completely:

Death,
Down with it,
In the coffins!
Together with God and other rubbish!
Close dauntless ranks—
And [take] fate by the throat—
Biologist, poet, and worker!

The next step in the biocosmic revolution—interplanetary travel and habitation of the cosmos—would be taken by humanity as a whole. And Iaroslavskii calls upon humanity: "To assail the Universe, brothers!"

Iaroslavskii's poem clearly attests to his unshakable belief in science, embodied in the very idea of anabiosis as a state of "neither life, nor death" that could be achieved by certain scientific techniques and manipulations. In the *Poem of Anabiosis*, the triad of life, death, and resurrection, which had been shrouded in biblical, mystical, occult, or supernatural terms and symbols in countless previous fictional, theological, and philosophical tracts, loses its mystique and is replaced by the power of science—symbolized by anabiosis—that would assure the abolition of death and the eternal life of humans on Earth and in Cosmos. Indeed, Iaroslavskii portrays science as the main means of his biocosmic revolution.[136]

Bakhmet'ev's metaphor of life, stopped and restarted as a pendulum clock, offered numerous possibilities for using anabiosis in fiction, first of all as a sort of a time machine, though of course going only in one direction—to the future.[137] This is how Vladimir Mayakovsky, the country's most acclaimed "futurist" poet, used "freezing and thawing" in his popular comedy, *The Bed-Bug* (1928). Its protagonist, "former worker" Prisypkin, loses his "proletarian nature" under the conditions of NEP. As a result of an accident, he is frozen and buried under the rubble. Fifty years later, he is found and "thawed out," only to be placed in a zoo as a specimen of *Philistinus vulgaris* not to be confused with *Homo sapiens* of the future (now present). Mayakovsky deploys the possibility of freezing and thawing humans only as a seemingly believable means of preserving a particular human type that differes profoundly from the conscientious proletarians who inhabit the future. There is no science or scientist involved in placing the protagonist in the anabiotic state, nor any discussion of anabiosis as such.

Alexander Beliaev in his novella "Neither Life, Nor Death" also used anabiosis as a kind of a time machine, but he did so in an entirely different manner, going to great lengths to explain the science behind anabiosis, its history, and its possible applications. Beliaev made his character Carlson repeat almost word for word Shmidt's explication of Bakhmet'ev's experiments and their import. It is this recounting of the actual scientific discoveries (along with the size of profits supposedly earned by Steinhaus the fish-monger) that finally convinces the capitalist Gilbert to buy into Carlson's "project." They form a joint venture as Gilbert agrees to finance Carlson's idea of building a Conservatorium—a special warehouse for people in anabiosis—in Greenland. But before they can start using it, they need to demonstrate that anabiosis is safe. They launch an advertising campaign, promising a reward to volunteers. Naturally, however, nobody wants to be "frozen to death." Finally, two individuals come forward to volunteer for the project. One is Edward Leslie, a professor of astronomy, mortally ill with tuberculosis. According to his doctors, he has only a few months left to live. Leslie wants to be placed in anabiosis so that he can be revived in ten years when a particular comet will again come close to Earth, enabling him to confirm his cosmological theory. The second volunteer is Meré, a starving young poet, who is mainly interested in receiving the promised reward for his participation in the experiment.

Leslie and Meré are frozen. After spending a month in anabiosis, they are successfully thawed in front of a huge crowd of onlookers. Physicians attending the "revival" discover that Leslie is completely cured of tuberculosis, thus—following Bakhmet'ev's earlier speculation—proving that anabiosis is not only safe but also therapeutic. The Gilbert-Carlson venture becomes a booming business, bringing the owners millions of pounds. Thousands of people sick with TB flock to the Conservatorium to be cured. In addition, the British government decides that it is "more humane" and, of course, cheaper, to put criminals in anabiosis than to keep them in prisons or put them to death.[138] Echoing another idea of Bakhmet'ev's, Gilbert & Co. also branches out its business to transport cattle in a state of anabiosis and thus is able to bring to its British customers fresh, instead of frozen, meat from Australia.

In time, Carlson's original idea of freezing the miners left out of their jobs begins to be realized as well. Gilbert & Co. offers five pounds a week to the family of every frozen worker and, driven by despair and economic woes of their families, many unemployed miners agree to go to the Conservatorium to be placed into anabiosis. Of course, "freezing workers does not solve the crisis of British capitalism." To the contrary, it fuels the class struggle and leads to a worker uprising. Brigades of workers fly to Greenland to revive their "sleeping" comrades, so that they, too, can take part in the uprising. Carlson and Gilbert send a cable ordering their underlings to destroy the Conservatorium. The storage is blown up and all the "sleepers" are killed. But the revolution succeeds!

Seventy-three years pass and a historian discovers in the archives the blueprints of the Conservatorium. On the blueprints, he notes one particular mine situated away from the main complex of underground galleries. He thinks that it might have survived the explosion that destroyed the Conservatorium. A special expedition sent by the government indeed discovers three undamaged "coffins." One holds the body of astronomer Leslie, another poet Meré, and the third Benjamin Johnson, a worker. The trio is revived in the year 1998. Alas, they do not fit into the bright future of the victorious proletarian revolution, despite all the technical and social advances it has brought to humanity. Johnson, though united with his descendants, is thoroughly disconnected from them. His wife is dead, and his son, who was only two years old when Johnson went to the Conservatorium, is a grumpy seventy-five-year-old stranger. Johnson feels like "Ahasuerus, perpetual wanderer, punished by immortality, alien to everything and everybody." He thinks of suicide, but decides instead to find his "comrades-in-anabiosis," Leslie and Meré.

The trio get together. All of them suffer from "the disjunction of time," feeling "lost in this new, incomprehensible world." Even Leslie has lost his purpose. His theory is made obsolete by the great progress of science. All the discoveries Leslie had only dreamt of have already been made. As a scientist, he is left far behind and has nothing to contribute to the new science. But he has an idea: "Since the fate tore us away from the present, let's live for the future." He suggests that the trio become observers of the progress of humanity. They will go into the "anabiosis sleep" and wake up every hundred years to see "what has happened on Earth and the heavens." "We will see things that the people who live for only a normal life span cannot even dream of," Leslie exclaims enthusiastically. "Is it not worth living for?" Johnson is at first horrified by the prospect of going back into the frozen sleep: "Again?!" But after a few minutes of gloomy contemplation, he replies apathetically: "It's all the same." End of story.

Like in "Professor Dowell's Head," the plot of Beliaev's new novella was definitely and directly inspired by science. Beliaev made extensive use of such "practical applications" of anabiosis envisioned by Bakhmet'ev—and popularized by Shmidt—as transporting live fish and cattle in a state of anabiosis and curing tuberculosis. But exploiting further the popular theme of the "failure" of science in an immoral capitalist society, he added a few "applications" of his own: capitalists' freezing of prisoners and workers in an attempt (unsuccessful, of course) to avert the inevitable proletarian revolution. Strangely enough, even though the anabiosis technology was still available after the revolution, Beliaev did not find any use for it other than to place back into the anabiotic state the trio of "wandering strangers," which presents the only example of moral and socialist applications of Bakhmet'ev's discovery in the story.

The idea of this application came—via Shmidt—from . . . John Hunter (1728–1793), a Scottish surgeon and scientist, who had been one of the first to study the freezing and thawing of living organisms. In his "lectures on the principles of surgery," Hunter described his studies on "the heat of animals" and particularly his unsuccessful experiment with freezing and thawing fish. He concluded:

> I had imagined that it might be possible to prolong life to any period by freezing a person in the frigid zone, as I thought all action and waste would cease until the body was thawed. I thought that if a man would give up the last ten years of his life to this kind of alternate oblivion and action, it might be prolonged to a thousand years: and by getting himself thawed every hundred years, he might learn what had happened during his frozen condition. Like other schemers, I thought I should make my fortune by it; but this experiment undeceived me.[139]

Shmidt used this very quotation in his 1923 popular treatise on anabiosis, and obviously it caught Beliaev's attention, providing him with a plausible way to employ anabiosis in the socialist future.[140]

But other Soviet writers found different uses for anabiosis in their own work. Bakhmet'ev's compatriot (who also grew up in the same Volga region and went to the same real school a few years after Bakhmet'ev) Aleksei Tolstoy found another application in his comedy, "The Factory of Youth" (1927).[141] By that time well known to Soviet SF fans for his novels *Aelita* (1923) and *The Hyperboloid of Engineer Garin* (1925–1926), Tolstoy portrayed freezing and thawing of humans as a means of their "rejuvenation."[142]

The play begins in a communal apartment in Moscow. Among its inhabitants are a couple, Sergei and Maria. Maria is six years older than Sergei and at forty-one she feels old and unattractive. She is afraid that Sergei will leave her for a younger woman, even though Maria has given her best years and all her energies to help Sergei become a successful artist and architect. Her fears prove justified: Sergei indeed leaves her for a young movie starlet. Another inhabitant of the apartment is an inventor, Prishchemykhin.[143] In his room, he experiments with rats and dogs, periodically frightening his neighbors with loud explosions. He claims "to have conquered death" and promises soon to be able to "resurrect the dead." Prishchemykhin freezes the animals at a "temperature of −8°" in "the atmosphere of pure nitrogen" and keeps them for three days between the poles of a huge electromagnet "with a current of 30,000 volts," while "electromagnetic irradiation" rejuvenates the frozen organism. His animal experiments are "85 percent successful," and he is trying to convince someone among his neighbors to volunteer for one of his trials. Understandably, nobody does, until, despairing after her husband's betrayal, Maria begs Prishchemykhin to "bring back her youth." She writes a note that would absolve the inventor of any responsibility should she not survive the rejuvenation. Prishchemykhin administers chloroform and starts his machinery. The experiment is a resounding success: after three days, Maria emerges from Prishchemykhin's apparatus looking better than she did at seventeeen. Her youth has returned! Now she can take "revenge" on her unfaithful husband and his paramour.

Tolstoy clearly used the anabiotic rejuvenation to reanimate the standard dramatic plot of a husband leaving an aging wife for a younger woman and to give a humorous commentary on the life and mores of NEP Russia. Yet his choice of this particular rejuvenation technique plus the specified temperature (the "magic" −8° invented by Bakhmet'ev) attests to the wide popularity of the idea of freezing and thawing humans during the 1920s.[144]

A short story with the telling title "The Affair of Death," which appeared in the *New World* just as Tolstoy's play was premiering in Moscow in February 1928, provides similar testimony. The story was authored by a very unusual pair, "Bor. Pil'niak and Prof. N. M. Fedorovskii."[145] The name of Boris Pil'niak (the pen name of a popular Soviet novelist, Boris Vogau) was well known to the readers of *New World*.[146] But for the second author, it was his first attempt at writing fiction.[147] Although the publication identified Nikolai Fedorovskii as a "professor," it did not mention that his scientific specialty was not biology or medicine, but mineralogy, nor that he was a prominent Bolshevik who had joined the Bolshevik Party in 1904 and, since the Revolution, occupied important positions in the party and state apparatus.[148]

The story opens with a lyrical discourse on a birch tree—a recognizable symbol of Russia, beauty, love, and sweet sorrow immortalized by generations of poets and artists—sitting in a corner of the asphalt-covered courtyard of an Institute of Life. Organized under the auspices of the "Moscow Communist Academy" and directed by an aged reputed scientist, academician Pavlishchev, the institute's main purpose is to study the biology of the cell to illuminate issues of life, death, and immortality. In addition to several laboratories located on the main floor of a "former princely villa" that houses the institute, its basement has several "crypts" equipped with refrigerators and other machinery to control the temperature, humidity, pressure, and chemical composition of the air.

One of the institute's talented young researchers commits suicide but leaves no explanatory note behind. Deeply puzzled, Pavlishchev sifts through the younger man's papers and the results of the autopsy, looking for a reason behind the suicide but unable to find one. He comes to the conclusion that the scientist's "fate was determined by his heredity." But he also finds among the deceased's papers a short note, written about a year prior to his suicide, which resonates with Pavlishchev's own thoughts and prompts him to contemplate the relationship between life and death. Pavlishchev admits that humanity knows very little about death or the "laws of life," but thinks that "the restoration of an individual life will be possible only when its form could be preserved." He states that "humanity has discovered methods of curing [diseases] by freezing" and suggests that the solution to "issues of immortality lie in the field of freezing and reviving a living organism." Pavlishchev is convinced that "at the very least, freezing will allow keeping the form until the time when human knowledge catches up." He believes that "science must master control over human life" and admits that he "wants to live" until such time will come. Pavlishchev goes down to the institute's basement and locks himself in one of its crypts. On the crypt's door he leaves a note: "Please, keep my body frozen until that time when science finds a method to revive me." The last paragraph of the story repeats its few opening lines, glorifying the beautiful birch tree still sitting in the institute's courtyard.

To its contemporaries, the story contained numerous recognizable and telling details. The description of the Institute of Life pointed unambiguously to the Institute of Experimental Biology, which at that time was indeed housed in a "former princely villa" of a wealthy Moscow merchant (see Figure 3.5). Its director, Nikolai Kol'tsov, was Russia's leading authority on the biology of the cell and was deeply involved in testing various rejuvenation techniques.[149] Kol'tsov was also the founder and president of the Russian Eugenics Society, which advocated for and conducted extensive studies of human heredity.[150] The mention of the Communist Academy—an institution created in July 1918

FIGURE 3.5 The "princely villa" depicted as the Institute of Life in Pil'niak and Fedorovskii's story "The Affair of Death." This building from late 1924 until 1952 housed Nikolai Kol'tsov's Institute of Experimental Biology. The well-known Moscow architect Ivan Bariutin built it in 1911 in the then fashionable "modern" style for the family of prominent merchants, the Bardygin brothers. After the Revolution, along with all of the neighboring buildings, which had also belonged to various Moscow merchants, the Bardygin mansion was expropriated. In the mid-1920s, this particular neighborhood (*Vorontsovo pole*) became a virtual "science city" in the center of Moscow that housed numerous scientific institutions, including the Institute of Biochemistry and the Institute of Social Hygiene. Today the "princely villa" belongs to the Embassy of India. *Photo by the author, December 2011.*

as an alternative to the "bourgeois" Russian Academy of Science—clearly linked the research conducted at the Institute of Life with the Bolshevik Revolution and "revolutionary dreams" of eternal life. With their close ties to high-level Bolsheviks, both Pil'niak and Fedorovskii undoubtedly knew of the project to freeze Vladimir Lenin's corpse to preserve it for "eternity,"—advocated by Leonid Krasin, a prominent member of the Bolshevik Party, but ultimately rejected in favor of embalming.[151] For many readers, Pavlishchev's final note clearly alluded to this project, providing an additional layer of meaning to the whole story.

Like Tolstoy, Pil'niak and Fedorovskii did not dwell on the technical details of freezing and revival. But unlike Tolstoy's, their story provided a pointed commentary on the very idea that freezing and thawing could offer a means of, if not abolishing completely, at least postponing, death. Expounded upon by Pavlishchev, this idea signified an unwavering belief in the progress of science, which would inevitably result (if not tomorrow, then in a decade or two, or in a century at the latest) in conquering death. Clearly contrasted to the story's title, "The Affair of Death," the very name of the Institute of Life implied the promise that the unstoppable advance of science would abolish death as an inevitable part of human existence. Yet the bracketing of the story by the discourse on the *green* birch tree sitting on the *gray* asphalt-covered courtyard—with its obvious reference to Mephistopheles's famous line from Goethe's *Faust*, "Grau, teurer Freund, ist alle Theorie, und grün des Lebens goldner Baum"[152]—was clearly meant to raise in the reader's mind the question of whether science and scientists are indeed capable of delivering on that promise and whether the "theory" of abolishing death itself is worthy of pursuit.

The internal secretions are the most hopeful and promising of the reagents for control yet come upon by the human mind. [. . .] A few hundreds of investigators are engaged upon their study throughout the world. That is one of the ironies of our contemporary civilization. A concerted effort at the task of understanding them, backed by the labors of tens of thousands of workers, would, without a doubt, accomplish as much for humanity as the vast armies and navies that consume the substance of mankind.
LOUIS BERMAN, *The Glands Regulating Personality* (1922)

I do not say that biologists as a general rule try to imagine in any detail the future applications of their science. The central problems of life for them may be the relationship between the echinoderms and the brachiopods, and the attempt to live on their salaries. They do not see themselves as sinister and revolutionary figures. They have no time to dream. But I suspect that more of them dream than would care to confess it.
J. B. S. HALDANE, *Daedalus* (1923)

4 "The Billionaire's Last Will": Hormones and Institutions

WHILE MANY RUSSIAN writers and readers entertained the revolutionary dreams of a new world, a new society, and a new eternal life, many Russian scientists actively pursued different dreams, the dreams of a new kind of science. For years, scientists around the world dreamed of large institutions filled with complex machinery, numerous assistants, and abundant research materials, of well-stocked libraries and publishing facilities, and of regular conferences with their fellow scientists at home and abroad, which together would enable scientists to conduct their studies simultaneously on a large scale and along different lines of research. One of the first among Russian scientists to articulate forcefully and eloquently these dreams of grandeur was Porfirii Bakhmet'ev.

As we saw in Chapter 3, circa 1900 Bakhmet'ev worked under conditions typical for "little" science: a tiny laboratory set up in his own house and equipped with "homemade" apparatuses. He had to catch or trap his own experimental subjects, be they insects or bats, with one or two students and friends helping him conduct experiments. He was always scraping together money to buy equipment and books, attend a conference, or publish his research results. But he had extensive plans for his research and pursued studies in various areas of physics, entomology, physical chemistry, and biophysics all at the same time. He clearly understood the need for his "little" science to grow.

In the fall of 1904, deeply frustrated by his inability to find support for his studies of anabiosis in warm-blooded animals, Bakhmet'ev published a lengthy futuristic novella entitled "The billionaire's last will."[1] Appearing on the pages of *Natural Sciences and Geography*, one of the most respected popular-science magazines in Russia, it presented Bakhmet'ev's vision of how scientific research ought to be organized in the future. The novella's title indicated the keystone of its plot: a certain American billionaire left his fortune (four hundred million dollars!) to create a scientific institute. The novella was clearly inspired by the founding two years earlier of the Carnegie Institution of Washington. Moreover, Bakhmet'ev named his billionaire "Andrei Karnedzhi," a Russified version of the Scottish American steel magnate Andrew Carnegie, who endowed twenty-two million dollars for the establishment of the Institution.[2] Yet the novella deviated considerably from the actual events surrounding Carnegie's will and its implementation. Most important, the "central international research institute" Bakhmet'ev envisioned, and the manner in which this institute was supposed to operate, differed drastically from its "prototype"—the Carnegie Institution—and presented a detailed description of what eventually would be called "big science."

Bakhmet'ev actually imagined a "science city"—a large complex of nineteen specialized research institutes engaged in the development of seven different fields of knowledge: mechanics, physics, chemistry, astronomy, mineralogy, botany, and zoology. He placed this science city in the picturesque valley of Grindelwald at the foot of the magnificent Jungfrau, Eiger, and Monch mountains, not far from the town of Interlaken, which he had visited on several occasions during his years in Switzerland. He envisioned a Central Scientific Council that would run the complex comprised of the directors of the institutes—each one an internationally renowned researcher in his field elected to the post of the director by a ballot cast by forty scientific academies from around the world. The council would make all administrative decisions, allowing each director complete freedom to follow whatever lines of research he wanted.

Bakhmet'ev produced very detailed and precise plans for the nineteen institutes included in the complex. He allocated a large building of 57,000 square meters and a staff of 207 people to each institute. He planned several dozen auxiliary buildings to house a printing facility, a library, workshops for making instruments, and warehouses for storing necessary materials. He did not overlook the living needs of the personnel, incorporating plans for apartment buildings, a theater, a hospital, a restaurant, and a school. He made exact calculations of the sums of money necessary for constructing and running the buildings, paying salaries, buying materials and equipment, printing the institutes' *Proceedings*, and even providing medical care to the city's inhabitants.

Even if they might have disagreed on the exact details, many Russian scientists shared Bakhmet'ev's general vision of "big science" and put considerable efforts into its realization.[3] As we saw in the previous chapters, they successfully courted such Russian "billionaires" as Khristofor Ledentsov, Fedor Riabushinskii, Alfons Shaniavskii, and Moisei

Gal'perin, asking them to become their patrons and partners in creating scientific institutes, mounting scientific expeditions, and training new generations of scientists.

The February 1917 revolution, which dethroned the Romanovs and created a liberal Provisional Government that was to ensure Russia's transition to democratic rule, provided a new impetus to scientists' institution-building efforts. The Russian scholarly community moved quickly to capitalize on the liberties and promises of the revolution. During the spring of 1917, a group of the most prominent members of the Russian Academy of Science, including Nobelist Ivan Pavlov, took steps to organize a Free Association for the Development and Dissemination of the Positive Sciences, modeled after the British and US associations for the advancement of science. The initiative received wholehearted support from Russia's scholarly community, the Provisional Government, Russian industrialists, and the public at large. The organizing committee of the new association included eminent scientists, government officials, and public figures such as popular writers Maxim Gorky and Vladimir Korolenko. The organizers envisaged "a scholarly institution whose first and most important task will be the development and improvement of the exact sciences and their methods."[4] They wanted to establish "well-equipped research facilities and laboratories, along with museums, libraries, and teaching auditoriums."[5] During the spring and summer of 1917, the organizing committee conducted a massive propaganda and fund-raising campaign that featured public lectures by distinguished scientists and enthusiastic speeches by Alexander Kerensky, Andrei Shingarev, and Pavel Miliukov, leading members of the Provisional Government. As Pavlov emphasized in his address, Russia needed "special research institutes of natural sciences with huge resources."[6]

The Bolshevik coup of October 1917 shattered scientists' budding alliance with Russian capitalists and put an end to their collaboration with the Provisional Government. But it did not extinguish their dreams of "big science." They soon found an enthusiastic partner and a generous patron for realizing their institutional ambitions in various officials and agencies of the newborn Bolshevik government.[7] In fact, Russian scientists managed to involve the Bolsheviks in a large-scale experiment to create "big science" institutions throughout the country. Within just a few years after the Revolution, the joint efforts of scientists and their Bolshevik patrons would create a complex of research institutes closely resembling the science city imagined by Bakhmet'ev (located, of course, not in Switzerland, but in Moscow): the State Institute of People's Health Protection (GINZ). Run, as Bakhmet'ev would have wanted it, by the council composed of the directors of each individual institute that came under its umbrella, GINZ was charged with addressing questions pertaining to the entire field of medicine and public health in the country. And it was not a "billionaire" but the People's Commissariat of Health Protection (Narkomzdrav) that provided funds, buildings, materials, equipment, and personnel for establishing this unique institution. Several areas of research within the overlapping fields of experimental biology and medicine benefited especially from the new partnership between Russian scientists and their Bolshevik patrons and found institutional niches within GINZ's elaborate structures. One of them was the new science of endocrinology.

One of the most quickly developing branches of experimental biology and medicine in the late nineteenth and early twentieth century was research on endocrine glands and their secretions. The discovery of specific substances produced by endocrine glands, which in minuscule quantities profoundly affected the development and functioning of the body, cured previously lethal diseases, and promised instantaneous rejuvenating effects, generated much excitement and enthusiasm among scientists and clinicians. This discovery fit perfectly with the central tenet of visionary biology: to control life, death, and disease. Indeed, in 1905, introducing the new term "hormone" that he had coined for the products of endocrine glands, the British physiologist Ernest Starling announced: "within a reasonable space of time, we shall be in possession of chemical substances which are normal physiological products, and by means of which we shall be in a position *to control* not only the activities but also the growth of a large number of organs of the body."[8] Some fifteen years later, Starling's followers were even more optimistic. In his book, tellingly entitled *The Glands Regulating Personality*, one of the most active proponents of endocrinology in the United States, Columbia professor Louis Berman asserted forcefully: "The physiology of the internal secretions teaches that human nature does change and can be changed. A *relative control* of its properties is already in view. The *absolute control will come*."[9]

In the 1920s, the new discipline of endocrinology devoted to the study of internal secretions quickly developed in many countries around the world, including Britain, France, Germany, Russia, and the United States.[10] In Russia, its development exemplified the origins, dynamics, and mechanics of an unprecedented "institutional" experiment conducted jointly by visionary biologists and revolutionary dreamers—an experiment that resulted in building "big science" institutions and making science a mass profession.

As might be expected, the transformative power of hormones on the human body and personality inspired literary works that probed the social implications of the newest "elixirs of life." Yet scientists' visions of "big science" necessary to find and produce such "elixirs," as well as their extensive labors to implement these visions, failed to instigate literary experiments examining the implications of "big science" for both science and society in Bolshevik Russia. Even though, as we shall see, the stories of institution-building had the stuff of a good novel, including ambitious characters, high stakes, byzantine intrigues, powerful passions, and behind-the-scene machinations, the most successful among the experiments conducted jointly by Russian scientists and their Bolshevik patrons went virtually unnoticed by the literary community.

The Science of Internal Secretions

For most of its history, science in Russia developed in close contact with, and under the profound influence of, its counterparts abroad. Indeed, until the mid-nineteenth century, the majority of Russia's premier scientists were foreigners and after that nearly all of

Russia's leading scientists spent time abroad studying, doing post-graduate work, attending conferences, and, occasionally, lecturing at various scientific institutions in Western Europe. Not surprisingly, until the Revolution, Russian developments in experimental biology and medicine in general—and in endocrinology in particular—closely paralleled those in other European countries.[11]

As in the rest of Europe, in Russia studies of internal secretions and their organs began in the mid-nineteenth century following Claude Bernard's pioneering research into the *milieu intérieur* of the living organism.[12] Probably the first Russian work on what today would be called endocrinology was a doctoral dissertation, "On the role of the suprarenal glands in the animal organism," defended in 1862 at the Medical-Surgical Academy (later renamed Military-Medical Academy) in St. Petersburg. The dissertation described the results of experiments on the removal of the suprarenal glands (the adrenals) in guinea pigs, rabbits, and mice.[13]

Over the next twenty-five years, about fifty publications on internal secretions appeared in Russia. Some of them dealt with clinical and epidemiological aspects of various diseases of endocrine glands, including goiter, gigantism, Addison's disease, dwarfism, and diabetes. Others focused on anatomical, histological, and physiological investigations of the organs of internal secretion, including the thyroid, pancreas, adrenals, hypophysis, testicles, liver, and ovaries.

As elsewhere in the world, during the 1890s, a new direction in the studies of endocrine organs emerged in Russia—organotherapy. In 1889, Charles Brown-Séquard, a prominent French physiologist who had succeeded Claude Bernard, the "father" of experimental medicine, at the Collège de France, astonished the world with his announcement that injections of extracts prepared from animal testicles had "rejuvenating" effects on the human organism. He had tried it on himself and it had certainly worked: after the injections, he felt much younger. Following this sensational announcement, scientists and physicians around the world enthusiastically began to apply extracts of various animal tissues as a cure for a variety of diseases.[14] In Russia, several entrepreneurial pharmaceutical companies, including the leading firms of Alexander Pel' [Poehl] in St. Petersburg and Karl Ferrein in Moscow, immediately began producing and marketing extracts of various endocrine organs, primarily, of course, testicles.[15] They also initiated and funded research on the physiological and medicinal effects of such extracts in the laboratory and clinical practice.[16] In 1900, Pel' even added the word "organotherapy" to the name of the *Journal of Medical Chemistry* he had founded a few years earlier.[17] He also established a special "organotherapeutic institute" to produce and study various extracts, beginning with "spermin."[18]

As elsewhere in the world, by the turn of the twentieth century studies of internal secretions in Russia had diverged into three well-defined subfields: first, clinical work on the diseases of endocrine organs; second, laboratory research into physiology, anatomy, chemistry, and histology of endocrine glands; and third, pharmacological studies and therapeutic applications of the extracts of endocrine organs.[19] The three subfields differed not only in their particular focus—the disease, its mechanisms, and its treatments—but also, and most importantly, in terms of the institutional settings and workforce characteristics. Although there was a certain overlap in personnel, the first subfield was primarily the domain of practicing physicians; the second of pathologists, physiologists, and, to a lesser extent, chemists concentrated in the laboratories of various educational and research institutions (most notably Moscow and St. Petersburg universities, the Military-Medical Academy, and the Imperial Institute of Experimental Medicine); and the third of a group of pharmacists and chemists involved with the production and distribution of endocrine gland extracts and practicing physicians who used these extracts in the treatment of their patients. But, as we will see, early endocrinology also attracted a variety of scientists, clinicians, and entrepreneurs.

In the 1910s, the budding discipline of endocrinology found an unusual institutional setting under the auspices of the Russian Society for Goat Breeding. In early 1911, the society's Moscow Branch established a farm on the small estate of its patroness, Princess Sofia Shcherbatova, in the village Vsekhsviatskoe on the outskirts of Moscow. The farm's major goal was to provide goat milk—believed to be beneficial for TB sufferers—to sanatoria, clinics, and orphanages, as well as to paying customers in the Moscow region. The farm was created and run by one of the society's founding members, veterinarian Iakov Tobolkin (1870–1941).[20]

Born in 1870 in the small Siberian town of Tobol'sk, Tobolkin apparently came from a humble background.[21] Judging by his surname that denoted his town of birth, his father was probably a low-level priest in the Russian Orthodox Church and, in all likelihood, Iakov began his education at a local theological school (seminary). But he did not follow the clerical calling. Sometime in the early 1890s, Iakov entered a veterinary school established in 1874 in Kazan', a large industrial city in the eastern part of European Russia renowned for its university.[22]

Perhaps his initial ambition was to go to the university to study medicine, but Iakov could not enter a medical school—Russian universities required their entrants to have completed eight years of secondary education at a classical gymnasium—and the Kazan' veterinary school presented an acceptable alternative. Veterinary medicine was a rare and not very prestigious profession in the Russian Empire. At the turn of the twentieth century there were only four veterinary schools (institutes)—one each in Iur'ev, Kazan', Khar'kov, and Warsaw—which together produced fewer than one hundred graduates per year to serve the enormous, predominantly agricultural empire.[23] Veterinary schools were certainly second- (if not third-) rate educational institutions: professors were not required to have completed doctoral degrees, curricula were outdated, research facilities were lacking. But, unlike medical schools, veterinary ones welcomed seminary graduates, who constituted more than half of the student body at the Russian veterinary schools.[24] After four years of studies, graduates received the degree of "bachelor in veterinary medicine," and most of them found employment either with the *zemstvos* (local government bodies in the rural areas) or with military, state, and private horse-breeding establishments.[25] With Russia's enormous distances and slow tempo of industrialization and urbanization, the horse remained the main means of transportation, and horse breeding was patronized by the Russian royalty and nobility.

Tobolkin graduated from the Kazan' Veterinary Institute in 1897, but apparently the customary career of a zemstvo or a military veterinarian did not satisfy his ambitions. Some two years later he came to Moscow and joined the Russian Society for the Prevention of Cruelty to Animals to head one of the society's animal clinics.[26] He also established a small private practice and occasionally served as a veterinarian for the Moscow Zoo.[27] He obviously did well for himself, for the 1910 Moscow address directory lists him as the owner of a house in a prestigious neighborhood—Arbat—and the owner of a private veterinary clinic.[28] Around the same time, Tobolkin managed to attach himself to the Russian Society for Goat Breeding and proved instrumental in setting up its Moscow Branch. He was accorded a "membership for life" in the society and was very active in all its endeavors, including annual fairs, competitions, correspondence, and meetings.[29]

The society's goat farm was Tobolkin's pet project, and he personally oversaw all facets of its operations. As Tobolkin later reported, almost immediately after the establishment of the farm, physicians had begun bombarding him with requests for milk obtained from goats with excised thyroid glands, as well as for thyroid hormone (thyroidin) produced from the glands.[30] Most likely, these requests had been instigated by the demonstration of a goat with excised thyroids at the society's first exhibition held in late February through early March 1911. The goat attracted considerable attention as a "particularly interesting and appearing for the first time [in Russia]" specimen that might have important medical implications.[31] As a report on the exhibition published in the society's journal emphasized, milk from goats with excised thyroids might have a medicinal effect on patients suffering from Basedow's disease (hyperthyroidism).[32]

Tobolkin did all he could to accede to doctors' requests. On April 7, 1912, he delivered a report to a general meeting of the Moscow Branch of the Russian Society for Goat Breeding on "The thyroid gland in the animal organism, the technique of its extirpation in goats, and the demand for the milk from the goats with excised thyroids in the treatment of Basedow's disease in humans."[33] By September 1912, the farm had produced its first 180 glasses (about 45 liters) of "medicinal goat milk." Milk production steadily increased, and during the next year the farm produced 9,675 glasses (about 2,400 liters) of milk, satisfying the requests of sixty-one physicians and earning nearly two thousand rubles (quite a sum in those days) from the sales. With shrewd business acumen, Tobolkin also rented out his goats to private sanatoria in the region and sold thyroidectomized goats to doctors in the provinces, including Odessa, Kiev, the Crimea, the Caucasus, and even Siberia. In 1913 alone, he performed thyroidectomies on 178 goats for various private practitioners. The farm prospered and expanded considerably: in 1914 its milk production reached 11,511 glasses (almost 3,000 liters).

Although the demand for "medicinal goat milk" was huge, there was very little evidence of its actual medicinal properties.[34] Belief in the therapeutic effects of the milk from thyroidectomized goats was grounded in the contemporary understanding of the action of the thyroid glands and their secretions. Under normal conditions, the thyroid hormone was thought to "neutralize" a particular "toxin" circulating in the blood. In a patient suffering from hyperthyroidism, the glands excreted far more hormone than was "necessary," producing the characteristic symptoms of Basedow's disease: goiter, rapid pulse, nervousness and irritability, tremor of the outstretched hands and tongue, and "goggle-eyes." In goats with excised thyroids, and hence with no hormone, the "toxin" accumulated in the blood and made its way into the goats' milk. Thus, it was believed that drinking this milk provided patients with the "toxin" necessary to offset the excessive amounts of the hormone produced by their own overactive glands.[35] "Medicinal milk" certainly presented a very welcome alternative to the then standard treatment—the surgical excision of part or the entire thyroid gland to reduce the quantity of hormone in the patient's blood.

The farm had neither facilities for research, nor ways of controlling the actual uses of its product. So, with his characteristic entrepreneurship, Tobolkin got several researchers involved in studies of goat milk and its medicinal properties. Between 1913 and 1914, at Tobolkin's request, the director of the veterinary-biological laboratory of the Moscow Zoo, Nikolai Babushkin (also a graduate of the Kazan' Veterinary Institute), conducted comparative physical and chemical analyses of milk, blood, and urine in normal and thyroidectomized goats.[36] He discovered certain differences in the chemical composition and physical characteristics of the fluids from normal and operated animals but was unable to find and identify any substances that might have been responsible for the reputed therapeutic effects of the milk from goats that had undergone operation. Tobolkin also appealed to doctors who purchased the milk with requests to collect all possible information relating to the use of the milk and its effects on patients.

Among the early enthusiasts of "medicinal milk" was a professor at Moscow University's Medical School, Vasilii Shervinskii (1850–1941).[37] In contrast to Tobolkin, Shervinskii was born into a family of the old Russian intelligentsia.[38] His parents died when he was only three years old, and Vasilii was raised by his relatives, first his aunt and then his uncle, Pavel Chebyshev, Russia's leading mathematician. As was expected of a Moscow *intelligent*, Vasilii graduated from a classical gymnasium and enrolled into Moscow University. But he did not follow his uncle's passion for mathematics, instead choosing to study medicine. He was a diligent student. In 1873, Vasilii graduated from the Medical School and was appointed as an assistant to its department (*kafedra*) of pathological anatomy, a clear sign of his teachers' high regard for his abilities. He began working on a doctoral dissertation and six years later earned the degree Doctor of Medicine.[39] At the age of thirty, Vasilii Shervinskii became a professor in the department of pathology at his alma mater, and director of one of its clinics. He got deeply involved in various professional endeavors, becoming an active member of the Pirogov Society of Russian Physicians (as the chairman of its Tuberculosis Commission) and playing a key role in creating the Moscow Therapeutic Society.[40] Shervinskii also established a successful private practice and quickly earned a reputation as Moscow's leading specialist in internal diseases. He was one of the first

Russian physicians to grasp the significance of contemporary physiological studies of hormones for medical practice. In 1910, he published a long essay on the "clinical importance of internal secretions" in Russia's leading medical journal.[41] Shervinskii also chaired the Russian Anti-Tuberculosis League for many years. Perhaps it was through his involvement with the League that Shervinskii first learned about Tobolkin and his farm; after all, its stated purpose was the production of goat milk for TB sufferers.

Shervinskii became very interested in "medicinal milk," actually performing an autopsy of one of Tobolkin's thyroidectomized goats and investigating the anatomical and histological changes induced by the operation.[42] As the head of a Moscow University clinic and a therapist with extensive private practice, Shervinskii had a number of patients suffering from hyperthyroidism, and he began using the milk from the goats with excised thyroids on his patients. As he reported to the Moscow Therapeutic Society in 1913:

> Based on my personal experience, I can safely say that in some cases the milk of thyroidectomized goats certainly exerts a positive effect on patients with Basedow's disease. Although we do not yet have thorough research directed specifically to the elucidation of this question, the practice seems to justify this treatment.[43]

The outbreak of World War I in the summer of 1914 impeded the work of Tobolkin's farm. With the medical community's attention turned to urgent military needs, the farm—faced with widespread shortages—was all but abandoned during the war years. A new chapter in its history began after the Bolshevik coup of 1917, and the contact established between Tobolkin and Shervinskii over the use of "medicinal milk" proved momentous—not only for the two men, but for the entire field of endocrinology in Russia.

Hormones and the Commissars

On July 10, 1918, a Congress of soldier, peasant, and worker deputies adopted the first constitution of the Russian Soviet Federated Socialist Republic (RSFSR). The next day, the leader of the Bolsheviks, Vladimir Lenin, signed a decree establishing a special agency to "protect the health of the people" in the newborn republic: the People's Commissariat of Health Protection (Narkomzdrav). Issued by the highest government office, the Council of People's Commissars (SNK), the decree formed a ruling body—a Collegium composed of seven party members with medical backgrounds—and appointed Nikolai Semashko, a Bolshevik physician and a vocal advocate of social medicine and social hygiene, to head the new commissariat.[44]

The Commissar and his staff faced an enormous job. To implement the Bolsheviks' vision of "protecting the health of the people," Narkomzdrav's officials had to overcome the hostility of the majority of Russian physicians to the regime and to mobilize them to rebuild the country's public health system and medical services.[45] Naturally, one of the first priorities was to set up medical services for the newly created Red Army. With

World War I just ended (for Russia) and the Civil War rapidly escalating, the epidemiological situation in Russia was nothing short of catastrophic. Endemic and epidemic diseases, malnutrition, the collapse of sanitary infrastructure (from water supply to sewage disposal to garbage collection) in the cities and the total absence of such infrastructure in the rural areas, the acute shortage of even the most basic medicines, supplies, and trained personnel—all demanded urgent attention and immediate action. But, as we shall see, that was not all that was on the Commissariat's plate.

Just a few weeks after the establishment of Narkomzdrav, on August 1, 1918, Commissar Semashko received a note from SNK chief of staff Vladimir Bonch-Bruevich. The note introduced a short memorandum written by Iakov Tobolkin, which asked for the nationalization of his goat farm in order to "put the treatment of Basedow's disease on wide democratic principles."[46]

Why, we might wonder, did Tobolkin appeal to Bonch-Bruevich? How did he manage to get access to such a high-level official in the new Bolshevik government? And why did the SNK chief of staff decide that his appeal was worthy of support? Several documents preserved among Bonch-Bruevich's personal papers shed light on this mystery.[47] As it turns out, Tobolkin had a longtime association with several influential Bolsheviks. During his student years at the Kazan' Veterinary Institute in the early 1890s, Tobolkin befriended a fellow student, Nikolai Bauman. Bauman did not graduate, leaving the school in his fourth year of study. He turned to politics and chose a career as a professional revolutionary. He became one of the founding members of the Bolshevik Party and eventually came to head its Moscow organization.[48]

Tobolkin shared with many members of the Russian intelligentsia disdain for the monarchy and high hopes for a democratic revolution. In the early 1900s, he took part in the activities of the Socialist-Revolutionary Party, one of the most popular "agrarian" parties of the time. During the first Russian revolution of 1905–06, the Moscow police department kept Tobolkin under surveillance, and his name appeared in several police reports on revolutionary activities in the city.[49] The head of the police department's "special division" (*okhrannoe otdelenie*) characterized him as "politically unreliable."[50] During this time Bauman used Tobolkin's house as a drop for party propaganda materials and correspondence, as well as a place for clandestine meetings with fellow revolutionaries. At these meetings Tobolkin met a number of Bolsheviks, including the future SNK chief of staff Bonch-Bruevich. After the defeat of the 1905–06 revolution, Tobolkin, like many other members of the intelligentsia, abandoned his political radicalism and turned to advancing his professional career. But with the Bolsheviks now in power, Tobolkin's early flirtations with the revolution proved quite useful.

In 1918, his high official title notwithstanding, Bonch-Bruevich was first and foremost a personal assistant to his old-time friend, Vladimir Lenin, now the SNK chairman.[51] As it happened, Lenin's wife, Nadezhda Krupskaia, suffered from Basedow's disease, and Bonch-Bruevich's wife, Vera Velichkina, who had studied medicine in Switzerland, had served as a personal doctor for Krupskaia.[52] In the spring of 1918, after the Bolshevik government's move to Moscow, Krupskaia became a deputy head of the People's Commissariat of Enlightenment (Narkompros). She was working hard to revive and reform the country's system of primary and secondary education. This intense work seems to have led to a worsening of Krupskaia's condition. Perhaps Velichkina had heard about the "medicinal goat milk" that supposedly helped patients with Basedow's disease, or maybe her husband was familiar with Tobolkin's farm and its main product. Whatever the case, Bonch-Bruevich called Tobolkin to his Kremlin office and asked him to procure some "medicinal milk" for Krupskaia. Tobolkin obliged. He thyroidectomized one of the goats remaining at the farm and began supplying its milk to the Kremlin. Reportedly, the treatment was able to alleviate Krupskaia's condition and, according to Bonch-Bruevich, "literally, put her back on her feet" (see Figure 4.1).[53]

FIGURE 4.1 Vladimir Lenin's wife, Nadezhda Krupskaia, June 1919. Note her "goggled eyes," a characteristic symptom of Basedow's disease. In practically all surviving photographs, Krupskaia wears turtleneck sweaters or scarves to cover another visible symptom of her affliction—goiter. As this photpgraph clearly attests, the "medicinal milk" did indeed "put her back on her feet." This photo, taken less than a year after she had started using the "medicinal milk," shows Krupskaia aboard "Red Star"—a "propaganda steamboat." In the summer of 1919 "Red Star" took a long voyage along Volga to disseminate the Bolsheviks' plans for the country's future in the areas recently liberated from the White Guards in the course of the Civil War. During the voyage, practically every day, Krupsakaia delivered lectures and spoke at the meetings organized in numerous villages and towns along the route. *Courtesy of RGASPI.*

Tobolkin immediately capitalized on this success to revive his "medicinal milk" enterprise, and Bonch-Bruevich readily provided the necessary introduction to the head of Narkomzdrav. The endorsement by the SNK chief of staff was nothing short of a direct order. So, the next day after Tobolkin's memorandum had landed on Semashko's desk, on August 2, 1918, with the Commissar in the chair, Narkomzdrav's Collegium listened to Tobolkin's presentation of his case.[54] The farm was on its last legs: out of its original herd of seventy goats, only thirteen remained. There was no feed for animals, nor food, money, or fuel for personnel, yet there was a constant threat of looting or even torching of the farm by "revolutionary peasants" in nearby villages. Tobolkin asked the Collegium that the farm be nationalized and taken under the wing of Narkomzdrav as an important medical institution. This would both provide protection from the local "revolutionaries" and, through Narkomzdrav's subsidies, allow the farm to resume its operations. Whether all of the Collegium members knew of Tobolkin's deliveries of his "medicinal milk" to the Kremlin or not, they were clearly impressed by the presentation. They resolved that the farm should be "taken temporarily under Narkomzdrav's auspices in order to use it for the treatment of Basedow's disease" and asked Tobolkin to prepare an inventory of the farm's properties. It obviously helped that one of the Collegium's seven members was none other than Vera Velichkina.

A week later, Tobolkin presented the inventory and a preliminary operational budget to the Collegium. At its meeting on August 10, 1918, the Collegium resolved that the farm be nationalized and allocated funds for its daily operations, plus a special sum for the purchase of new goats.[55] Commissar Semashko personally signed the mandate indicating that the farm was now the property of Narkomzdrav. Although the mandate certainly helped Tobolkin protect the farm, the promised funds proved much more difficult to obtain. A month later, on September 16, Tobolkin again appealed to the Collegium for the pledged money to buy additional goats and feed for the farm.[56] This time, however, the Collegium forwarded the request to a newly created Scientific Medical Council—a consultative body comprised of eminent medical researchers and practitioners sympathetic to the new regime and chaired by the country's leading bacteriologist, Lev Tarasevich. The Collegium asked for appraisal not only of Tobolkin's request, but also of the entire project of maintaining and expanding the farm of thyroidectomized goats. It took the Council another month to get to the issue: on October 19, Tarasevich notified Semashko of the Council's inability to properly assess the project due to an absence of necessary information about the farm and its operations. Evidently, the Council members had never heard of Tobolkin or his farm.

Semashko immediately sent a message to Tobolkin asking him to prepare a statement and provide all relevant materials regarding the farm. Tobolkin promptly presented to the Council an eight-page memorandum that outlined the farm's history, plans for its expansion, and even a "program of future studies."[57] The memorandum demonstrated that Tobolkin had not been waiting idly while the Narkomzdrav officials had been pondering over his initial appeal. In September, he had delivered a report about the farm and its work to the Institute of Experimental Veterinary Science, recently (in January 1918) created under the aegis of the People's Commissariat of Agriculture (Narkomzem). Tobolkin claimed that he had secured an agreement with members of the institute for cooperative research on the properties of the blood, milk, and urine of thyroidectomized goats. The memorandum also listed several eminent physicians familiar with the farm's work, who, in Tobolkin's opinion, should be invited to take part in any discussion of the farm's future. The first name on the list was Vasilii Shervinskii.

Tobolkin's memo obviously made the desired impression on the Council members. They indeed invited Shervinskii to attend the Council's meeting on November 1 to discuss the issue. As a result of this discussion, the Council recommended that the farm be expanded. The Council also suggested that a special commission be set up to study the medicinal properties of the milk from thyroidectomized goats and to develop the production of several "hormonal" preparations. The list of preparations included "rodagen" (condensed milk), "antithyroidin" (the blood serum from the operated goats), and "thyroidin" (an extract from the excised thyroid glands). The Council forwarded its recommendations to the Collegium.[58]

The Collegium, however, was not satisfied. A week later, it returned the Council's recommendations, requesting additional materials. The Collegium members wanted a detailed justification for the farm's expansion: what exact size of the farm would be necessary for conducting scientific experiments; how the milk from the operated goats would be studied; and how the farm would manufacture the proposed preparations.[59] Perhaps the Collegium's request for additional details was prompted by the fact that a major promoter of "medicinal milk," Vera Velichkina, had died a month earlier. But most likely, the Collegium's dissatisfaction with Tobolkin's plans derived from concurrent discussions regarding the establishment of GINZ.[60] Just a few weeks earlier, on October 18, the Collegium had approved the founding of the first research institute that would become a part of GINZ—the Institute of Tropical Diseases. Apparently, compared to the detailed justifications for the establishment of this institute, which had been prepared by the Malaria Commission of the Pirogov Society of Russian Physicians that included Russia's leading specialists in tropical diseases, the Council's recommendations regarding scientific research at the farm—based on Tobolkin's memo—looked weak to Collegium members.[61]

To satisfy the Collegium's request, the Council appointed a special commission for "scientific control over the farm of thyroidectomized goats."[62] The commission was to supervise "experiments [with] and production of organotherapeutic preparations at the farm." It consisted of four members of the Council: its head Tarasevich, surgeon Aleksei Martynov, the dean of Moscow University's Medical School, Dmitrii Pletnev (the chairman), the eminent therapist, and the physiologist Mikhail Shaternikov. The commission also included one "invited member"—Vasilii Shervinskii.

Shervinskii was the only member of the commission who had previously dealt with the farm and who had had firsthand experience in the use of its main product—"medicinal milk." His opinions obviously carried considerable weight with the other members and were clearly reflected in the commission's report presented to the Council on December 20.[63] The report emphasized the general importance of organotherapeutic preparations as the main rationale for the preservation and expansion of the farm. It repeated almost verbatim Shervinskii's 1913 statement regarding certain "practical" evidence for the therapeutic effects of "medicinal milk," despite the lack of direct scientific research on the subject. Indeed, the report stressed that it was the need for such studies that necessitated the farm's expansion. The report went on to suggest that the farm should be expanded, at minimum, to its pre-war size to satisfy the demands of research and to supply milk to at least some of the patients suffering from Basedow's disease. The report also proposed that a special "manufacturing laboratory"—modeled after the pre-revolutionary establishments involved with the production of organotherapeutic preparations such as Pel's and Ferrain's laboratories—be set up at the farm to manufacture the proposed preparations. Shervinskii's presentation to the Council hit its mark: the Council approved the commission's report and forwarded it to the Collegium. Furthermore, that very day, Shervinskii was "elected" to the Council.[64]

At the end of January 1919, Tobolkin and Shervinskii could celebrate their first success: the Collegium had approved the farm's expansion and recommended that the patients for clinical tests be concentrated in one of Moscow University's clinics.[65] The Collegium's recommendation meant that control over the tests and over precious "medicinal milk" would remain in the hands of commission members Dmitrii Pletnev and Vasilii Shervinskii. Given the fact that Pletnev was one of the trusted physicians frequently invited to treat patients in the Kremlin,[66] the Collegium's recommendation indirectly supports other evidence indicating that Lenin's wife was one of the intended beneficiaries of the tests. A few months later, perhaps inspired by their success in establishing the farm on firm footing under Narkomzdrav patronage, Shervinskii and Tobolkin appealed to the Council with suggestion that the farm be transformed into a special organotherapeutic institute.[67]

It seems likely that the primary motive for this suggestion was the appearance of a formidable competitor. In mid-January 1919, a certain doctor, Nikolai Ishlondskii, had presented to Narkomzdrav a detailed project urging the "organization of a special institute for the production of organotherapeutic preparations."[68] The project promised to "resolve the current medicaments' crisis in the country" by establishing a manufacturing facility that would produce preparations from a variety of endocrine organs, including the adrenals, hypophysis, and pancreas. Appended with a list of the sixty-seven "most popular organotherapeutic preparations in Europe" and the names of the pharmaceutical companies that produced them, the project claimed that the "manufacture of the most important and valuable preparations such as adrenalin and pituitrin [an extract of the pituitary gland] could be organized immediately at very low initial cost." Ishlondskii estimated that approximately 300 liters of adrenalin (valued at 200,000 rubles) could easily be produced every month. He described the necessary personnel, supplies, and equipment and asked for 500,000 rubles to organize the projected institute.[69]

Ishlondskii had managed to get his project endorsed by the head of the Moscow Regional Sanitary-Military Administration, and Narkomzdrav's officials took it quite seriously. In the space of just two weeks, the project had reached Semashko's desk, and the Commissar forwarded it to the Scientific Medical Council for evaluation. As was its custom, the Council created a commission to address the issue. The chairman of this commission was none other than Shervinskii.

In evaluating Ishlondskii's project, Shervinskii faced the challenging task of undermining the competitor while endorsing his central idea. Needless to say, he accomplished this task admirably.[70] Shervinskii began his report to the Council by emphasizing the practical and theoretical importance of organotherapeutic preparations but noted the considerable difficulties of manufacturing processes, as well as the need for veterinary control of raw materials (animal glands) and for reliable, skilled personnel. He stated that establishment of the projected institute should be motivated not by financial gain—hinting that this was the primary motive of Ishlondskii's project—but by research and clinical needs. He further asserted that from the viewpoint of research and clinical applications, the creation of such an institute was indeed timely and desirable. Shervinskii then turned to the project itself, focusing primarily on its author's "commercial projections and estimates" and stressing that the market for organotherapeutic preparations was relatively small and unstable. He dismissed Ishlondskii's claim that creation of the institute would "resolve the current medicaments' crisis," for organotherapeutic preparations could not even theoretically replace other drugs. He recommended that the project be rejected as "unsubstantiated."

Yet this recommendation was not the end of Shervinskii's report. He proceeded to suggest that, as a first step, the manufacture of organotherapeutic preparations could be organized at "existing laboratories" (a veiled reference to Tobolkin's farm, which at that time was the only institution involved with organotherapeutic preparations under Narkomzdrav's wing), if their personnel and equipment were expanded. He projected that these laboratories "in time would grow into a large organotherapeutic institute, with all the required departments," including research and clinical facilities. He proposed that "the commission created for overseeing the production of preparations at the farm of thyroidectomized goats [e.g. Shervinskii himself] work out the issues of manufacturing organotherapeutic preparations and improving the existing laboratories created for this purpose."

On March 7, 1919, the Council approved Shervinskii's report—both its recommendation to reject Ishlondskii's project and its proposal to "work out" the issues of manufacturing organotherapeutic preparations—and forwarded its verdict to Narkomzdrav. Just a few days later, Shervinskii and Tobolkin presented their own appeal for the establishment at the goat farm of a special organotherapeutic institute to the Council.[71] It seems likely that in addition to the Ishlondskii affair, Shervinskii and Tobolkin were also encouraged by the rapid growth of GINZ. By March 1919, GINZ had expanded to include four separate research institutes—Microbiology, Sanitary Science and Hygiene, Vaccines and Sera, and Tropical Diseases—administered by a board composed of the director and leading researchers of each institute. Although at this time, Shervinskii-Tobolkin's initiative did not find much support in the Council—the Civil War was at its height and the Moscow medical community was busy fighting the horrendous epidemics of typhus and Spanish flu—the farm's "supervisors" continued their efforts to convert it into a full-fledged research institute.

Under the conditions of War Communism—with its shortages of literally everything and the Bolsheviks' strict administrative control over the distribution of food and goods—the farm offered attractive employment. Not only did the farm provide the opportunity to obtain daily food rations distributed by the government—consisting mainly of bread and pickled herring, barely sufficient for sustenance—but it also supplemented these meager rations with produce from the farm's vegetable garden and milk from its herd of unoperated goats. Not surprisingly, the farm's "supervisors" had no shortage of work applications.

In the summer of 1919, the biochemist Aleksei Bakh, a member of the notorious "People's Will" group, who had recently returned from political exile in Geneva,[72] approached Shervinskii with an offer to perform analyses of "the enzymatic constants in normal and thyroidectomized animals."[73] Bakh suggested identifying the median quantity of certain blood enzymes—a technique he had perfected during his studies in Switzerland—in normal and operated animals, and investigating how these numbers changed when animals were injected with thyroid preparations.[74] Obviously realizing what a valuable ally Bakh would make, Shervinskii immediately penned a recommendation and presented it to a meeting of the Scientific-Medical Council on July 4. He characterized Bakh's proposal as "extremely desirable," for these analyses could provide a "solid and positive basis for the evaluation of the action of organotherapeutic preparations," as well as improve the production of such preparations. The Council approved the recommendation and permitted Bakh to conduct his experiments at the farm.[75]

Tobolkin and Shervinskii persistently cultivated patrons and allies in the Narkomzdrav Collegium and the Medical Scientific Council. On July 10, 1919, Tobolkin reported to the Collegium "on the situation at the farm and on the course of scientific investigations into the influence of milk from thyroidectomized goats on patients suffering from Basedow's disease."[76] The Collegium reaffirmed the control by the farm supervisors over the farm's products: "only with a prescription" issued by members of "the scientific commission that studies the medicinal properties of the milk" could patients get access to "medicinal milk." At the same time, Shervinskii expanded his own influence in the Council. In August he came to chair a special "permanent pharmacological commission" instituted to "control the production, distribution, and use of various medicinal preparations" in the entire country.[77]

Thanks to its supervisors' untiring efforts, over the next year the farm steadily increased its milk production and expanded its "research agenda." In February 1920, Shervinskii reported to the Council that the first results of studies on the therapeutic effects of the milk from thyroidectomized goats on patients with Basedow's disease had proved very encouraging. He proposed that "clinical tests" be expanded to include patients with tuberculosis. The proposal made no sense from the medical viewpoint: what role could the "antithyroidin" supposedly contained in "medicinal milk" play in the treatment of tuberculosis? Yet apparently it made perfect sense from a "political" point of view. Several high-level Bolsheviks, notably the chairman of the fearsome "Cheka"[78] Feliks Dzerzhinskii and his deputy Viacheslav Menzhinskii, suffered from tuberculosis.[79] Given the close connections of the farm supervisors to the top level of the Bolshevik government (Shervinskii personally treated Menzhinskii),[80] it seems likely that Shervinskii's proposal meant that "medicinal milk" would now be administered not just to Krupskaia, but to other inhabitants of the Kremlin as well. The Council approved Shervinskii's proposal. Its members also recommended that Bakh broaden his research on the enzyme composition of "medicinal milk" by performing analyses of changes in patients' blood.[81]

In October 1920, Shervinskii proudly reported to the Council that "the farm's laboratory" had produced its first two preparations—thyroidin and antithyroidin.[82] But in fact, the "laboratory" existed only in Tobolkin's and Shervinskii's reports to their patrons. The farm itself had neither space nor equipment for production of, not to mention research on, these hormonal preparations. The preparations had been "homemade," as it were: the farm's technician had manufactured them in the basement of Tobolkin's house on Arbat in the center of Moscow. Not surprisingly, Tobolkin and Shervinskii regularly petitioned the Council to provide appropriate quarters and personnel for manufacturing and research.[83] At the end of the year, the duo also appealed to the Council to rename the farm the "Institute of Medicinal Milk."[84] They argued that the existing name—the farm—impeded research and even day-to-day operations due to confusion it created in various state agencies. Some bureaucrats evidently thought that the farm was an agricultural, not a medical, institution, and they refused to provide ordered supplies and equipment and even requested that the farm deliver milk as part of the obligatory "natural tax" imposed upon all agricultural producers during the Civil War. Although the Council did not explicitly approve the name change, Shervinskii ordered a letterhead stamp with the new name and began using it in official correspondence.[85]

Of course, Tobolkin's and Shervinskii's early engagement with Narkomzdrav could be construed as merely a means of survival for two "rotten *intelligenty*" (in the parlance of the day) under the brutal regime and the dire economic and social conditions of War Communism.[86] But their actions suggest that it was much more than that. The duo demonstrated tremendous personal initiative, persistence, and devotion in their efforts to build a special institute for endocrinology. We can only imagine the difficulties Tobolkin had to overcome simply to provide his goats with feed. During the period between 1918 and 1920, the country virtually disintegrated into a patchwork of autonomous regions. Peasants refused to provide the fledgling regime (or its opponents, for that matter) with their products, and the Bolsheviks had to send special military brigades to collect foodstuffs from the countryside. Whatever remained of the country's transportation system was mobilized to serve military objectives. Yet somehow, Tobolkin kept the farm going and even growing. For his part, Shervinskii, who turned sixty-seven at the time of the Bolshevik Revolution in 1917, had to walk almost five kilometers each way through the snowbound Moscow streets to attend the meetings of the Narkomzdrav Scientific-Medical Council and endure hours of discussions, with the temperature in the meeting room barely reaching 3 degrees Centigrade.[87] The tantalizing vision of a large specialized institute to pursue their scientific interests obviously provided a critical motivation for the duo's perseverance.

"Big Science" in the Making

By 1921 the Civil War had spent its fury: the Bolsheviks regained control over the country. The introduction of NEP and the subsequent rapid restoration of the national economy greatly facilitated the discipline-building efforts of Russian scientists. Scientists could start implementing their grand visions of "big science" in earnest, using the considerable resources the Bolshevik state now commanded.

But for Shervinskii and Tobolkin, this appeared to be a mixed blessing, since in January 1921 they lost one of their most valuable assets—Bakh. Using his reputation as an "old revolutionary" and his personal contacts with influential Bolsheviks, as well as his affiliation with the farm as a jumping board, Bakh managed to convince Narkomzdrav officials to establish a separate Institute of Biochemistry for his own research under the GINZ umbrella.[88] Perhaps inspired by this institutional success of one of the farm's associates, throughout 1921, Shervinskii and Tobolkin continuously bombarded the Council with requests that a special "institute for the manufacture of organotherapeutic preparations" be established.[89] But unlike Bakh's, their requests proved futile: Council members duly discussed their presentations and formed several commissions to address the issues, but in the end could not make up their minds.[90]

Faced with the reluctance of their patrons in the Council to accommodate their institutional ambitions, Shervinskii and Tobolkin adopted a different tactic in their efforts to transform the farm (aka Institute of Medicinal Milk) into a full-fledged research and production facility. Having failed to establish a separate institute, they attempted to put the farm—as an independent department—under the aegis of an existing institute. The Institute of the Physiology of Nutrition established as part of GINZ in late 1920 seemed an ideal candidate for such a move. The institute's director, physiologist Mikhail Shaternikov, was a member of the Narkomzdrav Scientific Medical Council, as well as a member of the original 1918 commission created for the "scientific control over the farm." Busy with his own research and institution-building, Shaternikov had not contributed much to the commission's work. Nevertheless, he was familiar with the farm and might look favorably at taking it under his wing, particularly because he had been keenly interested in the study of hormones. Indeed, Shaternikov had planned to establish—within his own institute—"a special department to study the influence of the organs of internal secretion on growth and nutrition."[91]

On September 22, 1922, Shervinskii presented the idea to the Council in a special report.[92] He argued that the farm's inclusion in Shaternikov's institute would help "expand its scientific research" and provide it with "necessary direction." As always, the Council formed a commission to assess the issue. Composed of the three individuals most interested in the entire project—Shervinskii, Tobolkin, and Shaternikov—the commission quickly came to an agreement. They recommended that the farm be included in the Institute of the Physiology of Nutrition, on one condition: "its budget, administrative subordination, and practical activities should remain fully autonomous." The institute was to provide only a "general direction for scientific work," through the cooptation of "the institute's representatives [read—Shaternikov and his deputies] into the existing commission for control over the farm [read—Pletnev, Shervinskii, and Tobolkin]." On November 17, after receiving consent of the GINZ board, the Council approved the commission's recommendations.[93]

Shervinskii's latest institutional maneuver was very crafty indeed. The "inclusion" actually meant that the farm would become a financially independent unit within the system of GINZ institutions, with only nominal affiliation to Shaternikov's institute. Such an arrangement satisfied everyone involved. It meant, first of all, that the farm supervisors could now claim a share of resources, most importantly space, personnel, and equipment, allocated by Narkomzdrav to the quickly expanding GINZ. It also allowed the farm to gain access to specialists, equipment, and techniques at other GINZ institutes and to farm out, as it were, the actual research on preparations it produced. At the same time, it provided researchers of GINZ institutes with access to the farm's products—"medicinal milk" and organotherapeutic preparations.

The arrangement worked very well. The researchers at Bakh's Institute of Biochemistry and Shaternikov's Institute of the Physiology of Nutrition began to utilize the farm's products in their own research.[94] Furthermore, in only a few months' time, the farm's "manufacturing laboratory" became fully operational: with new funding in hand, Tobolkin rented the former beer joint situated next to his own house on Arbat and hired several technicians.[95] The laboratory quickly moved its production beyond "medicinal milk" and its derivatives—thyroidin and antithyroidin. It began manufacturing a whole new range of preparations, including extracts of hypophyses, mammary glands, and, of course, testicles. Accordingly, Shervinskii and Tobolkin soon rechristened the expanded manufacturing laboratory the Institute of Organotherapeutic Preparations.

Having finally succeeded in creating a functional manufacturing facility and in organizing studies of its products at GINZ laboratories, Shervinskii moved to further advance his vision for the discipline of endocrinology. In November 1923, he made a presentation to Commissar Semashko on the work of the new institute, proposing that an outpatient facility to treat people with organotherapeutic preparations be established as a first step to creating a special "Organotherapeutic Clinic." Shervinskii argued that such a clinic, together with the farm's laboratory "for manufacture and pharmacological studies" of organotherapeutic preparations, would form what he named an "Institute for the Study of Organotherapy." The Commissar appeared convinced and decided that the Scientific Medical Council should discuss the idea.[96]

A month later, on December 21, 1923, Shervinskii addressed the Council, urging the establishment of an outpatient unit and its integration with the farm's manufacturing laboratory (aka the Institute of Organotherapeutic Preparations) into the Institute for the Study of Organotherapy.[97] He declared that the demand for organotherapeutic preparations had grown considerably, but the physicians who used such preparations had failed to report back to the laboratory on their actual therapeutic effects. The laboratory itself had been testing its products on animals, but these tests, though important in ascertaining pharmacological effects, remained insufficient to judge the clinical efficiency of preparations in humans. Shervinskii insisted that the creation of the Institute for the Study of Organotherapy, combining the manufacture, animal research, and clinical studies of endocrine preparations, was paramount for establishing organotherapy on a "solid foundation." After a prolonged discussion, the council members agreed that the creation

FIGURE 4.2 The three who made Soviet endocrinology. Left to right: Vasilii Shervinskii, Commissar Nikolai Semashko, and Iakov Tobolkin at the opening of the new Institute of Experimental Endocrinology in September 1925. None of them made any "brilliant" discoveries, originated an exciting new theory, or found an "elixir of life." Yet together they created something that would outlive them all: the discipline of endocrinology in Soviet Russia. *Courtesy of MIM.*

of an outpatient facility was indeed "desirable, in principle." But because the laboratory was currently part of GINZ—a research rather than a clinical institution—they could not reach a final decision without the input of the GINZ board.

By that time Shervinskii and Tobolkin obviously had learned quite well how to navigate the corridors of Narkomzdrav's growing bureaucratic apparatus. Shervinskii decided not to wait for the GINZ board's decision (which, he probably knew, would not be in his favor anyway—the GINZ board consistently avoided any engagement with the clinic, upholding the institute's status as an exclusively research establishment). Instead, he took a different route and expanded his institution-building activities in a very unusual way. He penned a proposal to establish a "museum of organotherapeutic preparations."[98]

Museums had for a long time provided an institutional springboard for disciplines, ranging from anatomy to zoology, which rooted their research methodology in collecting and examining particular specimens. Not surprisingly, in Soviet Russia several disciplines followed this trajectory in establishing their institutional base. For instance, Soviet anthropology began its institutionalization with the creation of an anthropological museum under Moscow University, which soon gave birth to a teaching department and then to a research institute.[99] Perhaps the Bolsheviks' active support for various museums as a vehicle both for the popularization of science and for antireligious propaganda also encouraged scientists to adopt this particular strategy in the institutionalization of their own interests. But creating a museum for *endocrinology* was quite unorthodox.

It seems likely that the field's champion drew inspiration for his latest institutional move from a template employed in the discipline building of Soviet social hygiene.[100] This discipline began its institutionalization in 1919 with the establishment of a State Museum for Social Hygiene in Moscow. The museum provided a foundation for a teaching department of social hygiene organized in late 1922 at Moscow University. A few months later, in April 1923, by order of Narkomzdrav, the museum became a State Institute of Social Hygiene. Perhaps Shervinskii hoped that by emulating the discipline-building template of the field that was Commissar Semashko's personal fiefdom, he would increase the chances of successful institutionalization of his own field. Whatever his thoughts, Shervinskii first circulated the draft of his proposal among the members of "the commission on the manufacture of organotherapeutic preparations" and then put the issue on the agenda of the Council's first meeting in 1924. On the eve of the meeting Shervinskii got sick, and it fell to Tobolkin to present the proposal to the Council on January 15.[101]

Tobolkin outlined the structure and goals of the new institution in an hour-long report. The museum was to include five separate departments—anatomy, experimental physiology, pathological anatomy, clinical applications, and organotherapy—with each department collecting materials related to all endocrine organs. All told, the museum would amass materials on the anatomy, physiology, and pathology of all endocrine organs, as well as on the manufacturing techniques and clinical effects of organotherapeutic preparations. Tobolkin described the museum's dual goal: first, to provide consultations to the manufacturers of organotherapeutic preparations and to the physicians who used them;

and, second, to disseminate knowledge of endocrine physiology and pathology among practicing doctors and the general public. Tobolkin's report unambiguously indicated that the envisaged "museum" was in fact to become the country's clearinghouse for the entire field of endocrinology.

Shaternikov immediately supported Tobolkin's presentation, noting that the creation of such a museum was "extremely desirable from both scientific and practical perspectives." The Council's chairman, Tarasevich, however, appeared much more skeptical. He concurred that, in theory, no one could object to the project. The issue was whether, in practice, the creation of the museum was feasible and whether the necessary funds, equipment, and space were available. He noted that Tobolkin had addressed none of these issues in his presentation and suggested that the decision be put on hold until Shervinskii and Tobolkin provided at least a preliminary budget for the organization and maintenance of such a museum. The Council voted in favor of Tarasevich's suggestion.

Shervinskii and Tobolkin were not discouraged by the failure of their latest attempts to squeeze the Council into backing their institutional plans. They moved to widen their support base, managing to instigate an article in the country's major daily *Izvestiia*.[102] On April 11, the newspaper published a short anonymous laudatory article on "the Moscow organotherapeutic institute," which stated that "a museum of organotherapeutic preparations is being organized at the institute according to Prof. Shervinskii's project" and that "an outpatient facility for the treatment of endocrine diseases is being planned." The article lamented "the lack of funds," which impeded these plans. A few weeks later, Shervinskii also presented his ideas about the museum of organotherapeutic preparations to the Seventh Congress of Internists held in Moscow under his chairmanship.

At the same time, Shervinskii began creating an endocrinological society. He drafted a charter for the society and spent several months gathering support among Moscow's leading clinicians and researchers. His initiative found a receptive audience. On April 18, 1924, following the established bureaucratic procedure, Shervinskii sent the charter of the Russian Endocrinological Society for approval to the People's Commissariat of Internal Affairs (NKVD).[103] Co-signed by several of Shervinskii's colleagues from Moscow University's medical faculty, the charter called for the "unification of all individuals working scientifically in the field of endocrinology in Russia."[104] After the usual bureaucratic delay, on September 5, 1924, the NKVD approved the charter.[105]

A month later, on the evening of October 16, the Russian Endocrinological Society held its inaugural meeting in the largest auditorium of the House of Scientists.[106] More than one hundred participants listened to the opening address delivered by the society's founder. Shervinskii eloquently defined the meaning of the event:

> At this very moment, a new collective entity, a new complex organism is being born and has to take its first, timid and unsure, steps; this young, not yet very strong organism needs, perhaps, [a] certain [amount of] support from the outside, but most important, [it needs] a harmonious unification of its forces on the inside.[107]

Shervinskii carefully crafted his speech to highlight and rally support for his vision of institutional structures necessary for establishing the discipline of endocrinology in Soviet Russia. After a brief excursion into the history of endocrinology with a tribute to Claude Bernard and Charles Brown-Séquard, he proceeded to describe his plan for a specialized endocrinological clinic. Shervinskii emphasized that only a large clinic could provide the appropriate setting for multifaceted research on endocrine organs and their pathology, surgical and therapeutical treatments for various diseases of the endocrine system, and clinical testing of available organotherapeutic preparations. He admitted that establishing such a clinic as an independent institution seemed at the moment impossible: "it would require such a huge amount of resources that the very thought of it kills all hopes for any possibility of its realization." But, clearly targeting the numerous Moscow clinicians gathered in the auditorium, he suggested that one of the existing Moscow hospitals might house such a clinic on its premises. Shervinskii also outlined the three main directions for endocrinological studies—laboratory research, production of endocrine preparations, and clinical investigations—and underlined the need for rigorous coordination among them. He emphasized that only the combined efforts of specialists in different fields—biology, surgery, physiology, pathology, therapy—could advance endocrinological knowledge. The main goal of the society, he reiterated, was to unite all individuals involved with endocrinological research. Starting in November, the society began to hold regular monthly meetings to discuss research papers and particular clinical cases. To make sure that the society's work reached as wide an audience as possible, Shervinskii managed to publish accounts (and even the minutes) of the monthly meetings on the pages of *Izvestiia*![108]

While busy organizing the society, Tobolkin and Shervinskii had continued to pester the Scientific Medical Council with requests for funds "for the expansion and additional equipment for the [manufacturing] laboratory at the Institute of Organotherapeutic Preparations."[109] Their persistence paid off: during the year of 1924, the manufacturing facility grew quite substantially and both the quantity and the variety of its production increased considerably. At the end of November, Tobolkin and Shervinskii proudly presented to the Council a list of thirty-nine (!) preparations currently produced by their institute.[110]

In addition to its "signature" thyroidin and antithyroidin, the Institute of Organotherapeutic Preparations now manufactured extracts from the adrenals, hypophysis, ovaries, thymus, testicles, liver, pancreas, placenta, spleen, prostate, and mammary glands. The institute obtained the glands of cattle, pigs, and sheep from Moscow slaughterhouses and produced most preparations in both liquid and dry (tablet and/or powder) forms. In a report accompanying the list, Tobolkin stated that "members of the scientific commission that supervised the institute" had tested and studied all the preparations in their own clinical practice with very favorable results. Tobolkin also submitted the protocols of successful animal tests of the two most important preparations—adrenalin and pituitrin—conducted at the Chemical-Pharmaceutical

Institute and the Institute of Experimental Veterinary Science. On December 19, 1924, the Council discussed Tobolkin and Shervinskii's presentation.[111] Its members approved all but two preparations (the extracts of spleen and liver) for medicinal use throughout the country and recommended that they be included in a new edition of the Soviet Pharmacopoeia.

The events that unfolded over the next few months demonstrate that Tobolkin and Shervinskii's presentation of their institute's products impressed not only the Council, but also the Narkomzdrav Collegium and Commissar Semashko himself. The successful production of Soviet adrenalin constituted a major achievement. Adrenalin had been the first hormone in the world to be industrially manufactured in its pure form, and it was widely employed in medical practice around the world, primarily as a heart stimulant. Since the early 1910s, an American company, Parke, Davis & Co., had virtually monopolized the adrenalin market,[112] and the hormone had never before been produced in Russia.[113] Adrenalin was also the most expensive of organotherapeutic preparations and Narkomzdrav had had to import it, spending large sums of precious hard currency. Achieving Soviet independence in the production of adrenalin undoubtedly testified to the viability of Tobolkin-Shervinskii's endeavors and justified their repeated appeals for expansion and additional funding. As a result, in just a few months, Tobolkin and Shervinskii saw their dreams beginning to be fulfilled.

In early 1925, Narkomzdrav ordered a separation of the Institute of Organotherapeutic Preparations from GINZ, spatially, financially, and administratively and transformed it into a new, full-blown establishment named the Institute of Experimental Endocrinology. Narkomzdrav allotted a large building to the new institute and expanded its budget and personnel considerably, appointing Shervinskii as the director "for scientific affairs" and Tobolkin as his deputy "for administrative affairs."[114] Narkomzdrav also founded a new journal, *Herald of Endocrinology*, with Shervinskii as editor-in-chief and Tobolkin as a member of the editorial board. By September 1925, renovations of the building for the new institute were finished and its staff moved into the new quarters, an event attended personally by Commissar Semashko and proudly reported by *Izvestiia* (see Figure 4.2).[115]

But the discipline's founders did not rest on their laurels. With Shervinskii at the helm, in early 1926 the Russian Endocrinological Society instituted separate branches in Leningrad, Astrakhan', Odessa, Baku, and Voronezh, and began issuing its own periodical—the *Protocols*.[116] In the middle of 1926, Shervinskii established within his new Institute of Experimental Endocrinology a "museum of endocrinological preparations" and an "exemplary laboratory" that became responsible for providing guidelines and standards for the manufacture of organotherapeutic preparations in the entire country. Meanwhile, Tobolkin created a primate breeding station that toward the end of 1927 began breeding several species of monkeys to supply the Institute of Experimental Endocrinology (as well as other research establishments of Narkomzdrav) with the new experimental subjects (see Figure 4.3).[117]

FIGURE 4.3 Iakov Tobolkin at the Sukhumi monkey breeding station with its first orangutan "Bobby," ca. 1929. Although the station's early attempts to breed apes (chimpanzees and orangutans) failed, rhesus macaques and baboons proved much easier to "domesticate": by the end of the 1920s, a sizable herd of these two primate species inhabited the station. *Courtesy of MIM.*

Building on their success with adrenalin, the institute's researchers moved on to developing techniques for the production of another wonder hormone—insulin—with Shervinskii presiding over a special "insulin committee" created by Narkomzdrav to oversee the project.[118] By the beginning of 1928, the institute had succeeded in manufacturing the first samples of Soviet insulin, an achievement promptly announced on the pages of both *Pravda* and *Izvestiia* (see Figure 4.4).[119] A popular daily, *Evening Moscow*, proudly reported this "victory of Soviet laboratories" and claimed that it would "save the country 50,000 golden rubles a year."[120] Perhaps as a reward for this feat, at the end of December, the SNK bestowed on Shervinskii the honorific title of "Worker of Merit," an honor also announced in both *Izvestiia* and *Pravda*.[121]

This rapid disciplinary growth demanded skilled personnel, and the country's medical schools (starting of course with Shervinskii's own base, Moscow University) began to offer special courses on endocrinology. In 1928 Shervinskii compiled the field's first Russian textbook, *The Fundamentals of Endocrinology*. Written by twenty leading specialists, the six-hundred-plus-page compilation provided an overview of all major subjects in the field for both laboratory researchers and practicing physicians. The book came out in

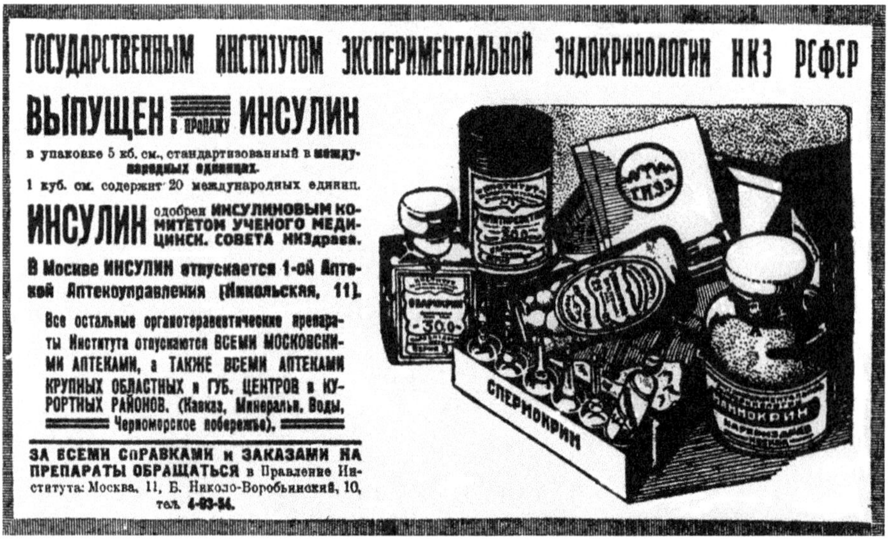

FIGURE 4.4 The announcement of the availability of Soviet insulin produced by the Institute of Experimental Endocrinology published in *Pravda*. It states that the quality of the preparation corresponds to "international standards" and is certified by "the insulin committee of the Narkomzdrav Scientific Medical Council" [read—Vasilii Shervinskii]. Interestingly, the drawing on the right side of the announcement depicts not insulin but other hormonal preparations produced by the institute, namely "spermatocrin" (testicular extract) in the center, "ovariocrin" (ovarian extract) on the left, and "mamocrin" (extract of the mammary glands) on the right. *From* Pravda, *July 31, 1928: 3.*

early 1929 with a first printing of 10,000 (!) copies.[122] A similarly voluminous manual for "students and physicians" entitled *Clinical Endocrinology*[123] came out the next year, and in 1930 the Institute of Experimental Endocrinology created a graduate program to train a new generation of endocrinologists. At the same time, Shervinskii finally realized his long-cherished plans for unifying production, laboratory research, and clinical investigations under one roof. At the end of 1929, the Institute of Experimental Endocrinology established an outpatient unit, and a few months later, a small clinic for the treatment of patients with various diseases of endocrine organs.[124]

By the end of the 1920s, thanks to Shervinskii's and Tobolkin's determined efforts, Soviet endocrinology had become a full-fledged discipline, with research laboratories, societies, periodicals, textbooks, teaching departments and graduate programs, clinics, manufacturing facilities, a standard-setting agency, a monkey breeding station, a museum, and a continuously growing number of researchers and clinicians engaged in endocrinological studies. The Tobolkin-Shervinskii tandem played the key role in the discipline's institutionalization across the country by providing other researchers with examples of a successful institutional template and legitimization strategy, as well as a coordinating center (the Russian Endocrinological Society) and a disciplinary forum (the journal *Herald of Endocrinology*).[125]

Elixirs of Life and Images of Science

As we saw in Chapter 3, the idea of anabiosis as a particular state of "neither life, nor death" inspired extensive literary experimentation, but did not generate much new research and failed to find an institutional niche in 1920s Russia. By contrast, the study of hormones—the newest embodiment of the old dreams of "elixirs of life"—resulted in an impressive amount of research and in the creation of elaborate institutional structures. Yet they found little place in fiction. Of course, scores of popular-science writers and journalists hailed the "wonders" of hormones in countless articles and brochures, describing "miracle cures" achieved by the applications of insulin, adrenalin, pituitrin, and other organotherapeutic preparations.[126] But, although Petr Bochkarev, a senior staff member of the newly established Institute of Experimental Endocrinology, entitled a popular account of the discipline and its flagship institution in Russia, *In Search of the Water of Life*,[127] the institutional aspects of this search did not attract any attention from the literary community.

Expectedly, the transformative and curative powers of hormones did inspire some literary experiments that explored the possible impact of the application of various extracts of the endocrine glands on human life, human relations, and human societies.[128] The most notable among these experiments was a novel entitled *The Man Who Lost His Face*, written by Alexander Beliaev and serialized in 1929 by a popular adventure magazine, *Around the World*.[129]

Like "Professor Dowell's Head," the novel is set in the United States. Its plot revolves around a famous, but extremely ugly, movie actor named Tonio Presto. Contrasted with the "patented" beauty of his female partner, Ms. Gedda Lux, Presto's repulsive appearance provides unlimited possibilities for various comedic situations, exploited in numerous movies, which bring the pair wide public admiration and, hence, fame and wealth. However, when Presto proposes marriage to Ms. Lux, she simply laughs at him. Angered by the rejection, Presto continues to make faces until Ms. Lux nearly dies of laughing. Presto thinks that she does die and flees the scene. Afraid of being accused of murder, he decides to change his appearance. But he does not turn to plastic surgery.

Instead, Presto comes to the clinic of a Russian physician, Dr. Sorokin,[130] who transforms people with various bodily deformities (giants, dwarfs, and assorted "monsters") into beautiful human beings by a careful application of mixtures of different hormones.[131] After several weeks of treatment, Dr. Sorokin is able to "cure" Presto's ugliness, transforming him into a "normal" and rather attractive man. Of course, this does not solve Presto's problems, but rather multiplies them. Trying to enter his own house, he is arrested and put in jail, since nobody recognizes him as his former self, "Presto the Ugly." With Dr. Sorokin's help, he proves to the court that he is indeed a "transformed" Presto. The court releases him from prison, but the district attorney (DA) convinces the judge and the state governor to freeze all of Presto's assets to make an example of him. According to the DA, the foundation of private property—and thus, capitalism—is ownership, which ultimately means an identifiable person. If people were to start changing their faces and bodies (and hence identities) as they please, this foundation would be thoroughly undermined.

Presto organizes a farewell banquet and invites everyone involved with his case to "make amends." He mixes different preparations he has stolen from Dr. Sorokin's clinic into the food and drinks served to his guests. As a result, all of his foes develop various deformities: Ms. Lux grows up to two-and-a-half meters tall, the DA shrinks to a lilliputian size, the state governor's skin turns black, and the judge becomes monstrously obese. Eventually, all of them go to Dr. Sorokin's clinic and he brings back their normal appearances. Instead of being praised for this feat, however, the doctor is expelled from the United States, since Presto's example makes Sorokin's science contrary to the established capitalist order. The novel ends on a ship sailing to Europe where Sorokin meets Presto, who has again changed his appearance by taking some of Sorokin's "hormones."

In addition to exploiting Beliaev's favorite theme of the "failure" of science in the "immoral" capitalist West, the novel contained lengthy popular expositions about the role that various glands of internal secretion, ranging from the thyroids to the hypophysis, play in the human body. Accompanied by a parade of Dr. Sorokin's patients "demonstrating" the actions of particular glands, the novel vividly illustrated the dream of control over the form and function of the human body spurred by the rapid advances of endocrinology during the 1920s, especially the discovery of the role that hormones play in such conditions as gigantism and dwarfism (see Figure 4.5).[132] What *The Man Who Lost His Face* did not illustrate, however, was the rapid growth of the discipline of endocrinology during the same decade.

Both scientists and writers saw hormones as a powerful tool for manipulating the shape and functioning of the human body. But, for scientists, hormones also became a convenient instrument for insinuating their own visions of "big science" into their patrons' revolutionary dreams and for securing funds to create large institutional structures for scientific research. Even though their Bolshevik patrons enthusiastically supported the development of science as a matter of principle, entrepreneurial scientists had to find ways to direct this enthusiasm to their own, not their competitors', disciplines and institutions. Yet this particular side of the "search for the water of life" did not fire up the literary imagination. The laborious process of courting patrons (who often had no idea about what scientists were actually doing in their laboratories), mobilizing allies and resources, convincing skeptical colleagues, and fending off competitors witnessed on the preceding pages did not attract attention from the writing and hence, reading, public. The emerging "big science," with its hierarchical structures, team efforts, distribution of labor, and "dirty" dealings with patrons, allies, and competitors did not lend itself to the romantic image of science as a single-minded pursuit of knowledge by a "lonely genius"—at most assisted by a single sidekick—living in an ivory tower and oblivious to all but his

FIGURE 4.5 The "parade" of Tonio Presto's foes who became Dr. Sorokin's patients, demonstrating various deformations of the body (such as gigantism, dwarfism, and obesity) produced by hormones that Presto mixed in their food and drinks. *From Aleksandr Beliaev,* Sobranie sochinenii v vos'mi tomakh *(Moscow: Molodaia gvardiia, 1963), vol. 4.*

research. Inspired by the actual practice of "little" science, this image had dominated the artistic renderings of science and the scientist—from Dr. Faust to Victor Frankenstein to Dr. Moreau—for centuries and, as illustrated by Prof. Persikov, Prof. Dowell, and Dr. Sorokin, it was still very much present in 1920s literature.

Indeed, someone like Shervinskii, whose efforts were pivotal for the establishment of endocrinology as a discipline in Soviet Russia but who did not make any "spectacular" discoveries, simply did not fit this image. To say nothing of Tobolkin, who never even conducted any experiments nor published a single scientific article but without whom Shervinskii perhaps would not even have engaged in the institutional development of endocrinology. Tobolkin certainly did not fit the scientist's customary image, even in the role of a sidekick. His roles as a manager, a procurer of resources, a "deputy-director for administrative affairs" had no place in "little" science.

Contemporary writers and journalists obviously saw very little glamour and drama in creating and managing scientific institutions. Not surprisingly, aside from the proud announcements of yet another "new," "the world's first," "Soviet" institute opened by various benevolent government agencies, the media did not pay much attention to the intense labors of both Bolshevik patrons and their scientific clients in building what would in just two decades become one of the world's largest research empires. Clearly, writers of "scientific fantasy" did not find the stories of science institution and discipline building particularly "fantastic" or inspiring.

Furthermore, entrepreneurial scientists like Shervinskii and Tobolkin themselves carefully hid the "dirty" side of their endeavors from the public eye. They certainly did not advertise their role in treating the various afflictions of the top-level Bolsheviks, including Krupskaia, Dzerzhinskii, and Menzhinskii. In their own "promotional" popular writings about science and scientists or in their accounts of the development of their particular disciplines, they omitted any mention of their social practices, which, unlike their investigative practices, produced not celebrated discoveries, fascinating ideas, and unique research methods, but careers, patrons, competitors, salaries, allies, students, awards, and institutions. They never publicly described and discussed the tactics and strategies they employed to achieve their institutional objectives. Accordingly, even when mentioning a research institute, a clinic, or a laboratory in their literary experiments, writers simply took for granted their existence without going into details of their origins or the social dynamics inherent to building, managing, and maintaining scientific institutions. This is exactly what Pil'niak and Fedorovskii did in their story of the "Institute of Life," even though Fedorovskii knew very well the mechanics and dynamics of the process: after all, he himself had created several scientific institutions, including the Institute of Mineralogy, in Soviet Russia.

Thus, even though the ultimate goal of the discipline of endocrinology—the control of life and human nature through the application of hormones—did lend itself to literary experimentation and scrutiny, the story of the discipline's actual construction remained obscure for both writers and readers.

Communism is the youth of the world,
And it is to be built by the young.
VLADIMIR MAYAKOVSKY, "*MIuD* [International Youth Day]" (1926)

Now—such is progress—the old men work, the old men copulate, the old men have no time, no leisure from pleasure, not a moment to sit down and think . . .
ALDOUS HUXLEY, *Brave New World* (1932)

5 "The Dog's Heart" and Monkey Glands: Rejuvenation

IF THE INTENSE efforts of Russian scientists and their Bolshevik patrons to build the discipline of endocrinology attracted little attention from litterateurs and the public writ large, another "visionary biology" project—connected to a certain degree to the studies of internal secretions—generated enormous public appeal and inspired numerous literary experiments: rejuvenation. In the 1920s dozens of physicians and biologists all over the country engaged in experimental attempts to "rejuvenate" animals and humans, publishing numerous articles on their research in professional periodicals, conference proceedings, and solid volumes. Dozens of journalists and fiction writers assessed these attempts in countless newspaper reportages, short stories, plays, and novels. As one might expect, the former physician Mikhail Bulgakov found this "rejuvenation craze" worthy of a close examination.

Soon after he had finished "Fateful Eggs" in late 1924, Bulgakov started working on its "sequel," which expanded on and elaborated the major themes and subjects of the first novella. This time, Bulgakov did not envision a future: he set the story, eventually entitled "The Dog's Heart," in his own present.[1] The main protagonist of "The Dog's Heart" is not a biologist, but an internationally reputed surgeon with a telling name, Preobrazhenskii (literally, "transfiguring," with its clear allusion to the Transfiguration of Jesus), working in NEP Moscow. Despite the "proletarian" revolution, Preobrazhenskii maintains his "bourgeois" lifestyle—a seven-room apartment, with a live-in maid and a housekeeper—by performing rejuvenations through the implantation of monkey sex glands: ovaries and testicles. He treats a select group of patients, who either pay him large sums of money or offer him protection from the excesses of the new proletarian militants, particularly their attempts to requisition rooms in Preobrazhenskii's apartment. But for Preobrazhenskii, the rejuvenation operations are primarily a means to continue his research under the new regime. Like Persikov, Preobrazhenskii is an experimentalist who studies the "mysteries of life and death," namely the workings of the brain and the glands of internal secretion. As he intimates to his assistant, he is concerned "about eugenics, about the improvement of the human breed. [. . .] You don't think I do it [rejuvenation operations] because of the money, do you? After all, I am a scientist" (126).

In search of the mechanisms of rejuvenation Preobrazhenskii conducts a daring experiment by transplanting the testicles and hypophysis (the pituitary gland) from a murdered petty criminal to a stray dog he has picked up on the street and named Sharik (ball). The dog survives the operation, but, instead of being rejuvenated, in the course of a few weeks it is transformed into a man. Assuming the name "Sharikov" (literally, a son of Sharik), this dog-man personality exhibits all of the negative traits of his human organ donor: he drinks, steals, lies, and swears. Furthermore, Sharikov quickly allies himself with the proletarian militants and makes the comfortable life of his creator miserable, whilst all of Preobrazhenskii's attempts at his "reeducation" fail utterly. The press gets wind of Preobrazhenskii's work and disseminates wild rumors about his experiment. The Moscow public eagerly follows the news and besieges Preobrazhenskii's apartment to get a glimpse of the dog-man. In the end, Preobrazhenskii manages to perform a reverse operation and Sharikov is transformed back into Sharik the dog, to the chagrin of the proletarian militants who had hoped to use him to get the professor arrested and evicted from his apartment.

In early 1925, Bulgakov read the novella to a group of his fellow litterateurs, but his efforts to publish it were squashed by a categorical verdict issued personally by Lev Kamenev, a leader of the Bolshevik Party and a member of its highest council, the Politburo: "This is a piercing satire on modern times; [it] can't be published under any circumstances."[2] In the Soviet Union, Kamenev's verdict was "honored" for more than sixty years: the first edition of the novella appeared only in 1987 during the heady days of Mikhail Gorbachev's *perestroika*. But ever since "The Dog's Heart" first appeared in print in Paris in the late 1960s, commentators have concurred with Kamenev's assessment and have praised the novella, much as they had "Fateful Eggs," as a parody on the Bolshevik Revolution and the Soviet "experiment."[3] However, much as they had with "Fateful Eggs," the commentators have also missed a major theme of "The Dog's Heart": the intertwining of the Bolshevik Revolution and the experimental revolution in the life sciences.[4]

If "Fateful Eggs" presented a veritable guide to the evolution of relationships among Russian experimentalists, the public, and the Bolshevik state, "The Dog's Heart" presented a snapshot of the major projects of visionary biology. The novella touched and commented upon the most prominent lines of research pursued by experimental biomedicine during the early 1920s—hormones, eugenics, brain studies, and organ transplants—all of which converged in the pursuit of rejuvenation. It identified precisely the major goals of visionary biology: control over life processes—and, by extension, human personality and, ultimately, human destiny—through scientific manipulation.

It demonstrated close links among various visionary biology projects with their common aim at rejuvenation and eventually immortality. It pointed to particular congruence of these projects to the Bolsheviks' revolutionary dreams of a new world, a new society, and a new kind of "advanced" human being they aspired to create within the one-sixth of the world under their rule.

Bulgakov's "fantastic" novella depicted quite accurately some of the major concerns, aspirations, and inspirations of biomedical research under the new regime. In the mid-1920s, Soviet scientists indeed performed transplantations, practiced rejuvenation, sliced brains, thought about eugenics, catered to powerful patrons among the Bolsheviks, and experimented on dogs.[5] In a literary form, Bulgakov reflected upon contemporary heated debates—spurred by the combined impact of experimental revolution in the life sciences and the Bolshevik Revolution—over human nature and the relative roles of nature (biomedical experimentation) and nurture (social revolution) in shaping human body, personality, and destiny.

More important, Bulgakov offered his own contribution to these debates by exposing what he perceived as the ultimate futility of both visionary biology and revolutionary dreams. Faced with the results of his experiment, Preobrazhenskii rethinks his aspirations and labors: "I spent five years extracting hypophyses from brains. You know what amount of work I've done—it boggles the mind. And now is the question—what for? To transform one day an amiable dog into such a scam that one's hair stands on end" (125). Preobrazhenskii readily admits the utter failure of his quest for rejuvenation, betterment of the human breed, and the more general vision of "controlling Nature." "Why?," he rhetorically asks his assistant, would one want to interfere with the "natural" way of things? Many Russian scientists thought they knew the answer to that question.

The Science of Aging

The possibility of stopping aging and thus preventing or at least postponing its final result—death—has always fascinated and inspired humankind. The quest for an "elixir of life," the "fountain of youth," "apples of immortality," or the "water of life and death" is a common theme in many cultures and a common trope of numerous myths, legends, and fairy tales, as well as literary and artistic works throughout human history, from the Greek myths of Hercules's labors to Juan Ponce de Leon's Florida adventures to Wolfgang von Goethe's *Dr. Faust* to Steven Spielberg's *Indiana Jones* movies. The rise of experimental biology and medicine in the last decades of the nineteenth century inaugurated a new era in this quest.[6] Armed with new experimental methods, scientists and physicians attacked the centuries-old mysteries of aging, and it seemed they would quickly succeed in finding suitable techniques to control and reverse the aging processes.

One of the first to approach aging from an experimental point of view was Charles Brown-Séquard, a prominent French physiologist.[7] In 1889, he reported the results of experiments (conducted on himself) with injections of extracts prepared from animal testicles. He claimed that the injections had clear rejuvenating effects on the human organism. It had certainly worked for him: after the injections he felt much younger and, as he demurely noted in one of his publications (alluding to his restored sexual drive), his wife felt it, too. Brown-Séquard's basic idea was an old one, namely that aging results from the decline in the activity of sex gonads. But this old idea received a new interpretation in Brown-Séquard's work, which indicated that aging resulted from the insufficiency of the "internal secretions"—a concept developed by his predecessor at the Collège de France, Claude Bernard—produced by the gonads.[8] Hence, with the new experimental methods, the aging processes could be reverted by supplying the aged organism with the secretions (extracts) of young sex gonads. Although initially greeted with much skepticism by the medical community, Brown-Séquard's experiments inaugurated the emergence of a new direction in therapeutics—organotherapy, treatments with various animal tissue extracts—which, as we saw in Chapter 4, paved the way for the discovery of hormones and the emergence of a new science of internal secretions, endocrinology.[9]

Another branch of the new, experimental medicine—bacteriology—also promised a solution to the problem of aging. In the early 1900s, Elie Metchnikoff, a Russian zoologist working at the Institut Pasteur in Paris, published his theory of senescence, which attributed aging to the "poisoning" of the organism with toxins produced by bacteria inhabiting its intestines.[10] Metchnikoff suggested that inhibiting the growth of noxious bacteria in the intestines by replacing them with healthy bacteria (such as the milk-souring bacteria present in yogurt) would prolong human life considerably.[11] One of the founders of the new science of immunology, Metchnikoff also advocated for the establishment of a special science of senescence and even coined a name for it: gerontology.[12]

Brown-Séquard's and Metchnikoff's work helped establish senescence as a legitimate subject of scientific inquiry and inspired other scientists to begin their own investigations. During the early 1900s, researchers working in two branches of experimental biology, cytology and biochemistry, offered yet another view of senescence as the result of the progressive accumulation in cells, particularly nervous cells, of certain "waste products." A number of investigators described changes in the appearance and chemical composition of old cells as compared to young ones. Some of them theorized that if there were a way to remove these "toxic wastes" from the cells, the organism could be "regenerated" and its aging would be delayed dramatically. Experiments on cultures of unicellular organisms, infusoria,[13] and, later, on tissue cultures, kept in the laboratory seemed to confirm this hypothesis: the cultures whose nutrient medium was regularly renewed lived much longer—as Alexis Carrel, a pioneer of the tissue cultures technique, claimed, forever—than the cultures whose medium was allowed to accumulate their waste products.[14] Some scientists suggested that a similar "renewal" of the milieu intérieur in a human organism could be achieved by means of blood transfusions from young to old.[15]

Since Brown-Séquard and Metchnikoff devised seemingly simple "remedies" for aging (injections of testicular extracts and yogurt drinking), their concepts attracted much public attention and raised expectations that soon, experimental science would be able to fend off old age, if not indefinitely, then at least for a long time. But it was the works of the Austrian physiologist Eugen Steinach and the French surgeon Serge Voronoff that in the early 1920s spurred an extraordinary obsession with rejuvenation around the globe, including Russia.[16]

In June 1920, Steinach, a respected physiologist and member of the famous Viennese Institute for Experimental Biology—the Vivarium—published a seventy-page article entitled "Rejuvenation through the experimental reinvigoration of the aged puberty gland."[17] The article summarized and expanded Steinach's investigations begun almost fifteen years earlier into the role the internal secretions of sex gonads—testicles and ovaries—play in the animal organism. These investigations had led Steinach to the idea that sex gonads consisted of two anatomically and histologically distinct parts. One was responsible for the production of sex cells (ova and sperm), the other for the production of sex hormones. He named the second part the "puberty gland." The 1920 article detailed Steinach's newest experiments with the ligation of seminal vesicles (vasoligature—an operation similar to vasectomy) in old male rats, as well as with the transplantation of sex gonads (both male and female) from young rats to old ones. Steinach claimed that both the vasoligature and the transplantation stimulated the puberty gland and produced clear rejuvenating effects in operated animals. Within a few weeks after the operation, a weak, bald, and sexually impotent rat was transformed into a strong, covered-with-new-fur, and sexually aggressive one. The operated animals also lived much longer than the controls; the operations extended the rat's lifespan by a third, from thirty months up to forty months. Supplemented by several photographs of rats before and after rejuvenation, the article ended with the description of three cases of rejuvenation in humans. Without their knowledge, three male patients (aged forty-four, sixty-six, and seventy-one) were subjected to the ligation of seminal vesicles. Reportedly, in a few weeks, all three individuals showed marked improvements in their health. The acute signs of the old age—face lines, crooked posture, and bodily weakness—began to disappear. The patients regained vitality, energy, appetite, muscle tone, and strength, as well as sexual potency.

In the fall of the same year, Serge Voronoff, a French surgeon of Russian extraction working at the Collège de France in Paris, published a lengthy monograph, under the enticing title *Vivre. Etudes des moyens de relever l'énergie vitale et de prolonger la vie*.[18] The book detailed Voronoff's experiments with the transplantation of sex gonads from young to old animals in sheep and goats. Illustrated by photographs of animals before and after the operation, Voronoff's book claimed that the transplantation led to rejuvenation of the old animals. Voronoff also included in his book a description of two cases of successful transplantation of the thyroid gland from apes to humans and concluded with an enthusiastic prophesy that the transplantation of sex gonads from apes to humans could become a certain means of rejuvenation.

Steinach's and Voronoff's publications caused furor in both academic and lay circles around the world. Among the many countries seized by the new rejuvenation craze, none seemed to provide a less likely locale for the experiments with sex gonads than Russia. Torn into pieces by the bloody Civil War, ridden with epidemics, and nearly starved to death, the country was on the brink of economic and political collapse; yet its medical and scientific communities, as well as the lay public, enthusiastically embraced the news of rejuvenation experiments.

One reason Steinach and Voronoff's research attained such enormous popularity in post-revolutionary Russia was the existence of a group of experimental biologists who had been developing similar ideas and techniques before the Revolution and who immediately seized upon the opportunities opened up by the possible applications of Steinach and Voronoff's works. As elsewhere in the world, in Russia experimental biology began to develop rapidly around the turn of the twentieth century in major university centers: Iur'ev, Kazan', Khar'kov, Kiev, Moscow, Odessa, St. Petersburg, Tomsk, and Warsaw. In Moscow, one man was principally responsible for its advance—Nikolai Kol'tsov (1872–1940).[19]

A scion of a large family of Moscow merchants, Kol'tsov graduated from Moscow University in 1894 with first-class honors and a gold medal for his research thesis on the comparative anatomy of "the hind legs of vertebrata." Slated for "preparation for the title of professor," over the next few years Kol'tsov conducted research in laboratories of various German universities, as well as at the biological marine stations in Naples and Villefranche. This research formed a core of his doctoral dissertation "On the form of the cell," completed in 1904. The same year Kol'tsov began a teaching career as an unpaid lecturer (*Privatdocent*) at his alma mater, Moscow University.

A brilliant experimentalist and theoretician, as well as an inspiring teacher and enterprising administrator, Kol'tsov was one of the most energetic proponents of the new, experimental biology in Russia. He laid a foundation for the development of experimental cytology, histology, embryology, zoopsychology, and genetics, creating a solid institutional base for his research through the establishment of several laboratories in various institutions of higher education, including Moscow University and the Women's Higher Courses. In 1908, Russia's first private university—the Shaniavskii Moscow City People's University—became his principal base. Here Kol'tsov created a laboratory and organized a special course on what at the time he called "experimental zoology." This course, named the "Big Practicum," consisted not only of lectures but also of a large number of laboratory exercises and experiments that students were required to perform under the guidance of their teacher and his assistants.[20] Kol'tsov recruited and trained a group of bright young researchers, skillfully directing their attention to, and encouraging them to specialize in, particular subfields of experimental biology (see Figure 5.1).

As part of student training, in the fall of 1911 Kol'tsov instituted a special weekly colloquium attended not only by his students and colleagues at Shaniavskii University but

FIGURE 5.1 Nikolai Kol'tsov with his students and assistants at Shaniavskii University, ca. 1913. Kol'tsov trained a large group of talented researchers who would dominate the field of experimental biology and medicine, with its quickly developing subfields, disciplines, and specialties, such as genetics, biophysics, zoopsychology, cytology, and embryology, for years to come. Seated left to right: M. Zavadovskii, R. Serebrovskaia, S. Skadovskii, N. Kol'tsov, Trofimov, and V. Efimov; standing: A. Serebrovskii, I. Kogan, V. Natali, and V. Savich. *Courtesy of ARAN.*

also students and researchers from other Moscow institutions, sometimes bringing together as many as two hundred people.[21] The colloquium aimed at keeping its attendees abreast of the newest literature on a variety of new experimental techniques such as organ transplantations, tissue cultures, and physicochemical analyses, as well as on the advances of numerous subfields in experimental biology, ranging from genetics and biochemistry to embryology and biophysics. It was at Kol'tsov's colloquium that Porfirii Bakhmet'ev first presented his research when he returned to Russia in March 1913.[22] The colloquium's sessions regularly featured reports on the latest publications in the field, but they also provided a forum for discussions of original research—often replicating experiments described in the literature—conducted by Kol'tsov and his students.

Among the various new lines of investigation in experimental biology of the time, Steinach's early studies on the "transformation of sex" and transplantations of sex gonads published in 1912 attracted the particular attention of Moscow investigators.[23] Kol'tsov assigned one of his older students, Ivan Kogan (1877–?), to examine numerous questions related to the determination of secondary sex characteristics in animals. Kogan quickly compiled a bibliography of publications on the subject and delivered several survey reports on the newest developments in the field at the colloquium. He also began his own experiments on the transformation of sex in amphibians. In early 1914, Kogan published a lengthy article on the "transformation of males into females" in the *Scientific Bulletins* issued by Shaniavskii University.[24] His article presented a detailed review of Steinach's 1912 work and briefly outlined a program of his own research on the subject. The Great War that broke out in the summer interrupted Kogan's work: he joined the army.

But sex transformations continued to intrigue Moscow researchers. Mikhail Zavadovskii (1891–1957), one of Kol'tsov's students who served as the scientific secretary of the colloquium, took up the subject: by the end of the 1910s it had become a primary focus of his own studies.[25] Graduated in the spring of 1914, Zavadovskii became Kol'tsov's assistant at Shaniavskii University, running one part of the Big Practicum and lecturing on his specialty—experimental embryology (developmental mechanics, as it was called at the time). By the end of 1916, he had assumed full responsibility for the Big Practicum, since at that time, Kol'tsov's long-cherished plans of creating a special Institute of Experimental Biology were finally coming to fruition. The Society of the Moscow Scientific Institute agreed to fund Kol'tsov's dream and he was spending most of his time attending to various administrative problems related to the establishment of the new institute. In the winter semester of 1917–18, while the Bolshevik coup d'état unfolded in Petrograd, Zavadovskii opened his own course on experimental biology at Shaniavskii University. Following in his teacher's footsteps, Zavadovskii engaged his own students in studies of his favorite subjects: in March 1918, one of them presented a report on Steinach's latest research to a "mini" colloquium that Zavadovskii organized at his own laboratory.[26]

The same month—March 1918—the Bolsheviks signed a peace treaty with Germany, ending Russia's participation in the Great War, but in just a few days the country was plunged into the Civil War. At the end of the month, threatened by the White Guard advancing on Petrograd, the Bolshevik government moved to Moscow, thus making the city the capital of a new, Soviet Russia. The Bolsheviks nationalized all research and teaching institutions on the territories under their control, but during the Civil War they had very limited resources to spend on science: the Red Army, not science, was their overriding priority. For the most part, scientists were left to fend for themselves. As Bulgakov vividly portrayed in "Fateful Eggs," until the end of the Civil War they had very limited access to fuel, food, or goods, to say nothing about equipment, publications, chemicals, laboratory animals, and other supplies necessary for their research.

Having barely survived the first terrible winter of the Bolshevik regime, in the late spring of 1919, Zavadovskii organized an "expedition" with a group of his students to warmer climates—to Askania Nova, the country's most famous nature preserve in the southern Ukraine.[27] In Askania Nova, Zavadovskii began extensive experiments on the transplantation of sex gonads in chicken and pheasants. But he was unable to complete his studies: the Civil War caught up with his expedition, transforming the nature preserve into a battlefield. Zavadovskii and his students were stranded in the Ukraine for almost two years. Only in the spring of 1921 was Zavadovskii able to return to Moscow. By that time, news of Steinach's latest research had reached Russia.

Rejuvenating Russia

Word of Steinach's newest work made its way to Russia with extraordinary speed. As early as October 1920—just three months after the publication of Steinach's latest article—Kol'tsov delivered a lecture on Steinach's rejuvenation experiments at the State Institute for People's Health Protection (GINZ) in Moscow. As we saw in the previous chapter, Narkomzdrav had created GINZ in the fall of 1918 as an association of research institutes to address issues pertaining to medicine and public health. Given the isolation of the new Bolshevik Russia from the rest of the world and the near total collapse of the country's publishing industry, one of GINZ's first tasks was to acquaint Moscow physicians with new foreign literature collected by the Narkomzdrav representatives abroad. For that purpose GINZ instituted a series of colloquiums, and it was at the first such colloquium held on October 27, 1920, at the House of Sanitary Education that Kol'tsov delivered his lecture, accompanied by a slide show made from the original illustrations included in Steinach's article.[28]

With partial restoration of scientific publications at the end of the Civil War, articles on Steinach's and Voronoff's experiments began to appear in Russian professional periodicals. At the end of 1920, the first issue of the just-established *Journal of Medical Abstracts* carried an extended summary of Steinach's 1920 article, along with abstracts of publications by several German physicians on Steinach's experiments.[29] A few months later, the same journal published abstracts of several articles written by various physicians in response to Steinach's experiments, as well as a summary of a 1918 article by Steinach and his collaborator, urologist Robert Lichtenstern, on the effects of castration and testicle transplants on homosexuality.[30]

As the country's economy began to recover under NEP, the number of publications on rejuvenation grew. In 1922, two different translations of a brochure entitled "Rejuvenation and prolongation of individual life," written by Steinach's ardent supporter and colleague at the Vivarium, Paul Kammerer, came out in Petrograd and Moscow.[31] The same year, Kol'tsov published a lengthy survey of Steinach's and Voronoff's operations in the first issue of his just-created journal, *Advances in Experimental Biology*.[32] Also in that same year, Semen Girgolav, a well-known Petrograd surgeon, published the first report on a Steinach operation he had performed at the Military-Medical Academy in Petrograd.[33] The news of Steinach's and Voronoff's latest works spread like fire throughout the country, with surveys and abstracts appearing in medical periodicals issued in Petrograd, Kiev, Khar'kov, and even in the faraway Central Asian capital, Tashkent.[34]

By the time Bulgakov began writing "The Dog's Heart" in late 1924, the country had witnessed a virtual explosion of publications on rejuvenation. In the space of three years, between 1923 and 1925, Russian biologists and physicians had produced several dozen books and more than one hundred articles on the subject.[35] Kol'tsov compiled two volumes, seductively entitled *Rejuvenation*, of translations with key works by

Steinach and Voronoff, as well as by other continental biologists and clinicians who had continued and expanded the research of the two pioneers.[36] He also published detailed surveys of the newest US literature on the subject.[37] Five different editions of Voronoff's books appeared in Moscow, Petrograd, and Khar'kov, edited and translated by eminent Russian biologists and physicians.[38] The well-known Petrograd publisher Practical Medicine also issued a translation of a monograph, entitled *The Theory and Practice of Rejuvenation*, by one of the most active German proponents of Steinach's operation, Peter Schmidt.[39]

Numerous Russian physicians and biologists attempted to replicate Steinach's and Voronoff's operations. Kol'tsov's Institute of Experimental Biology (IEB) became a major center of such attempts in Moscow. Ivan Kogan, who had been fortunate to survive the mincing machine of World War I and the Civil War, became Kol'tsov's point man in organizing a research group on rejuvenation at IEB. In 1922 Kogan resumed his pre-war studies on the transplantations of sex gonads and extended them from amphibians to chickens, guinea pigs, dogs, and then to humans. Since Kogan had no medical diploma and hence could not operate on human patients, Kol'tsov invited a surgeon, Alexander Zamkov, to join the institute's team.[40] In consultation with Kol'tsov, Kogan developed a program of research, while Zamkov performed the actual operations on dogs and humans. At the beginning of 1924, Kogan published the first report on "Transplantation of sex gonads in animals and humans" performed by his group.[41]

Naturally, IEB was not the only place in Russia where rejuvenation research commanded the attention of scientists and physicians. In the spring of 1924, a special collection entitled *Rejuvenation in Russia* came out in Petrograd.[42] The volume included six articles detailing the theoretical justifications for, laboratory research on, and clinical results of the Steinach and Voronoff operations performed by Russian investigators. Anton Nemilov, professor of histology at Petrograd University, opened the volume with a thirty-page analysis of "the physiological foundations of rejuvenation," detailing the theoretical underpinnings of Steinach and Voronoff's techniques, which he explained as providing an aged organism with sex hormones, produced either by the activated puberty gland (in the case of Steinach's operation), or by the transplanted sex glands (in the case of Voronoff's).[43] Nemilov paid particular attention to the histological transformations—his specialty—of testicles following the vasectomy or vasoligature, as well as to histological changes in the testicle transplants, illustrating them with the results of his own experiments on dogs. He concluded that neither Steinach's nor Voronoff's technique led to "true" rejuvenation but rather produced the "temporary refreshment" of the organism.

Nearly one half of the entire volume was taken up by an article written by Vladimir Gorash, a surgeon-urologist at the Petrograd Institute for Doctors' Continuing Education, who surveyed the results of clinical experiences with Steinach's and Voronoff's operations in Russia.[44] Gorash had collected data from doctors in Petrograd and the provinces, totaling 193 cases in three years, including twenty-one transplantations. Gorash himself had performed fifty vasectomies and vasoligatures. Although, like Nemilov, Gorash did not consider the results of the operations to be "true" rejuvenation, he claimed that after the operations most patients did experience a period of "reinvigoration," which included the restoration of the previously lost sexual drive, weight gain, and the overall improvement of physical condition.

The next article detailed research of Leonid Voskresenskii, a professor of physiology at the Pedagogical Institute in Tver'—a large industrial center situated halfway between Moscow and Petrograd. A former student of Russia's leading physiologist Ivan Pavlov, Voskresenskii created a physiological laboratory named after his teacher, where he experimented with Steinach and Voronoff's operations in humans and animals.[45] From 1921 to early 1924, Voskresenskii had performed thirty-seven vasoligatures and thirteen transplantations of testicles from human to human. To investigate the popular criticism (advanced by several researchers in Russia and elsewhere) that the rejuvenation observed after both Steinach and Voronoff's operations in humans was merely the result of suggestion and not of the operations as such, Voskresenskii supplemented his observations on humans with experiments on the transplantation of testicles in animals. He recorded temporary improvements in the general physical condition and sexual potency of operated animals, which slowly diminished in the course of two to three months, in parallel with the "dissolution" of the transplants. Although the results of his human experiments varied considerably from patient to patient, Voskresenskii endorsed the use of these operations as a means of rejuvenation.

The volume also included two short reports from doctors in the Ukraine on their experiences with Steinach and Voronoff's operations, along with a brief survey of data gathered by Moscow physicians, written by Lev Vasilevskii, a well-known physician-popularizer.[46] The volume was appended with an extensive bibliography (compiled by an intern-physician at Gorash's clinic) of more than 150 publications on the subject in Russian and in foreign languages available in Petrograd's libraries.[47]

Naturally, a single volume could not even begin to represent the entire scale of the contemporary rejuvenation craze in Russia. Indeed, not only in Moscow and Petrograd but also in Khar'kov and Smolensk, Baku and Kazan', Tashkent and Omsk, Arkhangelsk and Tiflis, and Irkutsk and Odessa, surgeons performed Steinach's and Voronoff's operations on humans, while biologists experimented on chickens, rats, guinea pigs, cats, rabbits, dogs, and horses.[48] Mikhail Zavadovskii, who had by that time become the scientific director of the Moscow Zoo, even performed rejuvenation experiments on a variety of exotic animals kept at the Zoo.[49] Between 1923 and 1925, practically all medical and biological periodicals, ranging from *Russian Physiological Journal* to *Archive of Clinical and Experimental Medicine* and from *Advances in Experimental Biology* to *Physician's Affair*, carried articles on rejuvenation.[50]

During the same period, meetings of various professional societies of Russian biologists and physicians hotly debated rejuvenation issues. At the Sixteenth Congress of

Russian Surgeons in May 1924, Steinach's and Voronoff's operations, and the transplantation of endocrine glands more generally, became a major subject of discussion.[51] Reports on rejuvenation figured prominently on the agendas of the Second Scientific Congress of Central Asian Physicians in Tashkent, regional conferences of Ukrainian surgeons, and meetings of physicians in Georgia.[52] The Seventh All-Union Congress of Internists in May 1925 held a special plenary session on the transplantation of endocrine glands with twelve reports.[53] The Society of Russian Surgeons in Moscow and the Pirogov Society of Surgeons in Petrograd regularly listened to reports on—and observed patients who had undergone—rejuvenation by Steinach and/or Voronoff's methods, as did the members of the popular forum of Petrograd physiologists, "Physiological Conversations."

Of course, replicating Steinach and Voronoff's experiments in Russia meant facing numerous obstacles, especially during the early 1920s. To begin with, finding old animals in a country that had just gone through nearly a decade of continuous warfare, famine, and epidemics posed a considerable problem. There was no system of supply of laboratory animals in the country: each laboratory and each researcher had to find their own experimental subjects. Voskresenskii solved this problem by entering into a collaboration with a local cavalry school of the Red Army: he performed testicle transplantations on the school's aged stallions and geldings. Other researchers appealed to the public in search of old animals and operated mostly on cats and dogs brought in by their owners with the hope of rejuvenating their beloved pets.

During the second half of the 1920s, however, the popularity of Steinach's technique began to diminish gradually: reports on vasoligature virtually disappeared from academic investigations and publications. Yet the transplantation of sex gonads, also advocated by Steinach but associated in the public mind most closely with Voronoff's name, continued to command considerable attention. Several Russian physicians even traveled to Voronoff's clinic to learn at the source his techniques of transplantation.[54]

The most difficult problem facing the proponents of rejuvenation á la Voronoff, however, was the availability of glands for transplantations. Human glands were not easy to obtain. Occasionally, as Bulgakov described in "The Dog's Heart," some surgeons used the glands extracted from cadavers.[55] Certainly, some volunteers were ready to sell one of their testicles for substantial sums of money.[56] But the major source of glands were the patients whose testicles had to be removed due to some medical condition, and, of course, there were very few such individuals. In the four years of his experimentation from 1921 to 1924, Voskresenskii was able to find only thirteen donors. Some surgeons used sheep, goat, or even cat testicles, but of course, following Voronoff's initial experiments, the preferred source of glands were monkeys. Alas, there were no monkeys in Russia and importing them from abroad for a large-scale research or treatment was prohibitively expensive. There was, nevertheless, a man who decided that this situation could be changed: Iakov Tobolkin.

As we saw in the previous chapter, Tobolkin, the deputy director for administrative affairs of the Institute of Experimental Endocrinology, played a key role in the early institutionalization of endocrinology in Soviet Russia. Thanks to his untiring efforts, by late 1925, the discipline's flagship institution had acquired a large new building, expanded its personnel, and become fully operational. It seemed that Tobolkin had achieved all of his goals and could now simply enjoy the fruits of his previous labors. But Tobolkin's inexhaustible inventiveness would not let him rest. Clearly building on the experience with his goat farm, he came up with a new idea—to establish a "monkey farm" that would supply the Institute of Experimental Endocrinology and other research institutions of Narkomzdrav with monkeys, including anthropoids. After canvassing leading biomedical specialists—including Ivan Pavlov—all of whom, of course, wholeheartedly endorsed the opportunity to have a domestic supplier of primates for their research, Tobolkin presented the idea to Semashko. The Commissar immediately bought it. The growing need for monkey glands for rejuvenation á la Voronoff certainly played a role in Semashko's willingness to support Tobolkin's plans.

Tobolkin spent the next two years building a primate breeding station. He traveled to Germany to assess the market for exotic animals and to confer with Carl Hagenbeck—a leading merchant of wild animals and founder of Europe's biggest zoo—about buying monkeys and about the problems of keeping them in captivity. He then went to France to observe the primate station Serge Voronoff had established on the French-Italian Riviera for his own operations. He also surveyed the shores of the Black Sea in search of a suitable locale, along the way courting local officials to support his endeavor. Tobolkin successfully marshaled the cooperation of countless scientists and administrators. As a result of his efforts—widely publicized by the media[57] —at the end of 1927, the Institute of Experimental Endocrinology's primate station in the town of Sukhumi on the Black Sea began breeding several species of monkeys, including rhesus macaques and baboons.[58] Along with supplying primates for physiological studies to research establishments around the country, the Sukhumi station became a major source of monkey glands for rejuvenation operations. Indeed, Leonid Voskresenskii, one of the pioneers of rejuvenation work in Russia, moved to Sukhumi to join the station's staff (see Figure 5.2).

The results of the early 1920s investigations of and debates over Steinach and Voronoff's operations in Russia were as diverse as those elsewhere. Some physicians and scientists rejected Steinach's or Voronoff's operation (or both), while others enthusiastically endorsed them.[59] Some reported complete success in achieving rejuvenation by Steinach's and/or Voronoff's techniques. Others claimed only limited achievements, describing the effects produced as temporary and arguing that the operations resulted not in a true rejuvenation, but only in a certain "refreshment" of vitality.[60] Still others announced absolute failures, recording "no effect whatsoever."[61] Neither the medical nor the scientific community could reach a consensus on the possible effects of the rejuvenation techniques. These disagreements only fueled the public fascination with rejuvenation, however.

FIGURE 5.2 The physiologist Leonid Voskresenskii (in the center) and two technicians holding young chimpanzees at the Sukhumi monkey breeding station, ca. 1929. A large group of people in the background apparently came to the station for an "educational excursion." *Courtesy of MIM.*

Sexy Science

The popular media, whose workings Bulgakov so sarcastically depicted in his novellas, widely disseminated the latest news on rejuvenation experiments. As one commentator observed as early as June 1922, "Rejuvenation is the latest fad of scientific fashion."[62] Indeed, the excitement and polemics over rejuvenation that raged within the Soviet scientific and medical communities spilled over from professional conferences and publications to the public scene instantaneously. A detailed account of Kol'tsov's first lecture on Steinach's experiments at the GINZ colloquium appeared in the next day's issue of the country's most widely distributed newspaper *Izvestiia* under the enticing title "Rejuvenation of organisms."[63] A few months later, Kol'tsov published the text of his lecture in the spring 1921 issue of *Priroda*—a leading popular-science journal in the country, which he had been editing since 1914.[64]

Kol'tsov's students followed their teacher's lead. In October 1921, the second issue of the just-established Bolshevik literary review, *The Press and Revolution*, carried a "survey of foreign and Russian literature on rejuvenation," written by the younger brother of Mikhail Zavadovskii, Boris (1895–1951).[65] His article reviewed more than twenty publications on the issues of aging and rejuvenation, including Steinach's and Voronoff's. A month later, Boris Zavadovskii also published a thirty-page-long essay on "the problem of old age and rejuvenation in light of the newest works by Steinach, Voronoff, and others" in the just-established first Soviet "thick" literary journal, *Red Virgin Soil*.[66] From 1921 to 1923, the country's most widely circulated newspapers, *Pravda* and *Izvestiia*, became engaged in the popularization of Steinach and Voronoff's work, publishing articles by and interviews with the leading Russian "rejuvenators" Nikolai Kol'tsov and Boris Zavadovskii.[67] Other newspapers, ranging from *Worker's Gazette* to *Whistle* (the newspaper for which Bulgakov was working as a staff writer), followed suit.[68]

From 1920 on, practically all popular-science magazines carried the latest news on rejuvenation research.[69] In 1922, *Man and Nature*, an influential popular-science monthly, printed a lengthy article on rejuvenation.[70] A similar article opened the first issue of a new magazine, *Science and Technology*, in February 1923.[71] In August 1923, *Book-carrier* published an extensive survey of the newest publications "on senescence, death, and the struggle against old age decrepitude."[72] *Hygiene and Health of Worker's Family*, another new magazine established in September 1923, followed suit.[73] Many other new weeklies that appeared between 1923 and 1925, such as *Flash*, *Red Panorama*, *Red Field*, *Worker's Journal*, and *Women's Magazine*, regularly published articles on the subject.[74] Even such literary journals as *Petersburg*, *Red Virgin Soil*, *Star*, *New World*, and *Young Guard* lent their pages to rejuvenation operations.[75]

All available media became engaged in the popularization of rejuvenation. During 1923 through 1925, a series of public disputes and lectures on the subject took place at Moscow's major venues for scientific propaganda: the Polytechnic Museum, the House of Sanitary Education, and the House of Scientists. In April 1923, Kol'tsov delivered a public lecture on "the newest works in the field of humans' surgical rejuvenation" at the House of Scientists. In February 1924, the Polytechnic Museum hosted a "public dispute" entitled "Is rejuvenation possible?" with Voskresenskii as the keynote speaker and Boris Zavadovskii as a moderator.[76] In December, the Polytechnic Museum held yet another "public dispute" on rejuvenation: this time with Boris Zavadovskii delivering the keynote address. In February 1925, a radio station in Moscow broadcast Kol'tsov's lecture on "Wondrous achievements of science" all over the country. In this lecture, Kol'tsov singled out current research on rejuvenation as one of the most impressive achievements of modern science.[77]

The printed and spoken word was soon supplemented with visuals. In the fall of 1923, two motion pictures about Steinach's research (produced a year earlier by a Berlin-based company, "Universum film,") were shown in Moscow, Petrograd, and other large cities.[78] One was a long documentary detailing Steinach's work for professionals. The second aimed at lay audiences and presented rejuvenations in animals and humans performed by Steinach and his collaborators in Vienna and Berlin. In June 1924, the House of Scientists organized a special viewing of Steinach's film "Rejuvenation" with extensive commentary delivered by Kol'tsov.[79] The same year, a widely publicized permanent exhibition on "the protection of health" established by Narkomzdrav in Moscow featured—along with exhibits on such subjects as "man-machine," "the origin of man," "first aid," and "the development of a human organism"—a special section on "rejuvenation."[80] At the same time, a state publisher issued a set of thirty-seven slides on rejuvenation for "cultural propaganda" in worker's clubs, schools, and libraries.[81]

Not to be outdone by their German colleagues, Russian proponents of rejuvenation also produced a motion picture about their research. Leonid Voskresenskii documented his work with Steinach's and Voronoff's operations on film.[82] Beginning in 1923, Voskresenskii filmed his research on both human patients and horses. Entitled "Who needs to be rejuvenated" the film was shown in Moscow in the fall of 1925, accompanied by lectures on rejuvenation and press interviews by its producer.[83]

But as Bulgakov indicated in his novella, the major source of public information on rejuvenation remained the daily press. In December 1923, with NEP in full swing, the Moscow City Soviet established a new daily, *Evening Moscow*, which became one of the most popular newspapers in the capital and beyond. Unlike other government newspapers such as *Pravda* or *Izvestiia*, *Evening Moscow* rarely published decrees or political propaganda. It was a typical "yellow" paper that brought to the Moscow public a variety of juicy rumors, crime chronicles, and short (usually humorous) fiction, along with information on sports and cultural events and happenings. For instance, one of its first 1924 issues carried a feuilleton written by Bulgakov about a mishap with a winning lottery ticket.[84] Among various subjects of popular interest, rejuvenation figured prominently on the pages of *Evening Moscow*. Between 1924 and 1925, the newspaper carried an item on the subject almost every month, reprinting "information" from foreign newspapers and publishing interviews with Russian proponents of rejuvenation.[85] The particular flavor of these publications could be seen through the following two examples. In September 1924, the paper reported that former British prime minister Lloyd George was planning to go to Paris to undergo rejuvenation in Voronoff's clinic. A few days later it informed readers that to date, Voronoff had rejuvenated nearly 1,500 individuals.[86] In 1925 the paper carried fifteen items on rejuvenation, ranging from a notice on the death of Baron Rothschild, who had been rejuvenated just a few months earlier, to a report on recent rejuvenation operations performed by Moscow surgeons.[87] It also became a major source of information for many provincial newspapers, which regularly reprinted articles first appearing in *Evening Moscow*.

In the second half of the 1920s, as the number of publications on rejuvenation in professional periodicals began to decline, the popular press became a major forum for debate, with *Evening Moscow* taking the lead and publishing no fewer than twenty-six items on rejuvenation in 1926 alone. On February 4, the newspaper carried a lengthy article by Nemilov under the telling title "One should be careful: sad consequences of rejuvenation á la Steinach."[88] According to Nemilov, "there is now a huge literature on Steinach's operation, which demonstrates quite convincingly that in humans vasoligation should be performed only in rare exceptional cases." Nemilov stated that Steinach's ideas underpinning his rejuvenation technique were mistaken.

Asserting that Steinach was "a brilliant experimenter, but a poor microscopist," Nemilov argued that his own and other researchers' histological studies did not support the Austrian scientist's postulate of two separate parts—one responsible for the production of sperm, another for the production of hormones (the puberty gland in Steinach's terminology)—coexisting in the testicles. Accordingly, Steinach's idea that by depressing (through vasoligation) the production of sperm one could increase the production of hormones and thus achieve rejuvenation was mistaken. In fact, the production of both sperm and hormones, according to Nemilov, is a function of the testicle as a whole and thus vasoligation completely upsets the "delicate mechanism" of the gonad. The short period of "reinvigoration" observed after Steinach's operation actually results from the "agony" of the gland, representing, as it were, its last burst of activity before dying. Nemilov warned that Steinach's operation in humans was not simply useless but dangerous, since it results in a de facto castration and the subsequent atrophy of the gonads. In Nemilov's view, it should be reserved exclusively for animals, whereby, for example, in an aged stallion, the vasoligation of one testicle would temporarily invigorate the activity of the other testicle, thus extending the reproductive ability of the valuable animal.

At the end of the month, *Evening Moscow* carried a reply to Nemilov's article written by Konstantin Viktorov, a well-known specialist in the physiology of domesticated animals and professor at the Kazan' Veterinary Institute, who completely disagreed with Nemilov's assessment.[89] Citing works by Steinach and his numerous supporters both in Russia and abroad, Viktorov refuted Nemilov's claims, insisting that Steinach's puberty gland does exist and that Steinach's operation does not result in "castration." He implied that "occasional negative results" observed after the operation were related to the insufficient qualifications of the surgeons, who did not follow Steinach's technique precisely. Viktorov maintained that a "one-sided vasoligation" is perfectly safe and "rejuvenating" and that, even if the Steinach operation fails to produce the desired result, one can always supplement it with Voronoff's transplantations.

On March 13, *Evening Moscow* expanded the polemic by publishing responses to a questionnaire distributed among leading physicians and scientists in Moscow and Leningrad.[90] Entitled "Is the return of youth possible?," the article cited the opinions of eight well-known physicians and biologists. The prominent surgeon Vladimir Oppel' stated that he "had considered the Steinach operation useless from the very start." His colleague Ivan Grekov was less categorical: "to answer the question of the usefulness of [both Steinach and Voronoff's] operations we need more investigations." The biochemist Efim London insisted that both operations had "only a temporary effect." Semen Girgolav, who had performed the first Steinach operation in the country, claimed that results of both operations were "not always beneficial," but in patients with "certain medical conditions such as arteriosclerosis" the positive effects were "indisputable." The histologist Anton Nemilov recapitulated his objections to Steinach's operation and endorsed the use of Voronoff's. The surgeon Vladimir Gorash rejected Nemilov's equation of Steinach's operation with castration, but cautioned against its indiscriminate use, joining Girgolav's opinion that rejuvenation should be reserved exclusively for patients with "certain medical conditions." Ruvim Gasul', a respected roentgenologist and TB specialist, warned that rejuvenation could activate latent diseases, particularly tuberculosis. The zoologist Petr Shmidt characterized vasoligation as useless but supported the limited use of transplantations.

Less than a month later, the newspaper carried yet another lengthy piece by Konstantin Viktorov, entitled "Professor Steinach's letter about the polemics in *Evening Moscow*."[91] Obviously irritated by Nemilov's claims that vasoligation was not only useless but harmful, Viktorov had decided to appeal to the ultimate authority—Steinach himself. Viktorov had forwarded to the Austrian scientist *Evening Moscow*'s clippings of Nemilov's and his own articles on the subject, along with an explanatory note. The maitre responded with a long letter, which Viktorov summarized for the readers. According to Viktorov, Steinach referred him to available statistical data, expressing regret that out of "many thousands of cases" only "few hundreds have been described in detail and published." Viktorov maintained that he did not want to "bother the readers with boring numbers"; it was sufficient to note that Steinach's method had been used "not only in large centers of Europe (Berlin, Vienna, London, Copenhagen, Madrid, and Moscow) but also in America and Japan" and that "no one among the numerous researchers from these centers had noted any negative effects of the operation." Furthermore, the complete absence of such negative effects was further demonstrated by the "data from America," where vasoligation "had been used for eugenic purposes" (such as the sterilization of criminals) on "more than 5,000 individuals," and in all these cases, it caused "no harm." In regard to Nemilov's reported observation of the complete atrophy of the testicles following vasoligation, Steinach suggested that the operation had been carried out by an inexperienced surgeon, who had obviously damaged blood vessels and nerves, which should have been carefully separated from seminal vesicles. "All further polemics are superfluous," Viktorov concluded.

It was, rather, Viktorov's conclusion that turned out to be superfluous, for the debates over the rejuvenation techniques continued unabated. The enormous publicity accorded

FIGURE 5.3 The 1923 satirical cartoons on rejuvenation á la Serge Voronoff, which appeared in the popular satirical magazine *Crocodile*:
a) The cartoon of August 1923 entitled "Monkeys' revolt." The caption says: "**Monkey with the rifle**: Enough is enough. You've been sucking our blood long enough." An inscription in a smaller font-type at the upper right corner provides the reason for the revolt: "Professor Voronoff has invented a method of transplantation of monkey glands to rejuvenate humans." The cartoon clearly implies that the monkeys rebel against capitalists and capitalism who exploit them to get rejuvenated. *From* Krokodil, *1923, 34: 1044.*
b) The caption to this cartoon that appeared a few months later in December says: "**Worker** (addressing the monkey on a leash): Ye-e-s, you've been a capitalist once. That's what you deserve! Don't get rejuvenated!"
This cartoon reflected a popular fear that the implantation of monkey's tissues and organs could lead to the transformation of the recipients into monkeys. In "The Dog's Heart" Bulgakov reversed the situation: the implantation of human organs into a dog leads to the dog's transformation into a human. This cartoon also serves as an illustration to the popular thesis (advanced in both Isai Lezhnev's article "Two grafts: Lenin and Steinach" and Andrei Irkutov's story "Immortality") that only capitalists (and capitalism) needed rejuvenation, while workers (and socialism) had no use for it. Note that this caption does not mention Voronoff's or Steinach's name and work (carrying only a notice "from newspapers"), indicating that the subject of rejuvenation became so familiar as to not require any substantiation or explanation by reference to the actual rejuvenation experiments performed at the time. *From* Krokodil, *1923, 46: 1237.*

FIGURE 5.3 a

FIGURE 5.3 b

to Steinach and Voronoff's rejuvenation works spurred a wide range of responses among the Soviet public. Some people were incredulous, others enthusiastic (see Figure 5.3). Some found the fulfillment of long cherished dreams, others nightmares. Some wanted to apply the new rejuvenation techniques immediately in order to achieve immortality, others warned of dangers inherent to such hasty applications.

Since Kol'tsov published a large number of popular accounts on rejuvenation (not to mention his numerous speaking engagements), while daily newspapers and weekly magazines regularly described his work, his Institute of Experimental Biology in Moscow became a center of particular attraction. Dozens of people besieged the institute in the hopes of being rejuvenated. Dozens more, from all corners of the country, sent letters and telegrams asking for advice, permission to come to the institute for the operations, or simply more information.[92] Who were all these people? Why did they want to undergo rejuvenation? And who among the hundreds of seekers was actually chosen for the operations?

In "The Dog's Heart," Bulgakov portrayed a male patient who had been rejuvenated and a female patient scheduled to receive monkey ovaries in Preobrazhenskii's clinic, indicating that both men and women sought rejuvenation. Indeed, both Steinach and Voronoff had developed certain techniques for the rejuvenation of women: Steinach suggested x-ray irradiation, and Voronoff the transplantation, of ovaries.[93] Yet despite a few highly publicized cases, the rejuvenation of women remained more of a theoretical possibility than a practical reality. Steinach and Voronoff's techniques of rejuvenation in men were relatively simple: both vasoligations and testicle transplants could be easily performed under local anesthesia in a doctor's office or even in a research laboratory. The transplantation of ovaries, however, was a major operation that required general anesthesia, a sterile operating room, and an experienced surgeon. Steinach's idea of using x-ray irradiation of ovaries as a means of rejuvenation did not inspire much following, since he had failed to develop a reliable protocol for the treatment: the dose, duration, and intensity of irradiation remained highly contested issues. Furthermore, the reported results of irradiation in women were not as encouraging as those of vasoligature in men. Nevertheless, women did appeal to Soviet rejuvenators. Some of them sought the implantation of ovaries because they lost theirs due to some surgical interventions. Others simply wanted to "regain their lost youth." Yet in all my searches through archival and published materials, I encountered only two reports on actual rejuvenation operations performed by Soviet investigators on women.[94]

The overwhelming majority among those seeking rejuvenation were men. Bulgakov's novella depicted two groups of patients who employed Preobrazhenskii's services: the "Nepman,"—a representative of the new bourgeoisie that emerged under NEP—who could pay the doctor large sums of money, and the new Bolshevik "ruler," who could offer the professor protection from the excesses of proletarian militants. Apparently, no other social groups needed rejuvenation. In a telling episode, when Preobrazhenskii's housekeeper entertains a suitor—a firefighter—in her room, she repels his particularly insistent advances: "'Just like a demon,'—mumbled Dar'ia Petrovna in the semidarkness,—'stop it!' . . . 'What's it with you, did you get rejuvenated too?' 'I have no use for rejuvenation,'—barely controlling himself replied [the man] in a hoarse voice" (96).

Some physicians did perform Steinach and Voronoff's operations for money. Under the NEP conditions that allowed private medical practice, rejuvenation offered a lucrative opportunity for enterprising doctors and an appealing possibility for the newly rich. Understandably, historical evidence of this kind of arrangement is scarce: in the "proletarian" state, neither party was interested in wide-scale advertising.[95] But Bulgakov's novella is not the only indication that the arrangement did in fact exist. The January 1924 issue of *Hygiene and Health of Worker's Family* carried an article tellingly entitled "Steinachs from Batum."[96] The article reported a court case of two doctors from the Georgian city of Batum who had been accused of profiteering by performing Voronoff's operations for money.[97]

On the other hand, the evidence that some of the new Bolshevik rulers did indeed undergo rejuvenation is ample. In November 1925, in a report on the recent rejuvenation operations performed in various clinics, *Evening Moscow* published the following case history:

> Citizen F. 75 years old, a professional revolutionary, spent 25 years in prison. Complains about the lack of strength and energy, the weakening of memory, etc. He is an emaciated old man, with a deeply lined face. Following the transplantation of a part of a gland (from a monkey in February 1924), after just one and a half months, noticeable improvement: less susceptible to colds; dark hair appeared on the back of his head.
>
> In May—walks more assuredly and freely, dark hair [appeared] on his head and in beard. Sexual drive re-emerged. In September, a mass of dark hair, a fresh-looking face, headaches and colds disappeared.
>
> At this moment the patient is under observation. How intense his mental capabilities have become is evident from [the fact that] he had delivered more than 40 public lectures in the course of the last year.[98]

The details provided in the account suggest that "Citizen F." most likely was Mikhail Frolenko (1848–1938), a member of the "People's Will" group, who had been arrested in 1881 in relation to the assassination attempt on Emperor Alexander II and had in fact spent twenty-five years in Russia's most infamous prison, the Shlisselburg Fortress near St. Petersburg.[99] After the Revolution, Frolenko came to Moscow and became an active member of the Society of Former Political Prisoners (*Obshchestvo byvshikh politkatorzhan*), lecturing and publishing on his revolutionary activities under the tsarist regime in the society's journal, *Prison and Exile*.[100]

Frolenko was certainly not the only "old revolutionary" to benefit from rejuvenation. It seems that the Society of Former Political Prisoners—composed predominantly of professional revolutionaries who had spent decades in prison and exile—became one of the major promoters and beneficiaries of rejuvenation operations in Russia. Indeed, one of the first Russian translations of Voronoff's 1923 book *Quarante-trois greffes du singe á l'homme* was initiated and prepared by Nikolai Morozov (1854–1946), another prominent member of the "People's Will" who also had spent twenty-two years in solitary confinement in the Shlisselburg Fortress.[101] After the Revolution, Morozov came to head the Petrograd Institute of Natural Sciences. Sometime in 1923, apparently intrigued by the numerous publications on the subject, Morozov initiated a correspondence with Voronoff, asking him for copies of his publications.[102] In early 1924, Voronoff sent Morozov his latest book on gland transplants from apes to humans. Morozov's wife translated the text into Russian, and Morozov carefully edited it.[103] In just a few months, a well-known publisher, Kniga, released the book.[104]

Although in a letter to Voronoff the sixty-nine year-old Morozov intimated that so far he himself felt no need for rejuvenation and would "wait for at least four more years,"[105] other "old revolutionaries" apparently were ready to take advantage of Voronoff's operation. Mikhail Arshenberger (1842–1926)—Morozov and Frolenko's fellow member of the "People's Will" and the head of its military organization—was slated for the Voronoff operation in October 1924. Furthermore, it seems that the "veterans of the revolution" had privileged access to rejuvenation. As a letter to Kol'tsov from the director of a "sanatorium for the veterans of the revolution" indicates, in 1924 Narkomzdrav had reserved several "slots" for the veterans at the busy rejuvenation department of the Institute of Experimental Biology.[106]

Among the numerous social experiments conducted by the Bolsheviks in their efforts to create a "new world" was a profound change in the societal regulations of, and public attitudes toward, sex and sexual relations, which many observers have described as a "sexual revolution."[107] Since both Steinach's and Voronoff's rejuvenation techniques involved certain manipulations of the sex gonads, in the public mind they became inextricably interlocked with the larger issues of sex. The main attribute of youth—and the major symptom of rejuvenation noted in numerous reports on Steinach and Voronoff's operations—was sexual prowess. In "The Dog's Heart" Bulgakov vividly portrayed this intertwining of sex and rejuvenation through the dog's observations of what was going on in Preobrazhenskii's "obscene apartment" (86). Thus in the public mind, the sexual revolution made Steinach and Voronoff's operations a preferred method of addressing all and any problems related to sex.

Available evidence indicates that the main group of individuals seeking rejuvenation was not the "old revolutionaries," but relatively young (aged between twenty and thirty) men who suffered from impotence. Russian rejuvenators, however, turned nearly all of them down. In the medical views of the day, impotence in young men was largely seen as the result of their own "bad" behavior—masturbation, sexual excesses, and venereal diseases. Doctors tended to dismiss their complaints and refuse "medical interventions," resorting to basic advice on healthy lifestyle as a "cure" for impotence.[108] In addition, from the viewpoint of researchers, young patients were of no particular scientific interest; they focused their attention on the old ones.

Rejuvenation and the promised restoration of sexual prowess certainly was a major, but not the only, reason for appeals to Russian rejuvenators from numerous individuals. An offshoot of Steinach and Voronoff's rejuvenation techniques was the possibility of "changing one's sex." Steinach's 1912 article described profound changes—both anatomical and behavioral—that followed castration and the subsequent implantation of male sex glands into females (or female sex glands into males), resulting in the appearance of the sexual characteristics of the opposite sex in the operated animal. Steinach himself clearly understood the clinical implications of his research. His follow-up article, coauthored with the eminent urologist Robert Lichtenstern, suggested that by boosting the puberty gland, both vasoligature and sex gland transplants could serve as a "cure" for homosexuality, viewed at the time as the result of inefficiency of homosexual individuals' sex gonads.[109] Steinach's works reignited the old debate on the mechanisms of sex determination and its possible manipulation by experimental methods.[110]

In Russia, one of the most active participants in this debate was Mikhail Zavadovskii. Within a year of his return to Moscow, in early 1922, Zavadovskii published the results of his experiments on sex transformation conducted in Askania Nova.[111] Entitled *Sex and the Development of its Characteristics*—lavishly illustrated with more than one hundred pictures and twenty color tables—the 250-page book described the "transformation" of roosters into hens and back (and vice versa) through castration and subsequent implantation of the gonads of the opposite sex. The state's major publishing house Gosizdat, which released the book, also arranged for Zavadovskii to deliver a series of popular lectures on his research at the Moscow Polytechnic Museum in the spring of 1922. The lectures were a huge success and the publisher decided to issue them in its series "Popular-Science Library" as a book, entitled *Sex in Animals and its Transformation*.[112] Published in the fall of 1923 and also richly illustrated, Zavadovskii's new book presented his old experiments conducted in Askania Nova, as well as some recent investigations conducted in his laboratory at the Moscow Zoo. Zavadovskii even exhibited several of his "hens-turned-roosters" and "roosters-turned-hens" at the Zoo. Zavadovskii's descriptions (and particularly his illustrations) of the changes in the sexual characteristics, which followed castration and implantation of gonads from the opposite sex, became a staple of popular literature on the "biological foundations of sex" and were replicated in practically every publication on the subject (see Figure 5.4).[113] Furthermore, Zavadovskii's lectures, books, and exhibits played an important role in popularizing the possibility of sex transformation as a viable surgical procedure that could be performed not only on animals, but also on humans. For certain individuals, such a possibility obviously held an enormous appeal.

Indeed, even before Zavadovskii published his books, the possibility of sex change had become one of the incentives for public appeals to Russian rejuvenators.[114] As early as January 1921, just a few months after the first publication of his lecture about Steinach's

FIGURE 5.4 One of the illustrations from Mikhail Zavadovskii's *Mechanics of Sex Development*, demonstrating the transformation of a hen into a rooster (right column) and a rooster to a hen (left column) following the castration (second row) and subsequent implantation of the gonads of the opposite sex (third row). This particular table was reproduced in practically every popular work on sex published in Russia during the 1920s. *From M. Zavadovskii,* Mekhanika razvitiia pola, *1923*.

rejuvenation technique on the pages of *Izvestiia*, Kol'tsov received a long letter from Chita—an administrative center in the depths of Siberia.[115] After having read the *Izvestiia* article, the letter's author—a fifty-year-old man—offered himself as a subject for experiments at Kol'tsov's Institute. He was not interested in rejuvenation, however. Without clearly stating that he was a homosexual, the man proposed that he be subjected to "castration, subsequent implantation of female sex gonads, and any further operations deemed necessary" to investigate his "certain psycho-sexual peculiarities." The man had obviously read diligently the available literature on the questions of sexual psychology and the role of sex hormones in an attempt to understand his own identity and its "peculiarities." He was ready to "give the last years or months of [my] life to the common good as a subject of scientific investigations." For himself, he wanted only to find out whether his "identity would remain unchanged [by the operations], or, if it did change, what would determine the change."

Zavadovskii's publications on the transformation of sex reinforced this trend, as evidenced by another letter preserved in Kol'tsov's files but apparently addressed to

Zavadovskii. Its author stated that he was not "a man in the fullest sense of the term but not a woman either, yet something in between," briefly describing his anatomical and psychological characteristics.[116] This man followed the publications on sex change with trepidation, lamenting: "Why am I not that lucky individual who is destined to become a real man?" Stating that "life in my current condition is awful, and if I die it won't be a big loss," he begged to be subjected to the "same experiment as the one you until now have performed on chickens."

The sex change operation also appealed to some women.[117] In July 1923 Kol'tsov received a letter written by a thirty-four-year-old female nurse from the Crimea.[118] Enticed by a public lecture on "the struggle against old age and sex change" she had attended, the nurse wanted to "be subjected to the operation of 'transplantation of sex organs,'" "in the name of Christ, for the progress of medicine, and for the good of humankind." But she apparently had other motives as well. Without clearly articulating why, she stated that she "wanted to become a man." She also put forward a condition: if she were to undergo the operation, she wanted to be enrolled into a medical school at the state's expense.

Another female correspondent was much more emotional and much more open regarding her desire to change her sex. In a long letter written in October 1925 and addressed to Ivan Kogan, a twenty-four-year-old woman from a small town in the Volga region complained that "nature played a cruel joke" on her.[119] "Physically, I am a woman," she explained. "I have female's external sexual organs and, as every woman, I menstruate and my breasts are well developed; but my inclinations, instincts are those of a man. I love women, love them the same way every man does." She stated that her "passion and instinct are directed to women only." Furthermore, she had met a "woman who shared" her passion, but "this mutual love brings [us] nothing but suffering," since "being physically a woman, I, of course, cannot satisfy neither myself, nor that woman." She provided a history of her "disease," explaining that from a very early age she felt and behaved like a boy, played boys' games, and "despised, as a boy does, dolls and other girls." When she reached puberty she became interested in other girls, but did not understand it. At the age of fourteen, she discovered from "reading medical books" that her "orientation" was "abnormal." She tried to hide it, forced herself into relations with men, but felt only repulsion. At the age of twenty she fell in love with a young woman. That woman loved her, too, but, in the end, "could not go against nature," broke their relationship, and married a man.

The correspondent detailed her attempts to cure her "disease." She went to see a psychiatrist, who of course "did not say anything of substance," reassuring her that "it will pass." But it did not: her passion for women only grew. Her friends who knew of her orientation recommended that she marry a man, but as she put it: "You know, doctor, no matter how strongly I want to get rid of my disease, I would rather die than do that." She went to see another physician, this time a surgeon-gynecologist, who proffered an explanation of her "disease"—the insufficient activity of her sex gonads—and offered a solution: "transplantation of ovaries from a healthy, normal woman." She refused, feeling that she "could not, did not want to, be a woman." She thought of suicide, but had to keep on living to support her aged mother: "I now live only with the hope that sooner or later, I will become a man." This newfound hope was grounded in "the experiments with sex change by Steinach's methods." "You cannot imagine," she exclaimed, "how happy I was to learn about this!" She read Mikhail Zavadovskii's 1922 book and was "amazed" and "excited," for she hoped that "similar operations could be performed on people." "This book I've read instilled in me the hope," she explained, "that since such operations have been conducted with great success on animals then I would stop at nothing, and, if you agree, would become your patient." She had only one question: "is it only the instinct, or also the external sex [organs] that change" as the result of the operation; "since I have the instinct, only sex [organs] need to be changed." "Please, tell me straight," she begged, "whether you can do such an operation on me."

Kogan himself certainly did not do any such operations on humans: he was not a surgeon to begin with. My search through the archives of Kol'tsov's institute failed to produce any evidence that even the surgeons affiliated with the institute—for instance Alexander Zamkov, who did perform both Steinach and Voronoff's operations on human patients—ever tried to perform a sex-change operation. But available evidence indicates that at least one Moscow clinic did practice such operations. As an *Evening Moscow* reporter informed his readers in December 1926, professor Il'ia Golianitskii, well-known for his work on plastic surgery and tissue transplantations, had successfully performed sex-change operations on both men and women.[120] Entitled "Underdeveloped people" and illustrated by a photograph of one of the patients, the reportage described five successful sex-change operations: four on women and one on a man. Other anecdotal evidence suggests that, following Steinach's example, some Russian surgeons attempted to "cure" homosexuals by implanting sex gonads.[121]

From Rejuvenation to Immortality

Bulgakov's "The Dog's Heart" was probably the best, but certainly not the only or even the first, literary assessment of the concurrent rejuvenation craze in Soviet Russia. The public fascination with Steinach's and Voronoff's techniques found an expression in a variety of literary works that employed rejuvenation as part of their plots, symbolic representations, metaphoric devices, or simply as a recognizable sign of the times. Significantly, many of these literary experiments quickly extended the hopes for rejuvenation into a definite possibility of immortality.

As one would expect, the Biocosmists became the first literary group to embrace wholeheartedly Steinach's works, conceiving of them as a first step on the way to their proclaimed goal—immortality. They insisted that Steinach's research "has shaken the biological foundations of death."[122] During the three years of the group's existence, its members regularly invoked Steinach's name and experiments in lectures and publications on their doctrine.[123] They appealed to Steinach himself, asking the maitre for his latest works to be translated and published in their journal, *Immortality*.[124] They even wrote poems about rejuvenation, lambasting the Viennese bourgeoisie "buying immortality" from Steinach.[125]

By 1923, both Steinach's and Voronoff's names had attained the status of "household words" that required no explanations for readers. A peculiar article, entitled "Two grafts: Steinach and Lenin," illustrates this perfectly.[126] Isai Lezhnev, a well-known journalist and the editor of *Russia* (who two years later would publish Bulgakov's first novel *White Guard*), authored the article, which appeared in the spring 1923 issue of his journal. Clearly alluding to Oswald Spengler's famous work *Der Untergang des Abendlandes* (1918–1923) that spurred heated discussions in Russia,[127] the article juxtaposed the "decline of the West" and the rise of a "new civilization" in Bolshevik Russia. Steinach represented the former, Lenin the latter. As Lezhnev put it, "If the West is getting comforted by Steinach's graft, we need a mass graft of civilization, of Leninism." Tellingly, the article lumped together Steinach and Voronoff by playing on Voronoff's technique—often referred to at the time as grafting—but using Steinach's name in its title and text.

One of the earliest literary experiments investigating the range of human reactions to the possibility of rejuvenation—and its natural extension, immortality—was a short story written by Mark Krinitskii, a well-known novelist.[128] Published in June 1923 on the front pages of *Flash*, a popular weekly, and misleadingly entitled "The Elixir of Immortality," the story was not about an "elixir," but about the discovery by "a Russian physiologist and surgeon" of a simple surgical operation that makes people "immortal."[129] Krinitskii did not mention Steinach's or Voronoff's name, but the story made it clear that the operation involved certain manipulations of the sex gonads.

To ensure the just access to immortality for everyone, the operation is "free of charge" (that is, the state covers the associated expenses), while the clinic performs the operation on patients in the alphabetical order of their last names. Everyone is lining up to become immortal. The story's main character, a man with a generic name, Ivan Ivanovich Dilektorskii, is a low-level clerk in one of the Soviet state agencies, where he "prepares papers." He enters his name into the waiting list, as does everybody else in the city. Since his last name starts with "D," he has to wait several weeks for his turn. Naturally, the waiting is a torment: he is afraid of falling gravely sick with typhus or influenza, or suffering an accident, and thus dying before he can become immortal. He observes with trepidation the funeral processions passing by, since in "people's minds death has become somewhat inappropriate and shameful."

Finally, the waiting is over and Ivan Ivanovich comes to the clinic for his operation. He is greeted by a nurse and two doctors, who accompany him to the operating table and help him undress. Before the operation begins, one of the doctors warns him that the operation is "irreversible," and once it is done, Ivan Ivanovich will live for "thousands and thousands or even millions of years, until our sun cools down, or a world catastrophe ends life on the planet." Neither old age, nor diseases could kill him. The only way to end his life is a suicide: "you can jump in front of the train, hang or shoot yourself," explains the doctor. "You can also cut your throat or take a poison," helpfully adds his colleague.

Suddenly, Ivan Ivanovich realizes that he is not ready to commit himself to immortality. Trembling, he asks the doctors to postpone the operation, while he thinks through its implications. He sits in the hallway and it slowly dawns on him what immortality would mean to him personally. In a train of thought clearly reminiscent of the first circle in Dante's "Inferno," Ivan Ivanovich imagines as day after day, year after year, century after century, millennium after millennium, he would have to go to the same office, prepare the same papers, suffer the same old jokes by the same annoying coworkers, and the only way to end this endless procession of sameness would be to kill himself! He watches other people go in and out of the operating room, every one of them "with an idiot's smile of happiness on the face." It occurs to Ivan Ivanovich that "all these candidates for a suicide" have been handed "a rope to hang themselves, instead of immortality." Death—that used to be unanticipated, unpredictable, unimaginable—would become a matter of his own decision, his own choice, and his own responsibility. "Shaking as if he has just escaped a deathtrap," Ivan Ivanovich puts his clothes back on. He does not want immortality, he is not an idiot. He will leave it to others to spend "billions and trillions of years in the office." He will go to a restaurant and order a shot of vodka. "Oh, a shot of vodka, when it's unknown whether it will be there for you tomorrow. This is something completely different from a shot of vodka day after day for a billion years," muses Ivan Ivanovich. "The sunlight is so intoxicating because tomorrow perhaps will come a perpetual night; love and women—because tomorrow will come old age." He realizes that life is so enjoyable exactly because it has an end. With a cry of "Long live cholera!," Ivan Ivanovich runs out from the clinic, under the accompaniment of a doctor's remark: "Finally, I see a smart man."

As Krinitskii's story indicates, some writers found the current rejuvenation craze hilarious, using the very ideas of rejuvenation and immortality for making fun of certain individuals, situations, and events. Various cartoons published by daily newspapers and weekly magazines illustrate this trend. By 1927, even Soviet surgeons began to make fun of rejuvenation operations, as witnessed by a cartoon published in *Not-New Surgical Archive* (the humorous supplement to a leading professional journal, *New Surgical Archive*) (see Figure 5.5). But these cartoons also demonstrate that the subject of rejuvenation had become so familiar to the Soviet public as to be used without explanation in any conceivable context (see Figure 5.6.).

Indeed, by the mid-1920s, rejuvenation had become a staple of popular imagination and popular fiction. Short stories, novels, and plays explored possible implications and ramifications of Steinach and Voronoff's techniques actively pursued by their Soviet followers.[130] Although the secret police confiscated the manuscript of Bulgakov's "The

Эндокринный завтрак в Боткинской больнице.

На шести столах одновременно оперировали больных с нарушениями внутренней серекции. По требованию хиґурга подавались для пересадки различные эндокр. железы от обезьяны, оперированной в соседней комнате. См. „Новое в медицине"—„Правда" № 287 11/XII—1926 г.

МЕНЮ

1. Oeufs de singe à la cocque.
2. Окрошка из щитовидной и зобной железы.
3. Мозги со сморчками, подливка инсулин.
4. Рагу из ливера и печенки.
5. Почки соус с адреналином.
6. Компот из надпочечников à la Oppel.
7. Пунш глясе гипофизный со спермином.

FIGURE 5.5 This cartoon is entitled "An endocrine breakfast at the Botkin hospital." It appeared in 1927 in a humorous supplement to the country's most authoritative surgical journal, *New Surgical Archive*. Underneath it carries a quotation from an article published in *Pravda* on December 11, 1926, under the title "News in medicine": "On six tables simultaneously [surgeons] operated patients with endocrine disorders. At the request of a surgeon various endocrine glands from a monkey operated upon in the next room were delivered to the operating table." A seven-course "menu" placed below lists "dishes" made from various endocrine organs (testicles, liver, thyroids, adrenals) and their products (adrenalin, spermin, insulin). *From* Ne-novyi khirurgicheskii arkhiv, *1927*.

Dog's Heart," making it unknown to Soviet readers for more than sixty years, in his play *Zoika's Apartment* (written in late 1925 and premiered at the Vakhtangov Theatre on October 28, 1926), Bulgakov revisited the subject in a brief but memorable episode. One of the play's main characters, Count Obol'ianinov, tells his lover, Zoika, not to call him "Count" anymore. He explains that on the previous day he had a visit from one of the new rulers, who informed him that now he is a "former count." When Obol'ianinov objected, "What does it mean 'former' count? Have I disappeared? I'm right here in front of you," the visitor told him that he should "be placed in the museum of the revolution." Furthermore, on his way to Zoika's apartment, Obol'ianinov saw an advertisement for an exhibit at the Moscow Zoo: "Today's demonstration—a former hen." Irritated by his previous conversation with the proletarian, Obol'ianinov went to the Zoo and asked a guard at the exhibit, "What *is* this former hen now, under the Soviet power?" The guard answered: "She is now a rooster." It turns out, Obol'ianinov continues, that "One of those bandits, some Communist professor,[131] did something horrible to the poor hen and, as a result, she turned into a rooster. Everything turned upside down in my head, I swear. . . . *Cauchemar*!" The count begs Zoika to take him away from Soviet Russia: "I cannot look at former hens anymore. [I need to get] away from here, whatever the price."

Aside from presenting a recognizable feature of its time,[132] in this episode Bulgakov clearly mocked the Bolshevik obsession with abolishing pre-revolutionary social hierarchies and reshaping the identities of people according to the goals of the Revolution. In Soviet Russia, the epithet "former" (*byvshie*) was typically applied to the members of so-called exploitator classes (the royalty, nobility, clergy, military, state bureaucracy, merchants, and capitalists) stripped of all their privileges by the Revolution.[133] Considered to be "remnants" of the tsarist past, these people had no place in the new social order, unless they denounced their past, shed their identities, and joined the proletariat in building socialism in Russia, which, according to Bulgakov, was somewhat akin to hens turning into roosters.

Naturally, the rejuvenation fad inspired writers not only in Soviet Russia, but all over the world.[134] During the 1920s, obviously responding to public demand, Russian publishers issued translations of several short stories and novels that dealt with rejuvenation, gland transplantations, and immortality.[135] *Le nouvel Adam*, written by the Swiss-Canadian novelist and journalist Hélène Dufour Pittard and published in 1924 under the pen name Noëlle Roger, appeared in Russian translation—just a few months after its publication in Paris—under the title *The Man of the Future*.[136] Two years later, a different publisher issued a new translation of the novel, under the title *The Coming Adam*.[137] The same year, 1926, the Russian reading public was treated to another French novel, *Osmant le rajeunisseur*, written by Marcel Roland.[138]

Not to be outshined by foreign competitors, several Russian novelists took up the subject. In 1928 Vsevolod Valiusinskii, a young aspiring writer, published his first novel, entitled *Five Immortals*.[139] The novel told a story of a future world divided into two competing

FIGURE 5.6 Cartoons on rejuvenation:
a) This 1923 cartoon is entitled "When carriages will be rejuvenated." The caption says: "**The horse** (escaping from the carriage's shafts): [How] could anyone catch up with this carriage! Look how it has become rejuvenated." An inscription in the upper right corner says: "In Tver', the head of a physiological laboratory professor L. N. Voskresenskii successfully conducts experiments on rejuvenating horses." Obviously, the target of this cartoon is not rejuvenation, but the dilapidated state of public transportation in Soviet Russia, particularly in large cities like Moscow, which was a point of special concern for both state agencies and the public during the early 1920s. *From* Krokodil, *1923, 38: 1098.*
b) This 1926 cartoon is titled "Rejuvenating rays." It accompanied an eponymous satirical poem that made fun of the current "rejuvenation craze" in Russia, portraying rejuvenation as a routine biweekly treatment available at every bathhouse or barbershop. Note that unlike the earlier cartoons of 1923 (see also Figure 5.3), this 1926 cartoon does not present rejuvenation as an exclusively "capitalist obsession" anymore. Apparently, by the mid-1920s rejuvenation has become acceptable for "proletarians" as well. *From* 30 dnei, *1926, 5: 92.*

FIGURE 5.6 a

FIGURE 5.6 b

parts: the Americas is the domain of capitalism, while the rest of the world forms the Great Union of Socialist Republics.[140] An international team of six scientists, working at a biological station in the Union and led by a brilliant Russian biologist with the telling name Kurganov (clearly alluding to death, since *kurgan* in Russian means a burial mound), discovers a method to manipulate—to speed up, slow down, and ultimately stop—the aging processes. According to Kurganov's findings, aging is the result of progressive accumulation of "poisons" produced by the organism, which is controlled by a special center located in the "beta-region of the middle brain." In numerous experiments on rabbits, dogs, and sheep, Kurganov discovers that implanting a small piece of brain from the beta-region to the same location into an animal of the same species but the opposite sex, leads to immortality. The transplantation performed between animals of the same sex greatly increases the speed of aging and leads to premature death. Furthermore, in the process of becoming immortal, the recipient clearly becomes "much smarter," but loses both primary and secondary sexual characteristics: its sexual organs atrophy. It becomes "immortal and sexless." And, of course, in order for the recipient to become immortal, the donor has to die. Kurganov conducts his research in total secrecy and nobody outside of his team knows what they do at the station.

Having failed to find another way of making experimental animals immortal, and faced with his own mortality—which could prevent him from finishing up this work—Kurganov decides that his team should go forward and perform the transplantations on themselves so that surviving members can continue the research unhindered by their own mortality. Since his team consists of five men and only two women, Kurganov recruits three more women to form five pairs for the operation. They decide by drawing lots which member of the pair will be the donor and thus die in order for the other to become immortal. Kurganov performs the operations on the first four pairs and disposes of the donors' bodies. The operations are successful, and the first four immortals, two men and two women, are created! After several months, they are sufficiently recovered to perform the last operation on Kurganov. Although Kurganov draws the lot making him the donor, his teammates decide that as the leader of the entire project, he must live. After Kurganov recovers, the five immortals blow up the station to cover up their tracks and disappear in the world.

For two centuries, they search for a way to achieve immortality without the horrible price of sacrificing one person for the immortality of another. In the course of time, their minds and bodies undergo dramatic transformations. In addition to the disappearance of sexual characteristics, which make "former" men and women look completely alike, their mental abilities increase dramatically. They become utterly rational, while their emotions wither away. Their skulls become 50 percent larger to accommodate the ever-growing brains, with the concurrent diminution of facial bones, particularly jaws, giving them a child-like appearance. In time, however, the immortals begin to notice certain signs of aging in themselves. Obviously, the effects of transplantation start to wear off. They need a boost, possibly another transplant, which makes their search ever more urgent.

In the end, they triumph. They discover a "death antitoxin" in the blood of immortalized animals and prepare a "serum of immortality," which injected into another animal renders the latter immortal. They prepare the serum from their own blood and test it on an unsuspecting man. The test is successful: the man's sexual organs begin to atrophy, which indicates that he is becoming immortal. Two of the immortals receive injections of the serum and, as expected, the signs of aging disappear.

The immortals are finally ready to present their discovery to the world. But alas, the world is not ready to receive it. It is still divided into two competing parts: the capitalist

Americas and the Great Socialist Union. The confrontation between the two is coming to a head: military conflict is imminent. The immortals' announcement of their discovery ignites the final battle, which becomes a fight for control over the immortals and immortality. Four of the immortals are killed, and the last one, Kurganov, having fallen into the hands of the Americans who use him to blackmail the Great Union forces into surrender, commits suicide. The Great Union wins the war, the last war on Earth. All of humanity comes together into the Great World Union. But the immortals are all dead and their secret is lost: "The immortality again became just a dream." The novel, nevertheless, ends with a resounding optimistic statement: "But everyone knew that it is possible and believed that sooner or later, the happy, crazy shout: 'There is no more death!' will roar again."

This ending summed up succinctly the major impression left by the rejuvenation craze on the Russian public: the promises of perpetual youth and eternal life were about to be fulfilled, and not through the benevolence of some deity, but through the efforts of scientists working in the land of victorious socialism.

Sag mir, was bedeutet der Mensch?
Woher ist er kommen? Wo geht er hin?
Wer wohnt dort oben auf goldenen Sternen?
HEINRICH HEINE, "Fragen" (1828)

Man is always concerned about his destiny—that is to say his position and role in the universe, and how he is to maintain that position and fulfill that role.
JULIAN HUXLEY, *Religion without Revelation* (1957)

6 "Quo Vadimus?": Human Biology and Human Destiny

DURING THE 1920S, scientists the world over actively partook in the quest for control over life, death, and disease, and biomedical developments in revolutionary Russia in many ways paralleled those in other locales in Europe and the Americas. Indeed, scientists everywhere conducted similar experiments: with isolated organs, blood transfusions, hormones, organ and tissue transplants, and the freezing and thawing of living organisms. The same new disciplines—biophysics, endocrinology, immunology, biochemistry, eugenics, hematology, and so on—emerged over the same time period in many countries, with specialized research institutions proliferating around the globe. Inspired, intrigued, or alarmed by the rapid growth of biomedical sciences, various European and American writers also produced similar SF stories, examining how new biomedical knowledge could shape human nature and human destiny.

Yet Russian biomedical developments differed significantly from those elsewhere. Certainly, in no other locale did biomedical sciences and scientists enjoy such a high level of state support, making the rapid institutional growth of Russian biomedical sciences a point of attraction and an object of envy for colleagues the world over.[1] Furthermore, unlike elsewhere in the world, the state was the sole patron of science in Bolshevik Russia, even though the multiplicity of government agencies and agents (as well as alliances and rivalries among them) during the 1920s provided enough room for maneuver for scientists looking for subsidies to pursue their research interests.

But what made the Russian case truly singular was the unique cultural landscape within which these developments unfolded: in particular, the revolutionary dreams of a "new world" and the militant atheism and materialism, which permeated the country in the aftermath of the Bolshevik Revolution and the decade of death. Around the world, scientists had to word carefully their public statements and to mold continually the public image of their research so as not to offend the religious beliefs and sensibilities of both the public and their patrons, as the highly negative assessments of Sergei Briukhonenko's experiments by certain Western observers as "the work of the devil," "abominable," and "diabolic" witnessed in Chapter 2 clearly indicate. In newborn Bolshevik Russia they could pursue their visions with impunity.[2] What is more, the banishment of organized religion, together with the militant antireligion campaign waged by the Bolsheviks, tore a huge hole in the very fabric of the country's cultural life, a hole that the new rulers aspired to mend with the help of science in general, and biomedical sciences in particular.

The intertwining of the actual and fictional experiments aimed at control over life, death, and disease witnessed on the preceding pages strongly suggests that esoteric, specialized knowledge generated by experimental biology and medicine was transformed into an influential *cultural resource* in revolutionary Russia. In this chapter I will examine the contexts and principles of this transformation, the various groups of actors who took part in creating this resource and shaped its content to their own ends, and the relative roles of science and literature in pursuing the dream of immortality in 1920s Russia.

Science and the Future: Speculative Science

Scholars often discuss the interrelations between science and literature as if the two were completely separate domains, using a variation of the simple formulae "life imitates art" and "art imitates life."[3] Yet ever since our ancestors acquired the abilities of abstract thought and of making art, various *art*ifacts of these abilities—both physical objects and "psychical organisms," as Konstantin Sluchevskii named them in his "Professor of Immortality," became integral parts of human life. And ever since our ancestors acquired these abilities, they asked themselves the same "big" questions: *Who are we? Where did we come from? What are we doing here? Where are we going? What is the meaning of life and death? Can we abolish death and live forever? Who (or what) defines our individual fates and collective futures?* In short, like the "young fool" in Heinrich Heine's poem, people have long puzzled over the questions of human nature, human place in the universe, and human destiny.

For millennia, human communities constructed coherent belief systems—we call them myths, sagas, legends, folklores, religions, philosophies, or ideologies—which provided answers to these existential questions, enshrining communal beliefs in various artifacts and actions (rituals). Music and monuments, dances and dresses, prayers and paintings, songs and sculptures articulated the specifics of a particular community's answers to the "big" questions. They served the critical function of introducing members of a community to a particular universe they inhabited by identifying its elements and forces, deities and demons, mores and morals, heroes and villains, structures and strictures, taboos and fashions. Along with educating and entertaining the community's members, these artifacts and actions formed the community's *culture* that shaped collective and individual identities and behaviors, as well as solidified intra-community relations and defined inter-community boundaries.

With the invention of writing, human communities recorded the particularities of their universes on clay tablets or papyri, bits of animal skin or tree bark, and pieces of cloth or paper carefully passed on from one generation to another. Embedded within the exposition of a communal belief system, these records contain (among many other things) both certain *practical knowledge* about the community's universe and various *artistic interpretations* of that universe. Eventually, the first gave birth to "science" as a specific form of human activity aimed at the systematic pursuit of knowledge about the universe (including, of course, the community's place within it). The second gave birth to "literature" as a specific form of human activity directed to artistic reflection upon that universe in written language. The two intertwined and informed each other at every step of human history, and both contributed to the construction and propagation of the community belief systems by providing certain answers to the "big" questions. There are myriad examples of such intertwining, ranging from Lucretius's *De Rerum Natura* (1st century BC) to Dante's *Divina Commedia* (1308–21) to Cyrano de Bergerac's *L'Autre Monde: où les États et Empires de la Lune* (1657) to Wolfgang von Goethe's *Faust* (1808) to countless travelogues, such as Charles Darwin's *The Voyage of the Beagle* (1839).[4] But the relative positions, and hence the authority, enjoyed and exercised by each "science" and "literature" within the larger system of a community's culture did not remain static. Indeed, they underwent profound transformations in the course of the history of each particular community.[5]

In nineteenth-century Russian culture, literature held an indisputable upper hand in its relation to science.[6] In the course of that century, literature became much more than a means of entertaining the leisured class.[7] Men of letters enjoyed a dominant position in educated Russian society as the "rulers of thoughts" (*vlastiteli dum*) and the "teachers of life" (*uchitelia zhizni*).[8] They provided authoritative answers to the burning questions of Russian "communal" life, such as Alexander Herzen's *Who Is to Blame?* (1847), Nikolai Chernyshevskii's *What Is to Be Done?* (1863), Nikolai Mikhailovskii's *What Is Progress?* (1869), and Leo Tolstoy's *What Is Art?* (1897) and *What Shall We Do?* (1906). They tackled the mysteries of human nature in numerous works of fiction, from Ivan Turgenev's *Fathers and Children* to Fyodor Dostoevsky's *Crime and Punishment* to Leo Tolstoy's *War and Peace*. Based on their artistic insights, they articulated ideals, norms, and values of human existence, which shaped the thoughts and behaviors of educated Russians for several generations. They became the guardians of the country's spiritual life and custodians of its future.

Russian litterateurs were aware of the concurrent advances in various fields of science, which infringed on the "big" issues they addressed in their writings—whether Charles Darwin's theory of biological evolution or Ivan Sechenov's concept of brain reflexes as the foundation of human thought and behavior. Indeed the same "thick" journals wherein they published their literary works—*Russian Thought*, *Herald of Europe*,

Russian Wealth, *The Contemporary*, *Russian Herald*, and many others—had special sections devoted to science, which regularly carried accounts of the latest scientific theories and discoveries. Some writers even used facts and concepts of science to support and illustrate their own views, as did Turgenev in *Fathers and Children*.[9] But, as Sluchevskii's "Professor of Immortality" demonstrates, many of them summarily denied the ability of science to provide meaningful answers to the "big" questions of human life and death, past and future, good and evil, goals and means. For many writers (and their readers), literature—not science—was the primary instrument for explaining human nature and exploring human destiny. Consequently, some popular nineteenth-century writers—for instance, Leo Tolstoy—were quite vocal in their negative assessment of the growing influence of science in contemporary Russian culture and society.[10]

Furthermore, some Russian scientists overtly acknowledged this supreme cultural authority of literature and, along with their scientific pursuits, actively dabbled in writing fiction and poetry. Thus, Nikolai Vagner (1829–1907), an eminent zoology professor at Kazan' and, later, St. Petersburg universities, published seven volumes (!) of highly acclaimed literary works, while his younger colleague, Nikolai Kholodkovskii (1858–1921), produced the best Russian translations of Wolfgang von Goethe's *Faust* and John Milton's *Paradise Lost*.[11] As the nineteenth century drew to a close, however, science and scientists began to challenge the indisputable authority of literature and litterateurs in Russian culture and society.[12]

In the decades around the turn of the twentieth century, the rapid accumulation of knowledge about the physical characteristics of the universe reduced the Earth to a tiny speck of dust lost in the boundless Cosmos among billions upon billions of stars, each one of which had a beginning (at least several billion years ago) and would eventually have an end. It made space, time, and matter infinite and multidimensional. It raised the possibility of the existence of life, intelligence, and human-like civilizations on other planets within the solar system and beyond. And as a consequence, it pushed humanity from its privileged central place in the universe to an uncertain but almost assuredly subordinate position—witness numerous Martians figuring in contemporary fiction from H. G. Wells to Alexander Bogdanov—within the imaginary hierarchy of sentient beings inhabiting the Cosmos. At the same time, the rapid growth of experimental biology and medicine offered some plausible answers to the questions of human life and death, origins and evolution, body and mind, health and illness. Furthermore, experimental means of control over life, death, and disease made evident that, unlike other animals, humans were no longer passive subjects to God's will or Nature's inexorable laws: armed with new scientific knowledge, they could "conquer Nature" and become masters of their own individual fates and communal destinies.

Scientists themselves were the first to realize the potentially profound impact of their specialized knowledge on the answers to the "big" questions that haunted humanity for millennia. Some experimentalists took it upon themselves to explain to an unsuspecting humanity what their discoveries of certain facts and regularities in Nature actually meant. In addition to publishing the results of their research in specialized journals and monographs addressed to their colleagues, they presented them to a broader audience, delivering public lectures, organizing scientific exhibits, and publishing somewhat simplified accounts of their work in popular periodicals and books aimed at the general public. They *popularized* knowledge obtained through their research by making it accessible to nonspecialists.[13] As we saw, both before and after the Revolution, numerous Russian scientists and physicians, including Porfirii Bakhmet'ev, Nikolai Kol'tsov, Boris and Mikhail Zavadovskii, Petr Shmidt, and Lev Vasilevskii, actively popularized their scientific studies.

Some of them went even further, however, engaging in what almost a century earlier the German philosopher Wilhelm Friedrich Hegel had termed "speculative science" (*spekulativen Wissenschaft*). Hegel did not use the term in a pejorative sense; quite the contrary. According to Hegel, speculative science adopts and uses the empirical facts, laws, categories, and classifications of science, but it also expands them into other categories. Preserving "the same form of thought and the same laws and objects" advanced in "regular" science, speculative science is simultaneously "remodeling and expanding them with wider categories"—most often categories that previously constituted the exclusive domains of religion, philosophy, and literary fiction.[14] As one commentator in the influential "thick" journal *Russian Thought* perceptively observed in an 1893 article with the characteristic title, "Biologists on the Women Question," "more and more, science takes over those areas which until now have been considered to be the undisputable possessions of poetry, philosophy, mysticism,—of anything but science."[15]

Porfirii Bakhmet'ev's 1901 "Recipe to live on to the twenty-first century" illustrates the principles of, and aspirations behind, a particular subset of speculative science that gained prominence around the turn of the twentieth century: "visionary biology."[16] A comparison of Bakhmet'ev's "Recipe" with his previous popularizations of his own research, such as, for instance, his 1898 popular article "The temperature of insects," reveals profound differences, clearly reflected even in the titles. In all of his publications (both in professional and in popular periodicals) prior to "Recipe," Bakhmet'ev had limited himself to the presentation of his methods, along with facts and regularities discovered by the application of his techniques and apparatuses. The only difference between the articles he had published in specialized versus popular periodicals was that in the latter, he simplified the descriptions to make his work understandable to a general audience. Although preserving the technical apparatus of a professional publication (such as exact referencing of previous research on its subject), "Recipe" went far beyond making Bakhmet'ev's discoveries accessible to the lay public.

In "Recipe," adopting the format of "scientific fantasy," Bakhmet'ev quite deliberately engaged in speculation regarding exciting possibilities that his discoveries of concrete facts in the physiology of insects (a "jump" in body temperature, the anabiosis "curve," the "point of death," etc.) could create not only for understanding, but also for manipulating, human life and death. Such speculation allowed him to leapfrog over numerous gaps in his experiments and his knowledge of the phenomena under study, opening up the tempting possibility of answering the "big" questions around life and death through his investigations of "small," experimentally answerable questions of insects' responses to the lowering of temperature. In all of his speculative publications, and particularly in his public lectures on "the theoretical and practical consequences" of his research on anabiosis, Bakhmet'ev used facts, laws, classifications, and categories from his biophysical investigations (of insect temperature, crystal structure, or droplet behavior) and expanded them to the much broader categories of human life, death, and disease.[17] As we saw, in his 1902 discussion of "the seat of the soul," Bakhmet'ev effectively expanded his *biophysical* analysis (of the behavior of droplets) to the understanding of a *metaphysical* subject (the soul).[18]

Bakhmet'ev was not alone in this endeavor: many Russian naturalists actively engaged in speculative science. Just two years after Bakhmet'ev had published his "Recipe," the soon-to-be Nobel Prize winner zoologist Elie Metchnikoff wrote an unusual work, tellingly entitled *Études sur la nature humaine: Essai de philosophie optimiste*. The two-hundred-page book appeared almost simultaneously in English, French, German, and Russian versions, becoming an international bestseller that in the next few years went through several editions.[19] Much as Bakhmet'ev had, Metchnikoff applied knowledge accumulated in various subfields of rapidly developing experimental biology and medicine to the "perpetual" questions of human life, aging, and death (see Figure 6.1). Metchnikoff's *Études* incorporated the facts, laws, classifications, and categories of his own zoological, morphological, anthropological, bacteriological, and immunological research and extended them to issues of human nature and philosophical categories of pessimism and optimism.

Even more important, Metchnikoff forcefully asserted the ability of experimental science not only to provide definitive answers to the "big" questions, but also to formulate the ideals, values, goals, and norms of human life—a task previously taken up by religion, philosophy, and literature. The concluding sentences of his treatise summed up his views:

> If one could think of an ideal able to unite people into a kind of religion of the future, this ideal could not be based on anything but scientific data. And if it is true, as has been declared quite often, that it is impossible to live without a faith, such faith could be nothing else but the faith in the omnipotence of knowledge.[20]

Like Bakhmet'ev, Metchnikoff followed almost to the letter Hegel's description of the operation of speculative science by preserving the same form of thought, the same laws and objects he deployed in his own research, while at the same time remodeling and expanding them with wider categories—those of faith, religion, morals, and philosophy.

Even Russia's premier physiologist Ivan Pavlov, who had become the icon of "objective" science, was seduced by the temptations of speculative science.[21] In December 1904, Pavlov delivered a lecture at the Nobel Prize ceremony in Stockholm, recounting the research—which had won him the award—on the nervous mechanisms regulating the

FIGURE 6.1 A widely reproduced portrait of Elie Metchnikoff in his laboratory at the Institut Pasteur, ca. 1908. Perhaps it was a reproduction of this very portrait that Sharik the dog broke into pieces in Professor Preobrazhenskii's apartment. This particular reproduction appeared on the cover of a special issue of *Knowledge for Everyone*, a Petrograd-based popular-science monthly, devoted to "The Struggle Against Old Age." *From* Znanie dlia vsekh, *1915, 12.*

digestive glands in the animal organism. But he did not limit his report to past achievements. At the end of the lecture, he also introduced a new direction in his studies—conditional reflexes—which, Pavlov hoped, would illuminate the nervous mechanisms of the "psychical" regulation of digestion. Although at the time the actual facts discovered by the application of this new approach were limited to a few regularities in the action of dogs' salivary glands, Pavlov concluded his lecture in the following manner:

> Essentially, only one thing in life really interests us: our psychical nature.[22] Its mechanism, however, was and still is shrouded in profound mystery. Man had employed all of his resources—art, religion, literature, philosophy, and the historical sciences—in the attempt to shed light on this mystery. But man also has at his disposal one more mighty resource: natural science with its strictly objective methods. This science, as we all know, is achieving gigantic successes every day. The facts and concepts, which I have presented at the end of my lecture, constitute one of numerous attempts to employ a consistent, purely scientific way of thinking in order to

study the mechanism of higher life processes [read—the psyche] in the dog, this representative of the animal world so [biologically] close and friendly to man.[23]

This concluding paragraph of Pavlov's otherwise strictly nonspeculative, though somewhat simplified for his nonspecialist audience, account of his research exemplifies an explicit expansion of the "scientific way of thinking" presented in experiments on dogs to the issues and categories of the "human psychical nature." At the same time, it illustrates the implicit claim of science's ultimate superiority over "art, religion, literature, philosophy, and the historical sciences," all of which, according to Pavlov, had struggled with these issues and categories for centuries, but apparently without any, let alone "gigantic," success that his own "strictly scientific" approach promised in the nearest future.[24]

Visionary biology was most certainly not an exclusively Russian preoccupation. In the early twentieth century, it captivated many scientists the world over. They applied facts and regularities discovered in their research on particular subjects (be it hormones, blood transfusions, heredity, tissue cultures, brain structures, behavioral patterns, or organ transplants) to the "eternal" questions of life and death, asserting the role of science as the only source of true knowledge about human nature and the only instrument of effectively controlling human destiny.[25] This trend became particularly prominent after World War I.

For many intellectuals, the horrors of the Great War signaled the end of the world as they knew it and demonstrated the utter impossibility of life as it once was, making humanity's future an overriding concern.[26] The perpetual question "Quo vadimus?" resounded through numerous writings by litterateurs, philosophers, politicians, theologians, and . . . scientists. In the general atmosphere of disillusionment with the old order, old values, old institutions, and old ideals of Western civilization, epitomized by Oswald Spengler's monumental *Der Untergang des Abendlandes* (1918–1923) and clearly reflected in the fiction of the post-war decade (from Henri Barbusse to Ernest Hemingway to Erich Maria Remarque), science seemed to offer a beacon of hope. Although also implicated in the horrors of the war by supplying new technologies of mass extermination, from airplanes dropping bombs to poisonous gases, in the aftermath of the war, science claimed the mantle of humanity's new Savior.[27] It promised wonderful new technologies (from space flight to atomic energy) and exciting new techniques of manipulating life (from artificial foods to rejuvenation), which could solve all of humanity's problems, past, present, and future, and which would forever banish the four horsemen of the Apocalypse: War, Pestilence, Famine, and Death.

Many scientists throughout the world shared the view of science's vital role in shaping humanity's future. To give but one example, in 1924, a British publisher established a special series on "To-day and to-morrow," which explicitly addressed the question "Quo vadimus?"[28] Indeed, one of the first books written for the series—by E. E. Fournier d'Albe, a well-known physicist—bore exactly that title.[29] Published simultaneously in Britain and the United States, the series famously opened with J. B. S. Haldane's *Daedalus, or Science and the Future*, which optimistically envisioned techniques of experimental biology and medicine as the main tools of both defying the laws of Nature and defining human destiny. Followed by Bertrand Russell's much more cautious and pessimistic *Icarus, or the Future of Science*, the series grew—over the seven years of its existence—to more than one hundred books (with most titles deriving from various characters in Greek mythology). It included visions of the future conceived by such eminent intellectuals, scientists, and writers as William McDougal, James H. Jeans, and André Maurois, as well as by a host of now-forgotten journalists, engineers, philosophers, physicians, activists, and educators.

These visions covered a panoply of subjects, including the arts, governance, marriage, urban planning, industry, and fashion. Visionary biology figured prominently among the series titles. In addition to Haldane, H. S. Jennings (who had extensively studied the "immortality" of unicellular organisms) contributed *Prometheus, or Biology and the Advancement of Man*, and J. D. Bernal furnished *The World, the Flesh & the Devil: An Enquiry into the Future of the Three Enemies of the Rational Soul.* These works offered the authors' own optimistic views on humanity's future inspired by the new experimental biology and medicine.[30] But regardless of the optimistic or pessimistic position of individual authors, taken as a whole, the series attests unambiguously to the rapidly rising public visibility and cultural authority of science in general, and biomedical sciences in particular, in the post-war era.[31]

Russia was no exception to this trend. Fueled by the horrors of the Great War, the disillusionment with the "old regime" in Russia found violent expression in the February 1917 revolution that deposed the monarchy, the subsequent Bolshevik coup d'état in October of that year, and the ensuing four-year-long bloody Civil War. Yet in contrast to the pessimistic outlook of many Western and Russian intellectuals regarding post-war prospects for humanity, the Bolsheviks' revolutionary dreams of a new world forged in the flames of wars and revolutions offered an optimistic answer to the question "Quo vadimus?": to a new civilization—communism. As Isai Lezhnev's 1923 article "Two grafts: Steinach and Lenin" indicated, this vision of a new civilization being born of death and destruction enjoyed considerable popularity in post-revolutionary Russia. The Bolsheviks' early programs did not define precisely the specifics of that civilization and of the bright future waiting ahead, beyond a general vision of abolishing private property, exploitation, injustice, and inequality through world revolution.[32] They thus provided considerable freedom for creativity and virtually unlimited space for social, cultural, technological, and biomedical experiments, which were supposed to bring the communist future ever closer and which were embodied, for instance, in the famous slogan, "Communism equals Soviet power plus the electrification of the entire country."[33]

For some Bolsheviks, the rapidly developing fields of experimental biology and medicine offered inspiration and insight into what that future might be like. Nearly a decade before the Bolsheviks seized power in October 1917, Alexander Bogdanov—at the time Vladimir Lenin's rival for the leadership of the Bolshevik Party—had written an SF novel with a telling title, *Red Star*. Bogdanov had imagined a future—on Mars—in which the workers of the red planet, liberated from capitalist exploitation, create a "physiological collective" that rejuvenates and unites its members through regular blood exchanges. After the Revolution, Bogdanov began actual experiments with blood exchanges on himself. In 1926 he became the first director of the Moscow Institute of Blood Transfusion and two years later died in an attempt to prove experimentally his vision of "physiological collectivism" and the possibility of using blood transfusions as a means of rejuvenation.[34]

After the end of the Civil War, another prominent Bolshevik, Leon Trotsky, prophesied in a similar fashion:

> Man will finally begin to really harmonize himself. He will put forward the task of introducing into the movement of his own organs—during work, walk, [and] play—the highest precision, expediency, economy, and thus beauty. He will want to control semi-unconscious, and then unconscious processes in his own organism: breathing, blood circulation, digestion, and reproduction—and will subjugate them to the control of reason and will. Life, even purely physiological life, will become collectively-experimental. Humankind, frozen *homo sapiens*, will again enter into radical reconstruction and will become—under its own fingers—an object of most complicated methods of artificial selection and psycho-physical training. [...] Man will put forward a goal [...] to raise himself to a new level—to create a higher socio-biological type, an *Ubermensch*, if you will.[35]

Even though few shared the Bolsheviks' political goals or approved of their methods, many Russian scientists found the revolutionary dreams actively propagated by the new rulers quite appealing, resonating strongly with their own long-held visions of control over life and death, health and disease, human nature and human destiny. Caught in the wave of revolutionary enthusiasm, many Russian experimentalists actively engaged in speculation regarding what their own scientific research could mean for making those dreams a reality in the new, Bolshevik Russia. They even aspired to explain, and thus eventually to control, the ultimate mystery of human nature: creativity and genius.[36] In 1922, Vladimir Savich, a respected coworker of physiologist Ivan Pavlov, expanded upon a trend first inaugurated in 1863 by Sechenov's "Reflexes of the Brain" and further developed by his mentor in the 1904 Nobel lecture. In the original Soviet "thick" journal, *Red Virgin Soil*, Savich published an unusual article, entitled "An attempt to elucidate the creative process from the viewpoint of reflexes."[37] Savich first dissected Miguel Cervantes's masterpiece, offering an explanation of Don Quixote's "strange behavior" as a combination of specific reflexes to specific stimuli. He then moved to a detailed analysis of Charles Darwin's biography, insisting that Darwin's "creative processes" that had led to the *Origin of Species* were nothing but "chains of reflexes."

Three years later, in 1925, a group of psychiatrists, endocrinologists, and brain anatomists even founded a special journal, *The Clinical Archive of Genius-ness and Gifted-ness*. As the journal's subtitle indicated, the group was interested in "questions of the pathology of the genius-endowed personality and also in questions of the pathology of creativity."[38] Along with investigating the "creative abilities" of mentally ill patients, the group produced "pathographies" of many famous Russian writers and poets, including Alexander Pushkin, Mikhail Lermontov, Fyodor Dostoevsky, Leo Tolstoy, and Maxim Gorky.[39] The particular focus of these studies on literary creativity and literary genius clearly responded to the centrality of literature in Russian culture and perhaps served to further emphasize the bid to cultural authority laid by science and scientists.[40]

Science/Fiction: From Popular Science to Scientific Fantasy

Of course, not all Russian experimentalists engaged in speculative science. Like Bulgakov's Professor Persikov, many were quite reluctant to go beyond the facts and regularities uncovered in their own research. As we saw, neither Sergei Briukhonenko nor Vasilii Shervinskii attempted to "remodel and expand" their findings with wider categories of human nature and human destiny. Characteristically, despite his deep involvement in building the discipline of endocrinology, Shervinskii was very skeptical about, and consistently stayed away from, various speculations regarding the transformative powers of hormones, as captured by Alexander Beliaev in *The Man Who Lost His Face*. Unlike his deputy Iakov Tobolkin, Shervinskii also avoided any association with rejuvenation à la Steinach and Voronoff, which involved manipulations of sex glands and thus must have been of particular interest for endocrinologists. Although, in the 1920s, the Institute of Experimental Endocrinology did produce testicular extracts (which since the time of Brown-Séquard's work were reputed as "rejuvenating"), at the height of the rejuvenation craze from 1924 to 1926, the Russian Endocrinological Society meetings presided over by Shervinskii did not feature a single report on the subject of rejuvenation!

Yet despite certain scientists' refusal to speculate on the broader issues raised by their research, the facts they had discovered and the concepts they had developed quickly became "public property" and thus ripe for appropriation. As we saw in the fictional "Fateful Eggs" and in the actual cases of research on mitogenetic rays, isolated organs, hormones, sex gland transplants, and anabiosis, popular-science writers, journalists, and newspaper reporters shouted at the top of their lungs on behalf of those scientists who shied away from speculation.[41] As one *Evening Moscow* reporter candidly admitted in his interview with Petr Lazarev, the director of the Narkomzdrav Institute of Biophysics, regarding the possibility of "reading minds": "the scientist's modesty *cannot stop us* from envisioning these—truly fantastic—perspectives that brain science is opening up [for humanity]."[42] Unlike many scientists, journalists in the 1920s had no qualms "expanding and remodeling" concrete investigations of particular biomedical subjects with the wildest of speculations regarding, for example, "rays of life" or science's imminent "victory over death."[43] Many were certain that thanks to science's interventions, "our children will live to [the age of] two hundred years."[44] Others firmly believed in the ability of science to "conquer death" and widely disseminated such beliefs. The media coverage unambiguously attested to, and in turn enhanced even further, the authority of science and scientists in post-revolutionary culture and society, challenging the supreme authority enjoyed by literature and litterateurs in the previous century.

An enormous campaign to popularize science orchestrated jointly by scientists and the Bolsheviks and conducted by journalists and popular-science writers in the aftermath of the Revolution contributed tremendously to the rising authority of science by widely disseminating facts and insights from new experimental biomedical sciences. The campaign employed all available media: radio, cinema, theater, newspapers, magazines, books, exhibits, lectures, posters, and public disputes (see Figure 6.2). The country's new rulers certainly hoped that this campaign would shape popular understanding and practices in various areas of life, ranging from personal hygiene to marriage, child rearing, and handling prisoners.[45] Numerous inquiries sent to popular magazines by readers suggest that the popularization of biomedical sciences did not automatically (or immediately) make their subject matter either understood or valued by everyone.[46] But the campaign certainly did enhance public awareness of what was going on in biomedical research, of scientific advances, and of scientific promises, planting in the public mind new expectations, new dreams, and new fears.

FIGURE 6.2 This 1920 poster, entitled "The Day of Soviet Propaganda" and subtitled "Knowledge to everyone!," exemplifies the Bolsheviks' literacy and science-popularization campaigns of the early 1920s. The titles of the books "sawn" by the man on the foreground range from "History" and "Class Struggle" to "The Arts" and "Agriculture," while the buildings in the background are identified as "University," "Academy" and "Library." *From* Pechat' i revoliutsiia, *1922, 5.*

Several letters between 1923 and 1924 addressed to Ivan Kogan and preserved among Kol'tsov's files illustrate the range of public responses to the new biomedical knowledge.[47] Their author was a twenty-four-year-old railroad worker named Faddei Pshonka from a small town some six hundred kilometers southwest of Moscow. Judging by the style and grammar of his letters, Pshonka had only recently learned how to read and write, but he clearly tried to put his newly acquired skills to good use. Pshonka had read articles about Kogan's experiments on the rejuvenation of dogs in the popular weekly *Flash* and got very excited. But when he relayed the articles' contents to his coworkers, they simply laughed at him. He then wrote to Kogan asking for the scientist's help in convincing the "unbelievers" among his fellow workers that "the discovery of rejuvenation" was not a "bunch of lies disseminated by the Bolsheviks," or "by the rotten intelligentsia," but was "the most powerful weapon that would overcome all obstacles [. . .], the weapon of SCIENCE," capitalizing and underlining the last word. Utterly convinced that "only through science is it possible to discover and learn everything," he offered to send Kogan a "twenty-year-old dog" he had found, so that the scientist could "rejuvenate" the dog and send back to Pshonka "some protocols" of the experiments, thus proving to the doubting Thomases both "how well medicine works in the USSR" and that "thanks to science we can triumph over anything."

As these letters suggest, the science popularization campaign did succeed in infusing post-revolutionary culture with certain knowledge about experimental biomedical research. Numerous satirical feuilletons and cartoons that employed and exploited such feats of biomedical sciences as hormones, organ transplants, artificial insemination, blood transfusions, and isolated organs offer similar testimony (see Figure 6.3). The

FIGURE 6.3 "Medical" cartoons that appeared in 1925 in *Crocodile*:
a) This 1925 cartoon is entitled "Rejuvenated Wilhelm."
The caption says: "So, your Majesty, how do you feel after the renovation?
–Wonderful, only there is nothing to sit on. It'd be great if my throne were renovated too."
This cartoon depicts the deposed German Emperor Wilhelm II, who had allegedly undergone a rejuvenation operation. It implies that no matter what rejuvenation the capitalist system undergoes, the fate of capitalism is sealed (by the proletarian revolution) and there is no return to the thrones of the "old regimes." This cartoon could be seen as yet another illustration to Isai Lezhnev's thesis that "the West is getting comforted by Steinach's graft." The Soviet press regularly reported in a very sarcastic manner on rejuvenation operations performed on members of capitalist elites, ranging from the former British prime minister Lloyd George to Baron Rothschild to the Great Duke Nikolai Nikolaevich (a brother of the late Emperor Nikolas II and the major claimant to the title of the Russian Emperor). *From* Krokodil, *1925, 30: 5.*
b) This 1925 cartoon, entitled "The advances of medicine in Western Europe," depicts a "blood-sucking capitalist." The caption underneath reads: "The latest method of blood transfusion, which uncultured communists call the Dawes Plan." The cartoon obviously has nothing to do with either medicine or blood transfusions, but satirizes a 1924 plan of collecting the German reparations debt by the Allies, named after Charles Gates Dawes (1865–1951), an American banker and statesman who presided over an international committee set up for that purpose. In 1925, Dawes received a Nobel Peace Prize for his work in the committee and became a vice president under Calvin Coolidge, which probably explains why his "plan" is being satirized in the Soviet media in late March 1925. *From* Krokodil, *1925, 12: 12.*

ОМОЛОЖЕННЫЙ ВИЛЬГЕЛЬМ.

Рис. К. Елисеева.

— Ну, как вы чувствуете себя после ремонта, ваше величество?
— Прекрасно. Только вот сесть не на чем... Трончик бы что-ли кстати уж отремонтировали!

FIGURE 6.3 a

Рис. О. Берн. УСПЕХИ МЕДИЦИНЫ В ЗАПАДНОЙ ЕВРОПЕ.

Последний метод переливания крови, некультурными коммунистами именуемый: *план Дауэса.*

FIGURE 6.3 b

metaphorical use of the new biomedical knowledge in such cartoons clearly demonstrates that their audiences required neither detailed explanations of, nor even references to, the actual research that generated the knowledge in question.[48]

The effectiveness of the science-popularization campaign in penetrating the public mind to a large extent derived from the wide use by popular-science writers and journalists of well-known folklore and literary images, ranging from the "water of life and death" to "Dr. Faust," to render the advances of modern science "familiar" and understandable to the lay public. A popular book on the physiology of sex life written by Konstantin Krzhishkovskii, a respected physiologist and a colleague of Petr Shmidt's at the Leningrad Agricultural College, exemplifies this technique. Krzhishkovskii characterized the results of research conducted by Western and Russian rejuvenators, including Steinach, Voronoff, Mikhail Zavadovskii, and Nemilov, as "the fairy-tale world of miracles," presenting them as the "realization of dreams [expressed] in legends and fables." Explaining the meaning of their experiments, he stated poetically:

> A German legend says that to return his youth Dr. Faust sold his soul to the devil, signing the [sale's] contract in his own blood. If Dr. Faust were living now, he would not need to deal with the devil and pay such a high price. It would have been enough to ask a surgeon, who would have returned him his youth for a much more suitable price, and perhaps even for free, just for the sake of an interesting experiment.[49]

But the impact of science popularization was not limited to turning modern scientific advances into common knowledge. One can argue that it also raised greatly the cultural authority of science and scientists by implanting in popular audiences certain expectations regarding the nature, meaning, and future of experimental biomedical research. Pshonka's letters captured the essence of these expectations: science and scientists were omnipotent and in due time would solve all of humanity's problems. They indicate that "the faith in the omnipotence of knowledge," which Metchnikoff had envisioned as "a kind of religion of the future" and of which he had only dreamt, was becoming a reality in Bolshevik Russia, profoundly affecting the thinking of scientists and their patrons, journalists and medical professionals, novelists and the general public.

These expectations, along with the aspiration of science and scientists to cultural authority, found expression not only in various works of popular and speculative science or in journalists' exalted and sensationalized praise of "science that conquers death." They were also clearly expressed in—and further magnified by—a new literary genre: scientific fantasy, as it was called in Russia, or scientific romance, scientific fiction, scientifiction, and science fiction (SF) in the English-speaking countries. The appearance of the words *science* and *scientific* as modifiers of such nouns as *romance*, *fiction*, and *fantasy* signaled unambiguously the recognition (first of all by the literati) of the growing role of science in the life of people in industrialized nations.

But what did the words *science* and *scientific* actually mean in all these labels? One answer is suggested by the popular sentiment—first explicitly presented by Mark Twain in his 1889 classic *A Connecticut Yankee in King Arthur's Court* and some seventy years later codified by Arthur C. Clarke in his "third law"—that any advanced technology and hence, its foundation, science, is indistinguishable from magic.[50] Indeed, many SF writers used science as a substitute for magic in developing further an old respected literary form: the "literature of wonder." In much of the new SF literature, science served as an instrument of estrangement, of amazement, and of awe (captured in the titles of such early SF magazines as *Amazing Stories*, *Science Wonder Stories*, and *Astounding Science Fiction*), just as magic had done in the wondrous fiction of previous centuries, be it fairy tales, ghost stories, or gothic romances. As early as 1907, Carl Grunert tacitly acknowledged this substitution by naming a lead character of his story "Dr. Magnus Magician," clearly indicating that in the new literature of wonder a physician, scientist, or engineer became a leading protagonist, assuming the role that had been played by a witch, sorcerer, demon, or magician in the literature of previous eras (see Figure 6.4).[51] Similarly, in "The Dog's Heart," Preobrazhenskii's patients, his assistant, and even Sharik the dog call the professor a "magician and miracle-maker" (83), a "sorcerer, magician, and conjurer" (94), and even a "Creator" (105).

In 1925 a Soviet journalist covering the newest rejuvenation experiments recorded this substitution in the following manner:

> Mephistopheles, who rejuvenated Faust, is no longer the hero of a medieval tale. 'Magical' has been removed from rejuvenation. Nowadays, Mephistopheles is just a dried-out, diminutive professor. Continuing to advance the Faustian possibilities, science has already put forward the questions of rejuvenation's regulation by the state and of its role in the future.[52]

In numerous SF writings, "wondrous achievements of science," to borrow the title of Nikolai Kol'tsov's 1925 radio lecture, made wonders of magic—be it the philosopher's stone, homunculus, or the elixir of life—pale by comparison. As Preobrazhenskii's assistant records in "The Dog's Heart": "A new area is being opened in science: without any Faustian retort, a homunculus has been created" (105). Furthermore, according to Bulgakov, discoveries of experimental biology—exemplified by Persikov's ray of life—made even such recent futuristic projections as H. G. Wells's 1904 novel *Food of the Gods* look like "sheer nonsense."[53]

Yet as we saw in various fictional experiments, science in the post-revolutionary SF literature often became much more than a mere prop, backdrop, or simple instrument of wonder and entertainment: it also became subject to close scrutiny. Writers began to question science's practices and practitioners, ideas and ideals, goals and means, promises and premises. They examined scientists' attempts to answer the "perpetual" questions of

FIGURE 6.4 Scientist as "Faust." This drawing illustrated Mark Volosov's 1925 story "Rejuvenation," depicting a certain Dr. Hakens, the discoverer of the "fountain of youth." Following Metchnikoff's ideas, Dr. Hakens found a particular microbe that rejuvenates the tissues of the human body. Despite the modern buildings seen in the background, the depiction of Dr. Hakens's laboratory in this drawing clearly harks back to a medieval alchemical laboratory, with large folios on the shelf and a "Faustian" homunculus sitting in a large retort. *From* Krasnaia panorama, *1925, 5: 11.*

human life and death, human nature, and human destiny, as well as scientists' aspirations to become the supreme cultural authority in contemporary societies. Literati critically assessed the "possible futures" offered to humanity by visionary biologists.

Bulgakov, for instance, did not simply "reflect" the current state of Soviet biomedical research in his novellas. He focused specifically on and examined closely the answers to the "big" questions supplied by contemporary biomedical science. He challenged the ability of science to answer those questions satisfactorily and mocked scientists' claim to the role of cultural authority that could provide suitable ideals, norms, and values of human existence. A small but very telling detail in "The Dog's Heart" illustrates Bulgakov's position.[54] Upon entering Preobrazhenskii's apartment, Sharik the dog immediately develops an inexplicable dislike of a particular object that decorates his new owner's office: a stuffed owl. At the first opportunity, Sharik destroys this centuries-old symbol of knowledge and wisdom. Along with disemboweling the owl, the dog also breaks into pieces a portrait of Elie Metchnikoff, the author of the highly influential *Études sur la nature humaine* that asserted the "faith in the omnipotence of knowledge" and its source—experimental science.

Some writers attempted to meet headlong the challenge posed by science to their own cultural authority and to negotiate a division of "the spheres of influence." During the early 1920s, Alexander Voronskii, a cofounder and the editor-in-chief of *Red Virgin Soil* (and a talented writer in his own right), published a series of essays on "Art as the cognition of life."[55] Voronskii sought to substantiate theoretically what his numerous predecessors and contemporaries saw as self-evident: literature was an instrument of cognition to no lesser extent than science was.[56] As Voronskii put it, "like science, art studies life," and "the [ultimate] duty of [both] the artist and the scientist is reducible to the exact, experimental cognition of life."[57] Thus, both the artist and the scientist study the same object: human life, human nature, and human destiny. They approach their "common subject" differently, however: "the specificity, particularity of the artist is that he thinks not with abstract categories [as does the scientist] but with images." The two also differ in the way the results of their studies are verified: "While one confirms the truthfulness of a scientific concept by its logical evaluation, [the truthfulness of] an artistic work [is confirmed] by its aesthetic evaluation."[58]

It seems likely that Voronskii's affirmation of the role of art in general, and literature in particular, as a legitimate source of knowledge about—and hence, values, ideals, and norms of—human life was to a considerable degree prompted by the concurrent, though contentious, assertion that science was the *only* source of any true knowledge. For example, in 1922, Leonid Vasil'ev, a well-known physiologist actively involved in the studies of telepathy and the science popularization campaign, stated that methods of art and science were fundamentally different: scientists searched for "the universal and repeatable," while artists focused on "the individual and unique." Furthermore, according to Vasil'ev, "artists are unable to assimilate advances of scientific thought," and thus the "so-called 'artistic truth' quite often has very little resemblance to the scientific truth," while artistic "creations of even top-rank masters" appear "naïve, primitive, if not bizarre" to a scientist.[59]

In a mirror image of the situation in the nineteenth century when scientists had accepted the supreme cultural authority of literature, many Soviet writers recognized and acknowledged the rising cultural authority of science. As Vsevolod Valiusinskii's *Five Immortals* indicates, they did not doubt the ability of science to deliver on its promises and to develop techniques of manipulating various life processes, applauding the coming wonderful era of "life without death." In 1927, in preparation for the tenth anniversary of the Revolution, *Thirty Days*, a popular literary magazine, asked Soviet writers for a short piece about the future as they saw it. None of the respondents had ever written SF, but out of fourteen responses in prose and verse published in the November issue under the general title "The future," three included rejuvenation and the elimination of aging as a definite feature of that future.[60] As Fedor Gladkov, a well-known author of "industrial novels," put it, "Metchnikoff's naïve dream will turn into a real everyday fact of great importance: rejuvenation of the elderly will become as widespread a phenomenon as, say, smallpox vaccination of children."[61]

What some litterateurs did doubt, however, was the ability of humans, primarily scientists themselves and secondarily those in positions of power—be they capitalists (as in Beliaev's stories) or proletarians (as in Bulgakov's novellas)—to use wisely and control effectively the tremendous powers that science could unleash. This doubt led to questioning the very necessity of conducting particular kinds of research and "interfering" with nature. Indeed, the inability of scientists to control the powers inherent in biomedical experimentation figured in numerous SF stories. Some writers even had scientists killed by various "monsters" created by their own experiments.[62] (See Figure 6.5).

What's more, some writers sensed the particular affinity of visionary biology with revolutionary dreams and were deeply troubled by possible monstrous hybrids such an "unholy union" could produce: from the assortment of monsters created by the Bolshevik Rokk using Persikov's ray of life or the dog-man Sharikov created by Preobrazhenskii and shaped by militant proletarians in Bulgakov's stories to a "snake with the head of Beethoven" in Ivanov-Razumnik's correspondence with Andrei Belyi. As we saw in certain fictional experiments, for some writers, science became not the spring of hope and the instrument of bringing a wonderful future ever closer, but a subject of ridicule (as in Krinitskii's "Elixir of Immortality") or a source of profound fears and a possible cause of unthinkable catastrophes (as in "Fateful Eggs"), much like "black magic" had been in the literature of previous centuries.

Numerous literary experiments depicted on the previous pages were certainly much more than mere "representations" (be they accurate or inaccurate) of concrete scientific experiments or "reflections" of particular episodes in the actual history of Russian biomedical sciences. Writers both tapped into and stimulated further public fascination with contemporary biomedical research spurred by the deafening science-popularization campaign. Building upon their own literary traditions, employing familiar imagery and symbols, and filling their writings with connotations of and allusions to classic works of fiction (by Dante, Goethe, Leo Tolstoy, or H. G. Wells), writers both questioned and enhanced the rising cultural authority of science. Their own literary experiments pinpointed precisely what kinds of scientific experiments fired up public imaginations, expectations, hopes, and fears. They clearly demonstrated what excited

FIGURE 6.5 Science-made monsters:
a) A monster created by the "rays of life" in the eponymous story by V. D. Nikol'skii, published in 1927 in the popular magazine *The World of Adventure*. *From* Mir prikliuchenii, *1927, 10:16*.
b) A monster created by the application of hormones in N. Dolin's story "Blood of the World," published in the same magazine. *From* Mir prikliuchenii, *1927, 6:10*.
In both stories, the "monsters" eventually kill their creators. Note also that both illustrations depict a "great scientist" and his "assistant."

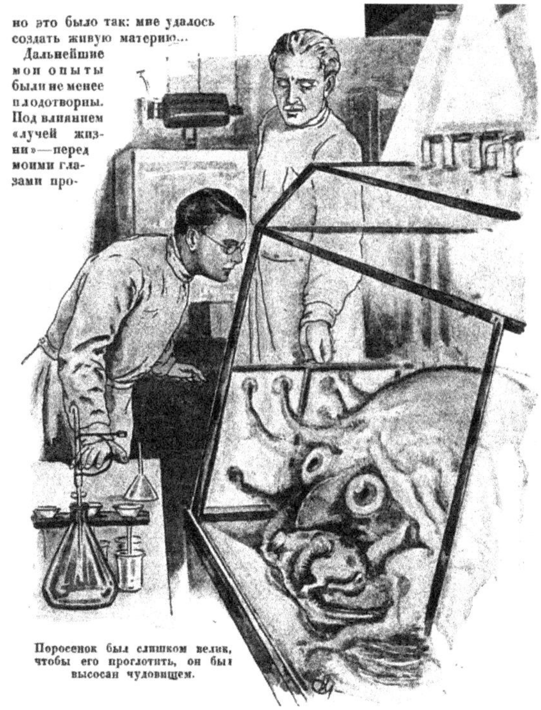

FIGURE 6.5 a

FIGURE 6.5 b

scientists, their patrons, and the general public and identified which particular features of biomedical research contributed to the shift in the public attitudes toward science.

Dreams and Realities: Big Science of Visionary Biology

Whether laughing at or lamenting about, lambasting or lauding *science*, practically all of 1920s SF writings focused on the *scientist*.[63] The image of a lonely—and often out of touch with the "real" world—genius assisted by a single sidekick (as in the cases of Dowell, Kern, Preobrazhenskii, and Persikov), or at most by a small group of disciples (as in the cases of Pavlishchev and Kurganov), dominated 1920s SF literature. This focus on the individual prompted some authors to question scientists' "noble" motives in conducting experiments aimed at controlling life and death, as did Beliaev in his portrayal of Professor Kern. Others obviously believed in scientists' benevolence and "childish innocence" during their single-minded quest for knowledge, as witnessed by Bulgakov's characterization of both Persikov and Preobrazhenskii.[64] Still others were incensed by scientists' "ethical deafness," as Ivanov-Razumnik called it, regarding the possible evil outcomes and immoral applications of their research. Yet the image of "mad" or "evil" scientists so popular in Western SF appeared in Soviet literature only in "capitalist" settings.[65] Scientists working in the land of socialism were neither mad nor evil. Even in Bulgakov's satirical writings, it is the world around scientists that is "gone mad" with the Bolshevik Revolution, while scientists themselves retain the ability to diagnose (even if not cure) this madness.

However, although many writers reflected upon the potential societal consequences of experimental revolution in the life sciences and the role of individual scientists in bringing forth a "new world," they did not really pay attention to or comprehend another scientific revolution unfolding at the same time: the emergence of "big science," which resulted in the appearance of large scientific institutions and which transformed the "lonely genius" into a mass profession (see Figure 6.6). One can hardly blame Russian writers for their inattention to the rise of "big science." Very few among them had ever set foot inside a scientific institution or had any first-hand experience with science in general or experimental biomedical sciences in particular. The majority of writers gained their knowledge of science and its workings from the science popularization campaign carried out by the media: daily newspapers, popular-science magazines, or radio and moving picture documentaries. In a 1926 survey of the contemporary literary scene, Voronskii pointedly commented that among Soviet writers, only "Zamiatin has substantial scientific 'luggage,' and also in Bulgakov['s writings] one feels that, for him, natural sciences are not some dark and impenetrable jungle."[66] But even for engineer Zamiatin and physician Bulgakov, the intricate maneuvers and "dirty" dealings involved in building, maintaining, and managing scientific institutions remained obscure.[67]

FIGURE 6.6 Scientist as a "lonely genius" versus science as a mass profession. A photograph of a general meeting of GINZ on August 30, 1920, (prior to the end of the Civil War!) on the occasion of a bacteriological conference. The photo was taken in the courtyard of the Institute of Vaccines and Sera decorated with a bust of Elie Metchnikoff. Seated in the first row are the leading lights of Soviet biomedicine (left to right): S. Dubrovinskii (GINZ's scientific secretary), A. Sysin (deputy director of the Institute of Hygiene), A. Bakh (director of the Institute of Biochemistry), N. Kol'tsov (director of the Institute of Experimental Biology), P. Diatropov (chair of microbiology at the Moscow University Medical School), Z. Solov'ev (deputy head of Narkomzdrav), N. Semashko (head of Narkomzdrav), L. Tarasevich (head of the Narkomzdrav Scientific Medical Council and director of the Institute of Vaccines and Sera), E. Martsinovskii (director of the Institute of Tropical Diseases), B. Slovtsov (director of the Institute of Experimental Medicine), V. Kedrovskii (director of the Bacteriology Institute), V. Barykin (director of the Institute of Microbiology), A. Berdnikov (director of the Saratov Microbiology Institute). Standing behind them are members of the scientific staff of various institutes under the GINZ umbrella. This photograph illustrates the transformation biomedical sciences were undergoing in Bolshevik Russia—from "a lonely genius" (symbolized by Metchnikoff's bust) into a mass profession (represented by the crowd of GINZ personnel). *From* GINZ, *1924: V.*

Yet after the Revolution, creating institutions was a major preoccupation of many Russian scientists. As we saw in the previous chapters, long before the Bolsheviks seized power, scientists had nurtured dreams of "big science"—of large specialized institutional structures staffed with numerous assistants and technicians that would enable them to pursue their subjects along various lines of investigation, produce necessary materials and research techniques, build and maintain an internal consensus regarding the subjects, methods, and goals of their studies, publish their results, and train new generations of researchers and practitioners. These dreams had particular currency among the proponents of emerging disciplines. The champions of new disciplines had to compete against the old established discipline-based communities who occupied existing institutional niches. They were constrained by the rigid institutional structures of the Russian Imperial science system and impeded by the general lack of interest and unwillingness of the tsarist government to provide support for scientific development. As did many of their colleagues elsewhere, Russian scientists turned to private patrons in search of such support, successfully courting wealthy merchants and industrialists, including Khristofor Ledentsov, Fedor Riabushinskii, Alfons Shaniavskii, Moisei Gal'perin, and many others.

The Revolution nipped the burgeoning alliance between scientists and capitalists in the bud. The Bolsheviks were set on demolishing the old Imperial system and building their own, new one. As the Russian-language version of the "Internationale" that became the official anthem of Bolshevik Russia boldly proclaimed:

> We will destroy this whole world of violence
> Down to its foundation, and then,
> We will build our [own], new world,—
> He who was nothing will become everything!

Science was not exempt from this dual process of destruction/construction, which presented to the scientific community both a threat and an opportunity. As we have seen in both fiction and reality, the Revolution liquidated private sources of support for science and stripped Russian scientists of the privileges they had enjoyed under the old regime, making the newly created Bolshevik state the one and only patron of science and scientists. But it also afforded entrepreneurial scientists an opportunity to insinuate their own visions of "big science" into the Bolsheviks' revolutionary dreams and to regain the high public prestige and social standing they had experienced before the Revolution. Scientists managed to involve their new patrons in a different kind of experiment—building large institutional structures exclusively with the state support to advance their own research interests.

The overlapping fields of experimental biology and medicine—at that very time quickly branching out into various specialized disciplines and subfields—were particular beneficiaries of this opportunity, finding willing and eager patrons in various agents and agencies of the fledgling government apparatus. Experimental science held a special appeal to the Bolsheviks. The very word *experimental* in the Russian language has connotations of something *new*, *never before attempted*, and *untried*. Given the Bolsheviks' determination to break free from everything "old," it is hardly surprising that they both supported the institutes of experimental biology and experimental medicine established before the Revolution and fostered the creation of new institutes of experimental veterinary science and experimental endocrinology (not to mention a variety of other cultural experiments such as Vsevolod Meyerhold's "experimental theater").

During the 1920s, nearly all top-level government agencies, including Narkomzdrav, Narkompros, Narkomzem, the Supreme Council of the People's Economy (VSNKh), the All-Union Central Executive Council (TsIK), and the Council of the People's Commissars (SNK), provided support for institutions involved with experimental biology and medicine. Thus, VSNKh funded the Chemical-Pharmaceutical Institute, TsIK—the Communist Academy Institute for Studies of the Brain, Narkompros—the Timiriazev Biological Institute, and Narkomzem—the Institute of Experimental Veterinary Science. Many proponents of experimental biology and medicine were able to take advantage of the destruction of old disciplinary hierarchies, the feverish construction of new institutional niches, and the multiplicity and variety of government offices and officials (as well as alliances and rivalries among them) able and willing to support their visionary experiments.

In the first decade of Bolshevik rule, Narkomzdrav was the principal source of funding for biomedical research. Under the enthusiastic administration of Nikolai Semashko, the agency built a huge system of research facilities, which by the mid-1920s included more than forty specialized institutes, laboratories, commissions, museums, and long-term expeditions aimed at "protecting the health of the people."[68] According to the conventional wisdom reiterated in the celebratory accounts of Soviet medicine and public health, the particular configuration of this new system reflected deliberate policies, ideological goals, and the proclaimed scientism of its patrons, the Bolsheviks.[69] One set of facilities was established to meet the challenges of concurrent health crises, and included institutes of microbiology and epidemiology, physiology of nutrition, vaccines and sera, sanitary science and hygiene, venereal diseases, tuberculosis, and tropical diseases. Another set was created to address the ideological commitments and health concerns of the "victorious proletariat" and was comprised of institutes of social hygiene, occupational diseases, the protection of maternity and infancy, labor hygiene, physiotherapy, and health resorts. And a third set was organized to advance basic biomedical research underpinning the progress of medicine and consisted of institutes of roentgenology and radiology, experimental endocrinology, blood transfusion, biochemistry, experimental biology, and biophysics.

The materials presented above suggest, however, that this perceived wisdom represents a later rationalization of rather haphazard and chaotic institution- and discipline-building processes, which unfolded under Narkomzdrav's patronage during the first post-revolutionary decade. The emergence of the large system of research establishments resulted not only and not even primarily from the deliberate policies of the Bolsheviks trying to address concurrent health crises or to fulfill the promises of the Revolution, but also and most importantly from the efforts of various scientists-entrepreneurs pursuing their own research agendas and institutional ambitions.

Consider, for instance, the case of Nikolai Kol'tsov's Institute of Experimental Biology (IEB).[70] Established in late 1916 with funds procured from the Society of the Moscow Scientific Institute, on the eve of the Bolshevik Revolution IEB occupied three rooms in a large building that had been purchased by the Society for its institutes in the center of Moscow. At this time, IEB had a staff consisting of the director (Kol'tsov) and three assistants, with a few of Kol'tsov's current and former students working on various projects without pay. The Bolshevik coup in October 1917 liquidated the endowments that funded the institute, and after the Bolshevik government moved to Moscow in the spring of 1918, Kol'tsov persistently courted the new rulers to support his institute. He offered his services simultaneously to Narkomzem, Narkompros, and Narkomzdrav and successfully tapped the resources of the three agencies, thus ensuring the survival of his students and colleagues during the harsh conditions of War Communism.[71] But only one, Narkomzdrav, eventually, in January 1920, took IEB under its wing.

A variety of reasons might have played a role in Semashko's willingness to accept an institution that at the time had at best only tangential connections to medicine or public health as a part of GINZ. Kol'tsov's reputation as Moscow's leading experimental biologist—in 1916 he had been elected a corresponding member of the Imperial Academy of Sciences—certainly was one of them. Kol'tsov's readiness to work on various projects that corresponded to Narkomzdrav's priorities was obviously of some importance: one of the first projects he offered to Semashko in 1919 was breeding rabbits, mice, guinea pigs, and chickens for Narkomzdrav's research establishments, which, given the total absence of a system of supply of laboratory animals in the country, was a very timely and appealing offer.[72] But probably the most important reason was Kol'tsov's longstanding association with Lev Tarasevich, head of the Narkomzdrav Scientific Medical Council and the de facto executive director of GINZ.[73] The two men first met in 1908 when Tarasevich, a one-time student and coworker of Metchnikoff, came from Odessa to Moscow and began teaching bacteriology as an unpaid lecturer (*Privatdocent*) at Moscow University. The two had much in common and quickly became good friends. At the time, Tarasevich was little known to the Moscow scientific community and it seems likely that Kol'tsov helped him obtain a paid teaching position at the Women's Higher Courses (where he himself had been teaching since 1903). Furthermore, when in 1914 Kol'tsov had come to head the country's leading popular-science journal, *Priroda*, he invited Tarasevich to join its editorial board. Obviously, when after the Revolution Tarasevich became Narkomzdrav's leading scientific expert, he extended a helping hand to his friend Kol'tsov by placing IEB under Narkomzdrav's patronage (while establishing his own Institute of Vaccines and Sera in the same building that at the time housed Kol'tsov's IEB).

During the early 1920s Kol'tsov's involvement with such highly publicized projects of visionary biology as rejuvenation and eugenics certainly played a role in consolidating IEB's position among Narkomzdrav's research establishments. In 1925 Kol'tsov's institute moved to a "princely villa" (vividly depicted in Pil'niak and Fedorovskii's "Institute of Life"), and its staff grew to more than forty full-time researchers (not counting technical personnel), working in seven separate departments on a variety of "hot" topics, including genetics, experimental cytology and embryology, zoopsychology, and eugenics.

Certainly scientists' involvement in the institution-building processes was to a considerable degree a matter of individual choice and temperament. For someone like Kol'tsov, creating and managing research institutions constituted a necessary (though not necessarily enjoyable) element of their engagement with science. For someone like

Iakov Tobolkin, it was the only scientific activity of interest. Many scientists, like Sergei Briukhonenko, preferred to leave institutional matters to somebody else, focusing exclusively on their research objectives. They "delegated" the difficult and often unpleasant tasks of dealing with patrons, allies, competitors, contractors, suppliers, publishers, and so on to their colleagues with a taste for administrative work. And, as did Briukhonenko in his "collaborative" publications with Oskar Steppun, these scientists often "repaid" the administrators by adding the names of these individuals to their own research publications.[74]

This critical role of scientists-entrepreneurs in building scientific institutions explains the paradoxical situation around the studies of anabiosis in Russia after the death of its champion, Porfirii Bakhmet'ev—with dozens of popular-science publications, but a nearly total absence of new research on the subject. Despite the wide popularization of, and numerous fictional experiments with, the freezing and thawing of living organisms, there was no scientist in Russia at the time willing and able to undertake the difficult task of establishing a special "anabiosis" laboratory, not to mention a research institute such as the "Institute of Life" created by Pil'niak and Fedorovskii in their story "The Affair of Death." Petr Shmidt, the main driving force behind the active popularization of anabiosis, had no connection to Narkomzdrav and was much more interested in pursuing his studies of fishes in the Russian Pacific than in building a special institution for studies of anabiosis. At the same time, Nikolai Kol'tsov, who had "inherited" what was supposed to be Bakhmet'ev's laboratory of low temperatures within the Biology Institute planned by the Society of the Moscow Scientific Institute (which eventually became the Institute of Experimental Biology), was preoccupied with his own research and had little interest in anabiosis.

Russian scientists and their Bolshevik patrons did not necessarily share the same values, interests, or dreams, even if the ultimate result of their interactions was one and the same: the establishment of large research institutions in particular fields of knowledge. Most of the time, the patrons did not really know or understand what exactly their clients were doing in the laboratories and institutes the patrons so generously funded. The state's resources were not unlimited and entrepreneurial scientists had to explain their research agendas and convince patrons to fund their, not their competitors', endeavors. They skillfully played upon their patrons' priorities and interests, translating the goals of their own esoteric research into a language understood and appreciated by the patrons. They successfully created an alliance with their Bolshevik patrons to advance their own research interests and institutional ambitions.

But the patrons had their own stakes in such an alliance. Their dreams of a new world notwithstanding, the Bolsheviks were hard-core realists and pragmatists. They saw science and scientists first and foremost as a suitable instrument for reaching their own objectives. Alexander Bogdanov's concept of "proletarian science"—science working exclusively on tasks defined by the needs of the proletariat, or more precisely, its "vanguard," the Bolshevik Party—formed a cornerstone of Bolshevik science policies, even though its author, Bogdanov, had been expelled from the Bolshevik Party and repeatedly criticized by no less a figure than its leader, Vladimir Lenin.[75]

The Bolsheviks' pragmatism was evident in their support for research that addressed the particular health problems of high-ranking members of their party. As Bulgakov's "The Dog's Heart" and the actual cases of endocrinology and rejuvenation demonstrate, enterprising scientists and physicians effectively exploited this "human" side of their new patrons—especially the widespread concerns over the health of the ruling elite and old party guard in the aftermath of the Civil War—to secure the support of those in power to advance their own agendas. As we saw, Nadezhda Krupskaia's affliction with Basedow's disease proved a decisive factor in the early stages of the institutionalization of endocrinology in Soviet Russia. Similarly, as I have detailed elsewhere, the creation in early 1926 of the Institute of Blood Transfusion in Moscow was to a considerable degree motivated by an "epidemic of Soviet exhaustion" (as Kremlin doctors named it) that plagued the top echelons of the Bolshevik Party.[76] One might suggest that if a Bolshevik leader had suffered from, say, diabetes, the story of early Soviet endocrinology would have looked quite different.

This instrumentalist attitude was also manifest in channeling Narkomzdrav's support for the institutional development of endocrinology primarily into the manufacture of various hormonal preparations such as "medicinal milk," thyroidin, adrenalin, and later insulin, rather than into basic or even clinical research. Indeed, all of the "endocrinological" institutions supported by Narkomzdrav—the farm of thyroidectomized goats, the institute of medicinal milk, the institute of organotherapeutic preparations, and finally, the institute of experimental endocrinology—were first and foremost manufacturing, not research or clinical, establishments. As we saw, Shervinskii had a hard time trying to convince his patrons of the need to establish a museum of endocrinology and a clinic for the treatment of endocrinological disorders.

Yet there was a third player that provided a critical nexus in the emerging alliance between visionary biologists and revolutionary dreamers—the media. During the 1920s Bolshevik agents and agencies had no set policy toward various subfields of experimental biology and medicine. As we saw in both the fictional case of Professor Persikov and the actual case of Sergei Briukhonenko, in the 1920s Bolshevik patrons often followed the lead of the media in identifying specific research subjects and directions as worthy of support. As the documents preserved in his personal archive indicate, Briukhonenko himself had never even tried to appeal to the powers that be in search of support for his studies. Instead, journalists and their publications lobbied, as it were, various government agencies on his behalf. At the same time, as we saw in the case of endocrinology, the role of the media as a critical nexus between science and state agencies certainly prompted such scientific entrepreneurs as Tobolkin and Shervinskii to use press interviews and publications to influence decision makers in the government apparatus. Shervinskii even used the pages of *Izvestiia* to publish the minutes (protocols) of the Russian Endocrinological Society. Many entrepreneurial scientists actively employed the media to promote their research agendas and institutional plans, as did Nikolai Kol'tsov in his live-audience and radio lectures and the countless articles he published in such widely circulated newspapers as *Pravda*, *Izvestiia*, and *Evening Moscow*.

Even more important, it was the popular media that effectively translated the esoteric knowledge generated by experimental biomedical research into a language understandable and appealing to state patrons and the general public alike. In media coverage, mitogenetic rays became "rays of life," hormones transmuted into the "water of life," anabiosis turned into the "fabulous century-long sleep," and experiments with severed dog heads transformed into "victory over death." It was the journalist Grigorii Grebnev who first christened Briukhonenko's apparatus an "artificial heart"; the inventor himself had always called it an autojector. This seemingly innocuous substitution of one name for another undoubtedly played a critical role in bringing Briukhonenko's work to the attention of state patrons, medical colleagues, and a general public. The simplified, colloquial, and often highly metaphoric language of media reports about science and scientists helped create a vocabulary that became an indispensible tool in the negotiations between scientists and their patrons over the research agendas, institutional growth, and cultural importance of biomedical sciences.

Science and Religion: Biology as a Cultural Resource

For centuries, religion constituted the dominant component of human cultures. All other elements of communities' cultures, such as music, literature, architecture, fashion, arts, science, moral codes, and aesthetic principles, developed in close connection to and evolved under the profound influence of religion. It was religion that for millennia provided answers to the "perpetual" questions of human life and death, human nature, and human destiny.[77] For most of recorded human history, religion remained the most influential cultural resource that inspired (and was heavily exploited by) generations of philosophers, artists, writers, musicians, and scientists. Yet that history also witnessed continuous attempts to challenge the supreme cultural authority of religion (and theologians) by certain philosophers, writers, and especially scientists.[78]

In the decades before the Revolution, the Russian intelligentsia, and scientists in particular, held somewhat dubious attitudes toward religion.[79] On one hand, like the fictitious Professor Preobrazhenskii, many Russian intellectuals had roots in the clergy that had long served as a recruitment pool for the country's professions and educated elites.[80] Bulgakov, Beliaev, Pavlov, Tobolkin, and Voronskii, to name just a few individuals covered in this volume, all came from the clergy, and the last four even began their education at theological schools (seminaries). Thus, many members of the Russian intelligentsia grew up in a religious milieu and received an orthodox religious upbringing, which deeply influenced their own intellectual endeavors.[81]

On the other hand, the Orthodox Church was one of the cornerstones of the tsarist regime, an important part of its oppressive state apparatus, and members of the Russian intelligentsia always prided themselves on (and defined themselves by) their opposition to the regime, including the state religion and its officials.[82] Furthermore, the Orthodox Church served as the supreme censor that watched over the country's "moral and spiritual health." Russian naturalists had had regular run-ins with censorship authorities over publishing their works, especially the ones written in the genre of speculative science and touching upon the theologically charged issues of human nature and human destiny, as happened, for instance, with the publication of Ivan Sechenov's "Reflexes of the Brain," or the Russian translation of Ernst Haeckel's *Die Welträtsel*.[83]

Circa 1900, as did their counterparts elsewhere, Russian theologians attentively followed contemporary scientific advances and published numerous works addressing the issues raised by modern, especially biomedical, sciences in their own periodicals and books.[84] But as the Russian biomedical community grew and matured in the decades leading up to the twentieth century, its members carefully avoided direct confrontations with theologians. In their professional publications, the majority of Russian naturalists, regardless of their personal religious beliefs, refused to address explicitly the theological dimensions of their own or their colleagues' works or engage in polemics with theologians. For instance, in the early 1900s, Vladimir Karavaev, a well-known Kiev entomologist, prepared a massive volume on the "comparative psychology of animals and humans," which included his translations of two major works on the subject by the eminent Jesuit entomologist Erich Wasmann.[85] Karavaev offered the manuscript to several science publishers, but all of them rejected it. One publisher explained his reasons bluntly: "This mixture of science and theology creates a very unfavorable impression and directly interferes with reading the book. Personally, I deeply regret this, for otherwise this is an excellent book with many interesting facts and a profound psychological analysis."[86] Eventually, Karavaev published the volume at his own expense. But although Russian scientists did review the book, they either completely ignored the religious aspects of Wasmann's works or merely mentioned them in passing, concentrating on his factual contributions to the studies of human and animal behavior.[87]

Similarly, the medical community was wary of the attempts of its members to mix science and theology (reminiscent of Abatulov's "theory of immortality" depicted by Sluchevskii). Thus, when Savelii Segal', a respected ophthalmologist, published (also at his own expense) a voluminous treatise under the enticing title *Questions of Life from the Viewpoint of the Natural Sciences* (1906), which included lengthy arguments about God, the soul, and religion, a reviewer in the leading medical periodical, *Physician's Gazette*, simply refused to broach a discussion of the treatise's theological aspects.[88]

Yet as Bakhmet'ev's "The seat of the soul," Metchnikoff's *Études*, and many other examples of speculative science clearly indicate, Russian naturalists continuously encroached upon the supreme cultural authority of, even while avoiding any direct dialogue (not to mention, confrontation) with, religion.[89] They consistently asserted their own authority over subjects and issues that through the ages had constituted the exclusive domain of religion (and theologians), ranging from the mechanisms of the psyche and the resurrection of the dead to human origins and human destiny. They actively propagated their own views on human life and death, health and illness, human nature, and humans' future evolution derived from their research, undermining in the process the millennia-old theistic views on these subjects. As one might have expected, the Bolsheviks actively capitalized on this materialist trend in contemporary biomedical research, making science their weapon of choice—and scientists their closest allies—in the militant propaganda campaign against religion waged in the aftermath of the Revolution.

Although they disapproved of the Bolsheviks' repressive actions against religion, especially the widespread arrests of clergy members and the demolition of churches, many Russian naturalists and physicians saw the process of replacing religious beliefs with scientific knowledge as a logical and necessary step in the progress of humanity (see Figure 6.7).[90] Following Metchnikoff, some of them even thought that science should

FIGURE 6.7 The reading room of the Scientific Library at the Timiriazev Biological Institute. Founded in 1920 (and renamed the Timiriazev Scientific Research Institute for the Study and Propaganda of the Scientific Foundations of Dialectical Materialism during the Great Break), this institution was a major base of militant "materialist-biologists" who marched "under the banner of Marxism." This 1931 photograph captures the militant antireligious stance of biology in the Soviet Union. The banner at the top left of the room reads: "The front of the struggle against religion," while the banner behind Vladimir Lenin's bust in the center says: "Materialism [is] our weapon." *Courtesy of ARAN.*

indeed become a "new religion." Many Russian naturalists were quite comfortable playing the role of "prophets" and "apostles" to the new scientific faith. For instance, in his 1921 presidential address to the Russian Eugenics Society that he had established a year earlier, Nikolai Kol'tsov proclaimed confidently: "Eugenics is the religion of the future and it awaits its prophets."[91] This "eugenic religion" was comparable, in Kol'tsov's opinion, to Christianity and Islam.[92] And as these "old religions" once had, it was now eugenics' turn to espouse an "ideal" that would "give meaning to [human] life and motivate people to [make] sacrifices and [impose] self-limitations," thus providing norms, values, and directions for human thoughts and behaviors. As the history of eugenics in Soviet Russia suggests, in the early 1920s, various Bolshevik agencies and officials (both Semashko and Lunacharskii supported the Russian Eugenics Society) actively encouraged scientists' aspiration to scientific "priesthood," while numerous journalists applauded the downfall of religious superstitions and the coming New Kingdom of scientific knowledge.[93]

There were, of course, some dissenting voices—as we saw, certain writers such as Ivanov-Razumnik and Bulgakov remained deeply apprehensive. But if Ivanov-Razumnik merely lamented scientists' "ethical deafness," Bulgakov actually mocked science's and scientists' pretensions to moral authority, even though, following an intelligentsia tradition, he also denied such moral authority to organized religion and its priests.[94] In "The Dog's Heart," Bulgakov repeatedly parodied science's role as a source of moral guidance and satirized scientists' role as high priests of the new religion.[95] In various places in the story, the advances of experimental biomedicine—encapsulated in statements like "there is no life after death" (77)—are presented as a foundation of and justification for abhorrent behavior. While working on his experiments, Preobrazhenskii continuously hums the same line from the aria of the king of Egypt in Giuseppe Verdi's *Aida*: "To the shores of the sacred Nile . . ." But if in the aria's Russian version the king continues the phrase with "Gods will show us the way, Gods will multiply our strengths!," Preobrazhenskii consistently leaves this ending out, thus clearly indicating that he doesn't need gods or their guidance and help in his research or his life. Bulgakov's portrayal of Preobrazhenskii (starting with the professor's very name and continuing through various details in his attire and his surroundings) often directly points to the professor's role as a "priest." Furthermore, throughout the entire story, Preobrazhenskii is continuously pontificating about various issues in life, from famously advising his assistant not to read Bolshevik newspapers before dinner to instructing the dog-man Sharikov on table manners. Even more important, Preobrazhenskii repeatedly justifies his wide-ranging opinions by references to his scientific research and his status as a world-renowned scientist. He even claims that he has conducted experiments on his patients, proving the harmful effects of reading Bolshevik newspapers on their digestion!

In contrast to the deafening antireligious campaign that thundered over the country and raised experimental science to the pedestal of a new religion, the dissenting voices remained effectively inaudible. Ivanov-Razumnik confined his to private correspondence, while Bolshevik censorship—no less vigilant than its tsarist predecessors—silenced Bulgakov's. Furthermore, the 1922 "philosophers' steamboat"—the forced exile of nearly two hundred eminent Russian intellectuals—made clear that the regime was prepared to take severe administrative actions against "idealists" and proponents of "religious philosophy." In this situation, Russian writers and scientists who might have objections to the militant antireligious campaign and reservations regarding the elevation of science to the status of "official religion" kept those objections and reservations to themselves. Thus, during the 1920s, Prince Aleksei Ukhtomskii (1875–1942), a prominent Russian physiologist and devout Christian, published numerous scientific papers, developing his concept of dominance as a basic physiological mechanism of the mind, but kept his theological musings on the subject in private letters and diaries.[96]

The forcible removal of religion created a "cultural vacuum" that was filled by science, contributing greatly to the rapidly rising cultural authority of science and scientists and profoundly shaping the cultural meanings of biomedical research in post-revolutionary Russia. The silencing of dissenting voices helped condition the Russian population to the idea that science—not religion or literature—was the only source of true ideals, norms, and values of human existence and that science was indeed capable of changing human nature and defining human destiny—up to and beyond fulfilling humanity's ultimate dream of immortality.

The power of biomedical knowledge as a cultural resource, however, was not merely arbitrary. It derived from scientists' demonstrated—and widely publicized—abilities to influence ("to control," in the parlance of the day) various life processes, such as metabolism, reproduction, growth, immunity, aging, cell division, and so on, through the application of certain physical, chemical, or biological agents. It was further enhanced by the proven successes of new diagnostic and surgical techniques, preventive vaccinations, and curative (vitamin-, hormone-, sera-, physio-, radio-) therapies grounded in experimental biomedical research and loudly trumpeted by the media as "science's victory over death." And, as Fedor Gladkov's likening of Metchnikoff's "naïve dream" of rejuvenation to the "widespread" smallpox vaccinations indicated, it was exactly these abilities and these successes that made the answers proffered by biomedical science and scientists to the "big" questions of human nature and human destiny plausible and believable.

Three major groups of actors lent a hand to the transformation of esoteric, specialized knowledge generated by experimental biomedical research into a powerful cultural resource. Scientists themselves initiated the process by actively popularizing their works and freely speculating on the possible intellectual, social, and cultural consequences of their studies. Excited, inspired, or troubled by scientists' discoveries and aspirations, journalists and writers eagerly joined in the process, affirming (even when questioning or ridiculing) scientists' claim to cultural authority and further molding public expectations, imaginations, and fears by bringing into play various literary symbols, images, and traditions. By virtue of their position as the one and only patron of science and through their tight control over the media, the Bolsheviks defined what kinds of knowledge would become culturally dominant by widely disseminating views and interpretations that confirmed and conformed to their own revolutionary dreams, while effectively silencing all dissenting and subversive voices (as they did with Bulgakov's "The Dog's Heart").

The historic coincidence of the Bolshevik Revolution with three major "scientific revolutions"—the experimental revolution in biomedical research that dramatically changed the understanding of life and death; the revolution of scale that led to the emergence of "big science" and to the transformation of science into a mass profession; and the "science/fiction revolution" that spectacularly increased science's public visibility and cultural authority—resulted in the continuous cross-examination, cross-pollination, and hybridization of various ideas and ideals generated by revolutionary dreamers, visionary biologists, and science/fiction writers. Inherent to the various projects of visionary biology, the motif of controlling human "biological destiny" resonated well both with the literary tradition of examining the "big" existential questions and with the Bolsheviks' firm belief in the ability of people to determine their own "social destiny" through a social revolution (forcefully expressed in the country's anthem: "No one will grant us deliverance: not a god, not a king, and not a hero. We will achieve liberation with our own hands"). This mutual cultural affinity certainly facilitated intense interactions—conducted in the dark shadow of the decade of death—among the three groups and profoundly shaped the actual content of the cultural resource they collectively created, giving the millennia-old dream of immortality new forms and new meanings in Bolshevik Russia.

Yet however strange, exotic, and fantastic the stories presented above may seem, in my opinion, they also illustrate certain general principles involved in the transformation of scientific knowledge into a *cultural resource*—principles that go well beyond Russian/Soviet contexts and whose actions can be seen in historical and contemporary public debates over evolution, race, eugenics, infectious diseases, genetic engineering, human origins, organ transplantation, transgenic foods, cloning, stem cells, and many other "hot" biomedical subjects.[97] The migration of scientific facts, ideas, and concepts into the public sphere entails much more than the creation of a convenient rhetoric that mediates negotiations between scientists and their patrons, or translates arcane scientific knowledge into a language comprehensible to a lay public. Such migration also creates a *zone of contest* between science and culture writ large, a zone that is opened for use and appropriation by any interested party, be it theologians, writers, politicians, public health officials, TV and movie producers, philosophers, scientists, journalists, business executives, jurists, artists, or state bureaucrats.

In the process of becoming a cultural resource, particular and often quite impenetrable scientific knowledge about concrete natural objects or processes is greatly simplified, fragmented, and reduced to a few, often disjointed, generalized statements, something like "men evolved from apes." In this process, the complexity of knowledge and the need for its continuous refinement and verification is eliminated and replaced by a "vivid picture" (to borrow Ludwik Fleck's astute characterization[98]), exemplified by numerous "human interest" stories about science and its advances, satirical cartoons, and science fiction writings. Scientific knowledge thus loses much of its conventionality, its embeddedness in a particular mode of production, its hypothetical, uncertain, relative, and provisional character. Instead it acquires dogmatic formulations (often in the form of various "isms" associated with a particular field of knowledge[99]) and doctrinaire imagery embodied in such popular cultural icons as "Frankenstein," "the brain in a vat," "Dolly the sheep," "the double helix," or the "mushroom cloud."[100] A cultural resource also obscures the historicity of scientific knowledge and masks the multitudes of scientists involved in its production, instead emphasizing one heavily fictionalized personality hailed as the "great discoverer," be it Isaac Newton, Charles Darwin, Ivan Pavlov, or James Watson.

Furthermore, once created, a cultural resource generates public debates whose forms, tones, and terms differ considerably from those used by scientists involved in the production of the scientific knowledge that gave birth to this resource. The arguments used in scientific disputes (logic, verifiability, controlled experiment, disciplinary consensus, technical apparatus, negotiated and agreed upon standards of measurement and presentation, and so on) lose their convincing power and are replaced by such "social" arguments as economic utility, political expediency, ideological conformity, traditional authority, and cultural affinity. A cultural resource acquires a life of its own, evolving in accord with its own environment (of which science per se constitutes only a small part). The transformation of scientific knowledge into a cultural resource thus undermines scientists' control over their "product" and its applications. It opens the scientific discipline from which this cultural resource emerged to invasion by "outsiders"—whether competing disciplinary groups or politicians—eroding its professional autonomy, disciplinary boundaries, and established scientific cultures. Yet it also helps set the agendas of future research and recruit new generations of practitioners by firing up people's imaginations with fantastic tales of scientific pursuits, be it a search for a cancer cure or a quest for immortality.

But that the dread of something after death,
The undiscover'd country from whose bourn
No traveler returns, puzzles the will
And makes us rather bear those ills we have
Than fly to others that we know not of?
WILLIAM SHAKESPEARE, *Hamlet* (1599–1601)

With a smile on his face, this man spoke of incredible things. With self-assurance that left no doubt, he called the wildest and most cherished dream of humanity—the dream of immortality—a scientific discipline.
GRIGORII GREBNEV, *Arctania* (1937)

Epilogue: An Unending Quest

NEARLY A DECADE after he had covered Sergei Briukhonenko's experiments for the *Komsomol Truth*, journalist Grigorii Grebnev decided to try his hand at writing science fiction. In the last three months of 1937, *Pioneer*, a popular children's magazine, serialized his first novel, titled *Arctania*.[1] Set in an undetermined but not too distant future, the novel depicted the world—joined in the Great Union of Socialist and Communist Republics—making great strides in developing science and technology and in conquering Nature. The people of this future constructed a huge flying science station and placed it in the Arctic (hence, the name "Arctania") near the North Pole to predict and control the weather. They are building a dam to close the Gibraltar Straits and to supply electricity to the entire Mediterranean. They use radium for lifting up underwater mountains and rockets for intercontinental travel. Combining the discoveries of Sergei Briukhonenko and Porfirii Bakhmet'ev (both of whom are directly mentioned in the text), they learnt how to "revive the dead" and treat senescence merely as a disease to be cured. They search for "the ways to make the life of every human born on this planet very, very long. So long [indeed] that it would be nearly equal to immortality."[2]

Yet all of these scientific and technological wonders serve merely as a decoration to the novel's main plot: a struggle with the remnants of "The League of Militant Christians"—a sort of fascist-religious sect that had led the capitalists' last assault on the Soviet Union and had been defeated by the world's proletariat. The League's leaders had managed to escape the capture and were hiding in an underwater base near the North Pole for several decades, until they were accidentally discovered by a young boy (the son of Arctania's director) and the final battle ensued. Even though Grebnev's story bore a subheading, "fantastic novel," and employed all the cliché of the genre (from a lost tribe of South American Indians to super weapons), it also clearly reflected a new reality of 1930s Russia.

To paraphrase Mikhail Bulgakov, everything in the world eventually comes to an end, and so did the "fantastic" 1920s. At the end of the decade, Russia plunged into a new revolution—the "revolution from above."[3] Joseph Stalin began to consolidate his personal power over the Bolshevik Party and that of the party apparatus over the nation, replacing nearly all top-level party and state officials, including the heads of all Commissariats (such as the well-known patrons of visionary biology, the head of Narkomzdrav Nikolai Semashko and the head of Narkompros Anatolii Lunacharskii) with his cronies.

The Great Break, as Stalin himself named it, however, meant much more than Stalin's ascendance to dictatorial powers. It also signaled that the multiplicity of revolutionary dreams—of a world revolution, a new world, a new society, and a new civilization—which had permeated the country during the first post-revolutionary decade, had been replaced with a singular vision embodied in a popular slogan: the construction of socialism in one country. The variety and vagueness of early Bolshevik programs for the country's future had crystallized into a specific plan of crash industrialization, the forced collectivization of the peasantry, and extensive militarization. The launching of the ambitious first Five-Year Plan inaugurated drastic changes in all facets of life. NEP was abolished, private initiative and the market suppressed, and a total state monopoly over resources, production, and distribution implemented, leading to the emergence of a system of strict centralized controls, administrative fiat, and greatly diminished local autonomy, accompanied by the creation of gigantic bureaucratic, military, and repressive (secret police) apparatuses.

Science was profoundly affected by the radical reorganizations of the new revolution: it was mobilized to serve the new policies of the state.[4] The grandiose plan for socioeconomic reconstruction reinforced the pragmatic, utilitarian attitude of the Bolsheviks toward science embodied in the famous motto "[let's put] science at the service of socialist construction" and in the title of a new journal, *Socialist Reconstruction and Science*, created in 1931.[5] The importance of scientific and, especially, technological developments to their new socioeconomic programs led the Bolsheviks to establish a complicated system of controls over all aspects of scientific activities: institutions, personnel, communications, and research agendas. During the 1930s, they greatly enlarged their support for science, vastly expanding the network of scientific institutions and their personnel and continuing to raise the public prestige and social standing of science and scientists, while simultaneously limiting the considerable autonomy enjoyed by the scientific community in the previous decade. Given their total dependence on state patronage, scientists were compelled to learn the "new rules of the game." They rapidly and skillfully adjusted to the new demands of their patrons by incorporating the new, Stalinist rhetoric and rituals into their own professional culture. Instead of using the media as a nexus of their relations to their patrons, scientists began to cultivate direct contacts with party agents and agencies via numerous letters that articulated their needs and wants in the new "party lingo." Indeed, to a certain degree scientists managed to turn the new system of controls into an instrument of advancing their own interests.

But, the demise of revolutionary dreams also spelled the end of visionary biology as a particular subset of speculative science and as a viable candidate for the country's new religion. The revolution from above severely restricted the creative freedom and social space available for cultural, technological, social, literary, and biomedical experiments, which had emerged in the aftermath of the Bolshevik Revolution. Only one "science"—Marxism-Leninism-Stalinism—was now allowed to "speculate" on the country's possible future and only within very narrow limits, often set up by Stalin personally and vigilantly enforced by his court ideologues. Visionary biology, which during the preceding decade had generated countless visions of what that future might be like if the techniques and discoveries of experimental biomedical sciences were implemented in various facets of human life, was stripped of its cultural authority and of its pretensions to attaining the status of a new religion. Marxism-Leninism-Stalinism became the country's official "religion," and party philosophers became its high priests and preachers. Official ideologues labeled attempts at applying biomedical knowledge to formulate the ideals, norms, and values of human life and to envision the country's (and, more generally, humanity's) future as the pernicious *biologization* of human nature, human place in the universe, and human destiny. They strictly limited the categories to which facts, regularities, laws, and classifications of biomedical research could now be extended and applied.

According to the rigorously enforced dogmas of Marxism-Leninism-Stalinism, human beings were first and foremost the products of their material conditions and social upbringing. Their personalities, individual fates, and collective destinies—not to mention their ideas—were all (pre)determined by the economic (and hence political and class) systems within which they lived. Thus only social—not biomedical—interventions, manipulations, and controls could effectively change this "economic" predestination and redefine human nature and human destiny.

The new emphasis on the social, as opposed to the biological, dramatically affected the interrelations of biomedical science and literature in Stalin's Russia. Ruthlessly controlled by the party apparatus (thanks to the complete suppression of private publishing and the market), literature and litterateurs were, nevertheless, once again elevated to a dominant position in the country's cultural life. In September 1934, at the First Congress of Soviet Writers, the party bosses bestowed on litterateurs the high status of the "engineers of human souls" and assigned them the task of propagating and inculcating new values, ideals, and behaviors consistent with the new vision of the country's future, a task that could be fulfilled by adopting "socialist realism" as the only permissible "method" of literary expression. Needless to say, science fiction writing suffered the most. The number of popular-science and adventure magazines published in the country dwindled down and the number of SF stories dropped dramatically. The entire genre was recast as a subset of children's literature. It quickly degenerated to a few standard formulas, depicting science's and scientists' role in "building socialism," "class conflicts," and "the inevitable military confrontation between socialism and capitalism." Grigorii Grebnev's 1937 novel presents a typical example of this new SF: it was geared to preteens and incorporated all of these officially endorsed formulas. Clearly echoing Stalin's famous statement that "there are no fortresses that the Bolsheviks could not take over," Grebnev boldly claimed that "there are no miracles that the science of communist humanity would not be able to perform."[6] Yet a year later, an acknowledged leader of early Soviet SF, Alexander Beliaev, felt compelled to call on his fellow writers: "Let's create Soviet scientific fantasy," as if the genre had not existed and had not flourished during the previous decade.[7]

With its cultural authority and public visibility greatly reduced, biomedical research in Soviet Russia nonetheless continued with great vigor and expanded at great speed. Manifested in the explosion of various research trends in the 1920s, the tremendous creative energy unleashed by the Bolshevik Revolution became "disciplined" and channeled into controllable institutional niches and predictable clinical routines during the subsequent decade. Thus, a "religion of the future"—eugenics—became transformed into medical genetics, while Briukhonenko's "fight against death" turned into studies of blood transfusions and the production of blood anticoagulants. In the 1930s, the utilitarian attitude toward science on the part of its sole patron—the Bolshevik Party and its leadership—defined the fates of particular research agendas and objectives. Some lines of research, which had been popular during the 1920s (such as, for instance, studies of telepathy, endocrine gland transplants, and sex change), were marginalized and abandoned, while certain research trends, which had been neglected in the previous decade (such as the studies of anabiosis), received a new boost.

Although the public infatuation with the dreams of rejuvenation and immortality subsided dramatically, overtaken by the daily struggle for survival under the oppressive regime, the Bolsheviks, of course, continued to support generously research on various rejuvenation techniques for use by the party's old guard. Thus, Alexander Zamkov, who in the 1920s had been deeply involved with rejuvenations á la Steinach and Voronoff at the Institute of Experimental Biology, in the early 1930s came up with his own new nostrum. A hormonal preparation Zamkov extracted from the urine of pregnant women (he called it "urogravidan") and injected in the buttocks of his patients became a treatment of choice for the aging members of the party elite.[8] Needless to say, the party apparatus immediately created a special institute for Zamkov's studies and for the manufacture of the new "water of life."

But all this is a different story. Future historians will have to untangle the complex interactions among intellectual and institutional developments in the biomedical sciences, socialist realist literature, and the new "religion" of Marxism-Leninism-Stalinism in Soviet Russia. They will need to figure out how these new contexts redefined the conditions and parameters of, but never did put a stop to, the quest for immortality.

Indeed, just as I was finishing this book in late May 2011, friends forwarded to me a copy of an open letter by the popular US movie actor, producer, and director Steven Seagal addressed to the then Russian Prime Minister Vladimir Putin.[9] An ordained Buddhist priest and a renowned master of martial arts (recognized in 1997 by one of the highest Tibetan authorities, His Holiness Penor Rinpoche, as the reincarnation of the seventeenth-century lama Terton Chungdrag Dorje), Seagal asked the most powerful man in Russia to support research currently being pursued in that country on the subject of... immortality. Specifically, Seagal urged Putin to extend a helping hand to a recently formed group of Russian scientists, physicians, writers, engineers, and media specialists, who call themselves "Russia 2045." Echoing its numerous 1920s predecessors, especially Biocosmists-Immortalists, the group advances an "ideology of the future, promoting technical progress, artificial intellect, 'multi-body,' immortality, and cyborgization." They strongly believe in the possibility and necessity "to eliminate aging and even death, and to overcome the fundamental limits of the physical and mental capabilities currently set

FIGURE E.1 When in the early twentieth century Elie Metchnikoff advocated for drinking soured milk as a "remedy" against old age, he of course did not anticipate that his idea would give birth to a multibillion-dollar yogurt industry. But his role is still appreciated, at least in post-Soviet Russia, as witnessed by this lid for a large container of yogurt I bought in Moscow in the summer of 2012. The top line reads: "Slavic traditions," while the bottom line says: "Metchnikoff's soured milk." *Photo by the author, July 2012.*

by the restrictions of the physical body." The group's manifesto promises that "before 2045 an artificial body will be created that will not only surpass the existing body in terms of functionality, but will achieve perfection of form and be no less attractive than the human body." According to the "Russia 2045" vision, "the new human being will receive a huge range of abilities and will be capable of withstanding extreme external conditions easily: high temperatures, pressure, radiation, lack of oxygen, etc." They foresee that "using a neural-interface, humans will be able to operate several bodies of various forms and sizes remotely." Most important, such "new bodies" will allow "extension of people's lives," "after the resources of the biological body have been exhausted," or, plainly speaking, immortality.[10]

Indeed, the dream remains alive, and not just in the new, post-Soviet Russia. It is enough to mention the Singularity and Transhumanism movements or Breathatarians, which gained extraordinary popularity in the West in recent years, to illustrate its continuing appeal in modern societies. Every time practitioners of the life sciences anywhere come up with new techniques, facts, concepts, and ideas, someone always attempts to adopt and adapt new biomedical knowledge to assuage humanity's inextinguishable fear of death or to make a career and/or a fortune by exploiting that fear (see Figure E.1).[11] And no matter how many doomsday scenarios, angry criticisms, or satirical parodies such attempts generate in contemporary cultures, someone will always be using the newly acquired tools of biomedical sciences in search of answers to the "big" questions of human life and death, human place in the universe, and human destiny, as well as in the pursuit of the ever-enticing but elusive dream of immortality.

Acknowledgments

As is always the case with any academic work, this book bears the mark of numerous individuals who helped me conceive, research, write, and, finally, publish it. Mark B. Adams's groundbreaking studies of "visionary biology" provided a major stimulus for this book and our continuous discussions (over the phone, e-mails, and in person) shaped its contents over the years it took me to complete it. My students at the University of Toronto willingly served as "experimental subjects" for testing the major ideas of this book, and their feedback helped hone and refine my arguments. In 2006, a group of graduate students at the Institute for the History and Philosophy of Science and Technology eagerly joined me in exploring the historical developments of experimental biology and experimental medicine at a research seminar on "The Biology of Death and Immortality." The next year, Valerie Watt, director of the human biology program at the Faculty of Arts and Sciences, invited me to teach a humanity course in the program, and when I designed a course on "Human Biology and Human Destiny: Science, Popular Science, and Science Fiction," she enthusiastically approved. For the last five years, teaching this course provided an invaluable opportunity to share some of my ideas with successive groups of extremely bright undergraduate students, who kept me excited about the whole project. In 2010, my syllabi for these two courses caught the attention of Christine Daum and Nelson Farber, an international team of documentary makers, who wanted to produce a film about the "pursuit of eternal life." We spent hours (the project is still ongoing) talking about science, immortality, Russia, and the Bolsheviks, as well as some recent "breakthroughs." The "visual perspective" brought in by Christine and Nelson provided a whetstone of sorts, helping to sharpen my thoughts on these subjects.

A major part of any historical study is gathering all possible information about its subjects: individuals, institutions, ideas, times, and places. Given the multitude of themes, characters, and locales examined in this book, I would have never been able to do it all by myself. The help of my friends and colleagues Aleksandra Bekasova, Dmitrii Derviz, Galina Savina, Aleksandr Sledkov, and Marina Sorokina in getting copies of materials from various archives, libraries, and museums in Moscow and St. Petersburg was indispensible. Other colleagues, including Dan Healey, Ganka Kamisheva, Mikhail Konashev, Mario Rovere, Susan G. Solomon, and Daniel P. Todes, generously shared with me their own precious archival finds. John B. Appleby, Yaroslav Ivanov, and Elena Maltseva provided much appreciated assistance in Toronto, carrying out the burden of surveying and copying relevant items from the secondary literature. The staff of numerous libraries and archives were extremely helpful, including the Archive of the Russian Academy of Sciences, the State Archive of the Russian Federation, the Russian State Archive of Socio-Political History, and the Manuscript Division of the Russian State Library (Moscow), the US National Library of Medicine (Bethesda, MD), the Libraries of the University of Toronto, the Library of the Russian Academy of Sciences (St. Petersburg), the National Library of Russia (St. Petersburg), the St. Petersburg Central State Archive of Literature and Arts, the Museum of the History of Medicine of the Sechenov First Moscow Medical University, the Museum of the History of Physics of the Bulgarian Academy of Sciences (Sofia), the Archive of Zurich University, and the State Archive of the Zurich Canton.

Numerous colleagues patiently heard at various scholarly gatherings and/or read pieces and bits of this book at various stages in its development and offered comments, criticisms, and suggestions. I am particularly indebted to Mark B. Adams, Anne-Emanuelle Birn, Michael D. Gordin, Yvonne H. Howell, Susan G. Solomon, and Daniel P. Todes, each of whom read the entire manuscript, for their thoughtful comments that helped improve the book in many significant ways.

Funding for research leading to this book came from the Social Sciences and Humanities Research Council of Canada and from the US National Institutes of Health (through National Library of Medicine grant, #G13 LM008632), for which I am extremely grateful. I have published different versions of Chapter 2 and Chapter 4 as articles (see Nikolai Krementsov, "Hormones and the Bolsheviks: From Organotherapy to Experimental Endocrinology, 1918–1929," *Isis*, 2008, 99: 486–518; and idem, "Off with Your Heads: Isolated Organs in Early Soviet Science and Fiction," *Studies in History and Philosophy of Biological and Biomedical Sciences*, 2009, 40: 87–100), and I thank their editors and publishers for permission to use the material from these articles in the present volume.

My wife and daughter stoically endured my long absences during frequent research trips to Russia and nurtured my obsession with this book in every possible way.

Naturally, I alone bear the responsibility for any mistakes and misinterpretations.

Notes

PROLOGUE

1. For a sample of this variegated literature, see Stephen G. Post and Robert H. Binstock, eds., *The Fountain of Youth: Cultural, Scientific, and Ethical Perspectives on a Biomedical Goal* (New York: Oxford University Press, 2004).

2. See, for instance, Nikolai Krementsov, "From a 'prominent biologist' to a 'Red Frankenstein': Il'ia Ivanov in Soviet and Post-Soviet biographies," in Thomas Lahusen and Peter H. Solomon, Jr. eds., *What Is Soviet Now? Identities, Legacies, Memories* (Berlin: LIT Verlag, 2008), 120–32.

3. D. Aleksandrov and N. Krementsov, "Opyt putevoditelia po neizvedannoi zemle: Predvaritel'nyi ocherk sotsial'noi istorii sovetskoi nauki," *Voprosy istorii estestvoznaniia i tekhniki* (hereafter—VIET), 1989, 4: 67–80; idem, "Sotsio-kul'turnye aspekty razvitiia sovetskoi nauki v 1920–1930-e gg.," *ibid*, 1990, 1: 166–68.

4. The classic accounts of the Russian revolutions by E. H. Carr, R. W. Davies, Sheila Fitzpatrick, Richard Pipes, Lewis H. Siegelbaum, and Robert C. Tucker focus almost exclusively on the political, economic, and social dimensions of the decade. Even the explorations of "Bolshevik culture" produced by such leading historians as Jeffrey Brooks, Katerina Clark, Abbott Gleason, Catriona Kelly, Peter Kenez, Stefan Plaggenborg, Richard Stites, and Nina Tumarkin pay very little attention to science and its place in the culture they examine. A voluminous introduction to "Russian cultural studies" does not have a single chapter on science; see Catriona Kelly and David Shepherd, eds., *Russian Cultural Studies: An Introduction* (New York: Oxford University Press, 1998).

5. A bibliographic database on the history of Russian/Soviet science created and maintained by Slava Gerovitch provides a vivid illustration of this disparity in historians' attention to various periods in the history they examine. See "Virtual guide to the history of Russian science and technology" at http://web.mit.edu/slava/guide/.

6. See Nikolai Krementsov, *Stalinist Science* (Princeton: Princeton University Press, 1997); and idem, "Big Revolution, Little Revolution: Science and Politics in Bolshevik Russia," *Social Research*, 2006, 73 (4): 1173–1204.

7. I am currently working on a book-length manuscript dealing with the other two themes.

8. For general assessments of the theme of immortality in Western science fiction, see George Slusser, Garry Westfahl, and Eric S. Rabkin, eds., *Immortal Engines: Life Extension and Immortality in Science Fiction and Fantasy* (Athens: University of Georgia Press, 1996); Vita Fortunati and Claudio Franceschi, "The Quest for Longevity and the End of Utopia," in Paola Spinozzi and Brian Hurwitz, eds., *Discourses and Narrations in Biosciences* (Goettingen: V&R unipress, 2011), 183–95; and Stefano Salvioli, "Sci-fi Quests for Longevity and Immortality," *ibid*, 197–209.

9. This is a paraphrase of a verse from a small 1940 poem, which became part of Akhmatova's famous cycle "Secrets of the [poetic] craft." It goes like this:

> If only you knew, from what rubbish
> Verses shamelessly grow,
> Like a yellow dandelion by the hedge,
> Like burdocks and goosefoot.

10. This asymmetry of attention is clearly reflected even in the order of words in the titles of many works seemingly dealing with the same issue—the interrelation of science and literature. While literary scholars usually put "literature" in first place in their titles, historians of science, expectedly, privilege "science." Compare, for instance, two collections: Ludmila Jordanova, ed., *Languages of Nature: Critical Essays on Science and Literature* (New Brunswick, NJ: Rutgers University Press, 1986); and Philip Coleman, ed., *On Literature and Science: Essays, Reflections, Provocations* (Dublin, Ireland; Portland, OR: Four Courts Press, 2007). For examples of the limitations of a disciplinary approach to the subject, see Katerina Clark, "The changing image of science and technology in Soviet literature," in Loren R. Graham, ed., *Science and the Soviet Social Order* (Cambridge, MA: Harvard University Press, 1990), 259–98; David P. Barash and Nanelle R. Barash, *Madame Bovary's Ovaries. A Darwinian Look at Literature* (New York: Delacorte Press, 2005); and I. V. Golovacheva, *Nauka i literatura: Arkheologiia nauchnogo znaniia Oldosa Khaksli* (St. Petersburg: SPbGU, 2008).

11. Although it is obviously connected to the themes examined in the present study, I do not address in any detail the equally interesting and important topic of the interrelation between the life sciences and philosophy in Russian culture, as well as their changes in the aftermath of the Revolution. For some preliminary observations, see Nikolai Krementsov, "Marxism, Darwinism, and Genetics in the Soviet Union," in DenisAlexander and RonNumbers, eds., *Biology and Ideology: From Descartes to Dawkins* (Chicago: Chicago University Press, 2010), 215–46; and Daniel P. Todes and Nikolai Krementsov, "Dialectical materialism and Soviet science in the 1920s and 1930s," in William Leatherbarrow and Derek Offord, eds., *A History of Russian Thought* (Cambridge: Cambridge University Press, 2010), 340–67.

12. See Krementsov, *Stalinist Science*.

13. See Nikolai Krementsov, *A Martian Stranded on Earth: Alexander Bogdanov, Blood Transfusions, and Proletarian Science* (Chicago: University of Chicago Press, 2011).

14. When historians and philosophers of science talk about "scientific revolutions," they usually mean drastic changes in the cognitive content of various disciplines, typically associated with

the name of a "great discoverer," such as a "Copernican revolution" in cosmology, a "Newtonian revolution" in physics, or a "Darwinian revolution" in biology. This approach—most clearly presented in Thomas S. Kuhn's 1962 classic *The Structure of Scientific Revolutions* and later elaborated by his numerous followers—treats science as a system of knowledge and investigates why and how that system changes over time. Science, however, can also be analyzed as a set of specific practices and as a social institution, and, hence, dramatic changes in scientific practices and science's social relations also deserve to be called "scientific revolutions." As we shall see later on, scientists themselves recognized the changes in their practices, in this particular case, the introduction of experimental methods into biology and medicine, as a revolution! For more on different kinds of "scientific revolutions," see below.

CHAPTER 1

1. For a classic biography of Bulgakov, see M. Chudakova, *Zhizneopisanie Mikhaila Bulgakova* (Moscow: Kniga, 1988); for an excellent recent rendering of his life, see Aleksei Varlamov, *Mikhail Bulgakov* (Moscow: Molodaia gvardiia, 2008).

2. Mikhail Bulgakov, "Belaia Gvardiia," *Rossiia*, 1925, 4: 3–99; 5: 3–82.

3. Mikhail Bulgakov, "Rokovye iaitsa," *Nedra*, 1925, 6: 79–148. For an English translation of the novella, see Mikhail Bulgakov, "Fatal Eggs," in idem, *Diaboliad and Other Stories*, eds. E. Proffer and C. Proffer, transl. C. Proffer (Bloomington: Indiana University Press, 1972), 48–134. The Russian adjective *rokovye* derives from the noun *rok* that means "fate." Fate (*rok*) plays an important part in the novella's plot and is even personified in one of its major characters whose last name is "Rokk" (Bulgakov added the second *k* at a proofs stage, apparently at his editor's insistence). Thus the commonly used English title of the story, "*Fatal* Eggs," is somewhat misleading, since the adjective *fatal* has entirely different connotations in English from those of *rokovye* in Russian. A more accurate translation of the title is "Fateful Eggs" or "Ill-Fated Eggs."

4. Mikhail Bulgakov, "Luch zhizni," *Krasnaia panorama* (hereafter—KP), 1925, 19: 1–4; 20: 3–4; 21: 5–9; 22: 4–6; ["Rokovye iaitsa"] 24: 4–6. On the publication history and contemporary reception of both the novel and the novella in Russia, see Chudakova, *Zhizneopisanie Mikhaila Bulgakova*. Unfortunately, no manuscripts of "Fateful Eggs" survived in the collection of Bulgakov's personal papers (*fond* 562) held at the Manuscript division of the Russian State Library (hereafter—NIOR RGB), making it impossible to trace the changes and edits Bulgakov made in the text at various stages.

5. Mikhail Bulgakov, "Rokovye iaitsa," in idem, *D'iavoliada. Rasskazy* (Moscow: Nedra, 1925), 44–124.

6. See, for instance, G. Gorbachev, "Tvorchestvo M. Bulgakova," *Krasnaia gazeta*, May 24, 1927, 5; A. Voronskii, "Pisatel', kniga, chitatel'," *Krasnaia nov'* (hereater—KN), 1927, 1: 237–38; Peter Doyle, "Bulgakov's Satirical View of Revolution in *Rokovye Iaitsa* and *Sobach'e Serdtse*," *Canadian Slavonic Papers*, 1978, 20 (4): 467–82; E. A. Iablokov, *Motivy prozy Mikhaila Bulgakova* (Moscow: RGGU, 1997); and many others.

7. See, for instance, V. Pravdukhin, "Nedra. Kniga shestaia," *KN*, 1925, 3: 287–89; N. Korotkov, "Nedra. Kniga shestaia," *Rabochii zhurnal*, 1925, 3: 156–57; A. Pridorogin, "Nedra. Kniga shestaia," *Knigonosha*, 1925, 6: 18; M. Lirov, "Nedra. Kniga shestaia," *Pechat' i revoliutsiia*, 1925, 5–6: 517–19; and many others.

8. I. M. Nusinov, "Put' M. Bulgakova," *Pechat' i revoliutsiia*, 1929, 4: 40–53, here–47.

9. See Walter Duranty, "Clowns Give Way to Russian Cynic," *New York Times*, September 25, 1925.

10. V. V. Maiakovskii, "Beseda s sotrudnikom gazety 'Zaria Vostoka': literaturnaia i kul'turnaia zhizn' Ameriki," in idem, *Polnoe sobranie sochinenii v 13 tt.* (Moscow: Khudozhestvennaia literatura, 1961), vol. 13: 229.

11. See, for example, A. C. Wright, "Animals and animal imagery in M. A. Bulgakov," *Zeitschrift für Slawistik*, 1991, 36 (2): 220–28; Edythe C. Haber, "The Social and Political Context of Bulgakov's 'The Fatal Eggs,'" *Slavic Review*, 1992, 51 (3): 497–510; L. M. Iur'eva, *Russkaia antiutopiia v kontekste mirovoi literatury* (Moscow: IMLI RAN, 2005), 79–180; and E. A. Ivan'shina, "O vozmozhnykh kontekstakh povesti M. A. Bulgakova 'Rokovye iaitsa,'" in *Tvorchestvo M. A. Bulgakova v mirovom kul'turnom kontekste* (Moscow-Iaroslavl': n. p., 2010), 25–35.

12. See, for example, Sigrid McLaughlin, "Structure and Meaning in Bulgakov's *The Fatal Eggs*," *Russian Literature Triquarterly*, 1978, 15: 263–79; Christine Rydel, "Bulgakov and H. G. Wells," *ibid.*, 1978, 15: 293–311; Vera Shamina, "H. G. Wells and Mikhail Bulgakov: Aspects of reception in the Island of Dr. Moreau, 'Rock's Eggs,' and the 'Dog's Heart,'" in Elmar Schenkel and Stefan Welz, eds., *Lost Worlds & Mad Elephants: Literature, Science and Technology, 1700–1990* (Berlin and Cambridge, MA: Galda und Wilch, 1999), 201–10; E. A. Iablokov, *Khudozhestvennyi mir Mikhaila Bulgakova* (Moscow: Iazyk slavianskoi kul'tury, 2001); idem, *Mikhail Bulgakov i mirovaia kul'tura: Spravochnik-tezaurus* (St. Petersburg: Dmitrii Bulanin, 2011), and many others.

13. If we accept the widely held opinion that Persikov is a stand-in for Lenin, then it seems likely that his surname Persikov (*persik* in Russian means "peach") was inspired by, and thus alludes to, the name of Aleksei Abrikosov (*abrikos* in Russian means "apricot"), a Moscow University professor of pathology who had conducted the autopsy of Lenin's corpse and whose signature appeared on the autopsy report published in every Soviet newspaper after Lenin's death in January 1924.

14. Bulgakov is practically unique in this respect. Neither contemporary observers nor later historians of early Soviet science have paid much attention to the role of the popular media—daily newspapers, weekly magazines, radio, and documentary films—in mediating the interactions between science and scientists, on one hand, and their state patrons and the general public, on the other. For more on this subject, see below.

15. For a highly readable account of the Russian Civil War, see W. Bruce Lincoln, *Red Victory: A History of Russian Civil War, 1918–1921* (New York: Da Capo Press, 1989). For a general discussion of the social, economic, cultural, and political aspects of the Civil War, see D. P. Koenker, W. G. Rosenberg, and R. G. Suny, eds., *Party, State, and Society in the Russian Civil War* (Bloomington: Indiana University Press, 1989).

16. For a general discussion of the social, economic, cultural, and political aspects of NEP, see Sheila Fitzpatrick, AlexanderRabinowitch, and Richard Stites, eds., *Russia in the Era of NEP* (Bloomington: Indiana University Press, 1991), and Lewis H. Siegelbaum, *Soviet State and Society between Revolutions, 1918–1929* (Cambridge: Cambridge University Press, 1992).

17. For details on the pre-revolutionary literacy and science-popularization campaigns, see Jeffrey Brooks, *When Russia Learned to Read: Literacy and Popular Literature, 1861–1917* (Evanston, IL: Northwestern University Press, 2003); for post-revolutionary campaigns, see James T. Andrews, *Science for the Masses: The Bolshevik State, Public Science and the Popular Imagination in Soviet Russia, 1917–1934* (College Station: Texas A&M University Press, 2003).

18. Bulgakov had spent the civil war years in the south of Russia—the stronghold of the White Armies—and came to Moscow only in the fall of 1921. Thus, he perhaps did not know of the

Bolsheviks' early efforts to preserve the nation's "scientific capital." In December 1919, while the Civil War was still raging throughout the country, the highest governmental body, the Council of People's Commissars (SNK), hoping to prevent Russian scientists from emigrating, issued a special decree, "On Improvement of Conditions for Scientific Specialists." The decree created a special commission "to improve scientists' living conditions" (KUBU), which provided scientists with food and fuel to survive the harsh conditions of War Communism. In January 1920, the Petrograd branch of KUBU (headed by Maxim Gorky) opened a House of Scientists in the palace of the Great Duke Vladimir Alexandrovich, with a canteen and a hostel for scientists. H. G. Wells visited the House of Scientists in 1920 and described it in his book, *Russia in the Shadows* (New York: George H. Doran Co., 1921), 48–51. For details on the functions and actual work of the Petrograd branch during the first three years of its existence, see *Kratkii obzor organizatsii PetroKUBU i ee deiatel'nosti za trekhletnii period* (Petrograd: Petropechat', 1923). The 1921 decree extended the purview of the commission created in 1919 and renamed it the Central Commission to Improve Scientists' Living Conditions (TsKUBU). On the early activities of TsKUBU, see *Shest' mesiatsev raboty Tsentral'noi komissii po uluchsheniiu byta uchenykh pri Sovete narodnykh komissarov* (Moscow: Mospechat', 1922).

19. See, for instance, Kendall E. Bailes, *Science and Russian Culture in an Age of Revolutions: V. I. Vernadskii and His Scientific School, 1863–1945* (Bloomington: Indiana University Press, 1990); Daniel P. Todes, "Pavlov and the Bolsheviks," *History and Philosophy of the Life Sciences*, 1995, 17: 379–418; and Vera Tolts, *Russian Academicians and the Revolution: Combining Professionalism and Politics* (New York: St. Martin's Press, 1997).

20. For details on the campaign for the popularization of science, see Andrews, *Science for the Masses*. On the history of the Soviet mass media, see T. M. Goriaeva, *Radio Rossii. Politicheskii kontrol' sovetskogo radioveshchaniia v 1920–1930-kh godakh* (Moscow: ROSPEN, 2000); and Kh. Giunter and S. Khengsen, eds., *Sovetskaia vlast' i media* (St. Petersburg: Akademicheskii proekt, 2006). Unfortunately, media historians have paid practically no attention to the role of the media in the popularization of science in Russia.

21. See Jeffrey Brooks, "The Breakdown in Production and Distribution of Printed Material, 1917–1927," in Abbott Gleason, Peter Kenez, and Richard Stites, eds., *Bolshevik Culture: Experiment and Order in the Russian Revolution* (Bloomington: Indiana University Press, 1985), 151–73.

22. See G. Bosse, "Populiarnaia biologicheskaia literatura posle oktiabria," *Knigonosha*, 1924, 26: 2; idem, ed. *Obzor nauchno-populiarnoi literatury po biologii* (Moscow: Timiriazev Institut, 1924); M. Gremiatskii and G. Bosse, eds., *Obzor nauchno-populiarnoi literatury po biologii* (Moscow: Izd. Kom. Un-ta Sverdlova, 1926). Given the fact that Russia's printing industry stood at a virtual standstill during 1918–1922, these figures are indeed very impressive and telling.

23. See F. Davydov, "Biologicheskie luchi (genial'noe otkrytie russkikh uchenykh)," *Izvestiia*, May 24, 1924, 7.

24. For Gurvich's original article, see A. Gurwitsch, "Die Natur des spezifischen Erregers der Zellteilung," *Archiv für Entwicklungsmechanik der Organismen*, 1923, 100 (1–2): 11–40. For a biography of Gurvich, see L. V. Belousov, A. A. Gurvich, S. Ia. Zalkind, and N. N. Kannegisser, *Aleksandr Gavrilovich Gurvich, 1874–1954* (Moscow: Nauka, 1970).

25. For Kravkov's original article, see N. Kravkow, "Über die Grenzen der Empfindlichkeitdes des lebend Protoplasm," *Zeitschrift für die gesamte experimentelle Medizin*, 1923, 34: 279–306; the next year the same article also appeared in Russian: N. P. Kravkov, "O predelakh chuvstvitel'nosti

zhivoi protoplazmy," *Uspekhi eksperimental'noi biologii*, 1924, 3 (3–4): 147–72. For a brief biography of Kravkov, see A. N. Kuznetsov, *N. P. Kravkov* (Moscow: Medgiz, 1948).

26. The worldwide interest in the "rays of death" was spurred by media reports on the works of the British engineer Harry Grindell-Matthew. For Russian reports on his works, see, for instance, "Neuzheli ne illiuziia," *KP*, 1924, 11: 17; Ia. G., "D'iavol'skie luchi," *ibid*, 1924, 18: 14–15; and Envish, "Diavol'skie luchi," *Krasnyi zhurnal*, 1924, 8: 605–09.

27. See, for instance, N. Karpov, "Luchi smerti," in *Biblioteka revolutsionnykh prikliiuchenii*, 1–3 (Moscow: Rabochaia Moskva, 1924); and idem, *Luchi smerti* (Moscow: Zemlia i fabrika, 1925). Also in 1924, there appeared a Russian translation of a lengthy novel, entitled "Death rays," by the German SF writer Hans Dominik; see Gans Dominik, *Luchi smerti* (Khar'kov: Kosmos, 1924).

28. A large segment of the film is available online at http://io.ua/vb3c4e4afe4433acb8bc0fb4b50789b0f, viewed on September 11, 2010.

29. For more on "The Dog's Heart," see Chapter 5.

30. See L. L. Vasil'ev, "Biologicheskie luchi," *Vestnik znaniia* (hereafter—VZ), 1925, 19–20: 1185–94. Furthermore, just one month after it had finished the publication of Bulgakov's novella, *Red Panorama* also carried a similar account of Gurvich's work under the title "The Rays of Life." See P. Iu. Shmidt, "Luchi zhizni. Otkrytie prof. A. G. Gurvicha," *KP*, 1925, 29: reverse of the cover page.

31. In the United States, this overlapping and interconnectedness of medical and biological research spurred by the experimental revolution was clearly manifest in the establishment of the Society of Experimental Biology and Medicine in 1903 and of the Federation of American Societies for Experimental Biology in 1912. On the early history of the new field, see Elizabeth Gaskins, *The Rise of Experimental Biology* (New York: Random House, 1970); and Garland E. Allen, *Life Science in the Twentieth Century* (New York: Wiley, 1975).

32. Of course, in certain areas, particularly physiology, this revolution had begun much earlier, in the middle of the nineteenth century, with the introduction of laboratory as a site of medical research. See Andrew Cunningham and Perry Williams, eds., *The Laboratory Revolution in Medicine* (Cambridge: Cambridge University Press, 1992).

33. This is readily evident in the title of a famous work by one of the leading proponents of the new biology in the United States, Raymond Pearl, *The Biology of Death* (Philadelphia and London: J. B. Lippincott Co., 1922). Philip J. Pauly has explored the general development of these ideas in the US context in his excellent monograph, *Controlling Life: Jacques Loeb and the Engineering Ideal in Biology* (New York: Oxford University Press, 1987), while Jon Turney has analyzed public responses to these ideas in the US and British contexts: "Life in the Laboratory: Public Responses to Experimental Biology," *Public Understanding of Science*, 1995, 4: 153–76; and idem, *Frankenstein's Footsteps: Science, Genetics and Popular Culture* (New Haven: Yale University Press, 1998).

34. See his groundbreaking analysis, Mark B. Adams, "Last Judgment: The Visionary Biology of J. B. S. Haldane," *Journal of the History of Biology*, 2000, 33: 457–91.

35. See, for instance, Mark B. Adams, "The Quest for Immortality: Visions and Presentiments in Science and Literature," in Post and Binstock, eds., *The Fountain of Youth*, 38–71.

36. For a history of this institution, see a jubilee volume issued for its centennial anniversary, N. P. Bekhtereva, ed., *Pervyi v Rossii issledovatel'skii tsentr v oblasti biologii i meditsiny* (Leningrad: Nauka, 1990).

37. For an analysis of the intertwining of the Bolshevik Revolution and the "big science" revolution, see Krementsov, "Big Revolution, Little Revolution."

38. For a vivid depiction of one particular instance of this process in Russia, see Daniel P. Todes, *Pavlov's Physiology Factory* (Baltimore: Johns Hopkins University Press, 2001).

39. On "big science" and its development in Western contexts, see Derek J. de Solla Price, *Little Science, Big Science* (New York: Columbia University Press, 1963); Peter Galison and Bruce Hevly, eds., *Big Science: The Growth of Large-scale Research* (Stanford, CA: Stanford University Press, 1992), especially the editors' introduction: Peter Galison, "The Many Faces of the Big Science," in *ibid*, 1–17; and afterword: Bruce Hevly, "Reflections on Big Science and Big History," in *ibid*, 355–63. What Western historians of science usually describe as "big science" did not emerge overnight or in one single place. As we shall see, the development of "big science" in Soviet Russia differed substantially from that in the West. In fact, many characteristics of what came to be seen as "big science" in the late 1940s and afterward—such as heavy government involvement in science funding and agenda setting, rapid institutional growth, the sheer size of scientific institutions and the ever growing number of people involved in scientific research and development, the transformation of science into a mass profession and the intricate division of labor within scientific research establishments, as well as their close ties to industry and the military—had begun to develop rapidly in Soviet Russia during the 1920s and 1930s. For some observations on the particularities of "big science Soviet-style," see Mark B. Adams, "Soviet Union: The Russian Research System," in *The Academic Research Enterprise within the Industrialized Nations: Comparative Perspectives* (Washington, DC: National Academy Press, 1990), 51–65.

40. In addition, such new disciplines as developmental mechanics (experimental embryology), dietetics, eugenics, experimental cytology and histology, experimental psycho-physiology, genetics, experimental veterinary science, immunology, the physiology of "higher nervous processes," "psycho-technology," social hygiene, and zoopsychology found institutional niches during the first post-revolutionary decade in Russia.

41. See, for example, the report on the 1921 annual meeting of the Federation of American Societies for Experimental Biology by the federation's executive secretary Chas. W. Greene in *Science*, 1922, 55: 379–80.

42. On the role of private foundations in the development of experimental biology and medicine in the United States, see Robert E. Kohler, *Partners in Science: Foundations and Natural Scientists, 1900–1945* (Chicago: University of Chicago Press, 1991).

43. On the Bolsheviks' technocratic and pragmatic attitudes toward science, see Kendall E. Bailes, *Technology and Society under Lenin and Stalin: Origins of the Soviet Technical Intelligentsia, 1917–1941* (Princeton: Princeton University Press, 1978); and Krementsov, *Stalinist Science*. For a detailed analysis of Bolshevik technocratic aspirations and practices more generally, see Don K. Rowney, *Transition to Technocracy: The Structural Origins of the Soviet Administrative State* (Ithaca and London: Cornell University Press, 1989).

44. The phrase comes from Tikhonov's popular "Ballad about nails" written in 1922. It goes like this: "if nails were made from these individuals, there would be no stronger nails in the world." Joseph Dzhugashvili's penname, Stalin, actually means "made of steel."

45. For an analysis of the role of these concerns in the development of Soviet "blood science," see Krementsov, *A Martian Stranded on Earth*.

46. For different estimates of this catastrophe, see Stanislas Kohn and Alexander F. Meyendorff, *The Cost of the War to Russia* (New Haven: Yale University Press, 1932); Frank Lorimer, *The Population of the Soviet Union: History and Prospects* (Geneva: League of Nations, 1946); V. A. Isupov, *Demograficheskie katastrofy i krizisy v Rossii v pervoi polovine XX veka* (Novosibirsk:

n.p, 2000); Iu. A. Poliakov and V. B. Zhiromskaia, eds., *Naselenie Rossii v XX veke. 1900–1939* (Moscow: Rosspen, 2000); A. G. Vishnevskii, ed., *Demograficheskaia modernizatsiia Rossii, 1900–2000* (Moscow: Novoe izdatel'stvo, 2006).

47. On forced migrations, see Nick Baron and Peter Gatrell, eds., *Homelands: War, Population and Statehood in Eastern Europe and Russia 1918–1924* (London: Anthem Press, 2004); and Peter Gatrell, *A Whole Empire Walking: Refugees in Russia during World War I* (Bloomington: Indiana University Press, 2005).

48. One should add considerable emigration to the human losses sustained as a result of the Civil War. Again, the available estimates are rather crude, and the numbers vary from 1 to 2 million people. See Petr E. Kovalevskii, *Zarubezhnaia Rossiia* (Paris: Librairie des Cinq Continents, 1971); Robert C. Williams, *Culture in Exile: Russian Émigrés in Germany, 1881–1941* (Ithaca: Cornell University Press, 1972), 126–35; James E. Hassell, *Russian Refugees in France and the United States Between the World Wars* (Philadelphia: American Philosophical Society, 1991); E. P. Chelyshev and D. M. Shakhovskoi, eds., *Kul'turnoe nasledie russkoi emigratsii, 1917–1940* (Moscow: Nasledie, 1994), vol. 1; and M. Iu. Sorokina, ed., *Rossiiskoe nauchnoe zarubezh'e: Materialy dlia biobibliographicheskogo slovaria* (Moscow: Dom russkogo zarubezh'ia im. A. Solzhenitsina, 2010).

49. For contemporary analyses of the demographic consequences of wars and revolutions, see P. A. Sorokin, "Vliianie voiny na sostav naselenia, ego svoistva i obshchestvennuiu organizatsiiu," *Ekonomist*, 1922, 1: 77–107; idem, *Sovremnnoe sostoianie Rossii* (Praha, 1922); and M. M. Gran, P. I. Kurkina, and P. A. Kuvshinova, eds., *Trudy komissii po obsledovaniiu sanitarnykh posledstvii voiny 1914–1920* (Moscow: GIZ, 1923), vol. 1.

50. V. Khodasevich, "Koleblemyi trenozhnik," in idem, *Koleblemyi trenozhnik. Izbrannoe* (Moscow: Khudozhestvennaia literatura, 1991), 203.

51. It would be interesting to compare the impact of the "decade of death" on Russian science and culture more generally with that in other European countries that also sustained large population losses during this period, such as France, Germany, or Belgium.

52. See Richard Stites, *Revolutionary Dreams: Utopian Vision and Experimental Life in the Russian Revolution* (Oxford: Oxford University Press, 1989). Unfortunately, this brilliant analysis of revolutionary visions and experiments by artists, poets, musicians, architects, writers, and social planners does not address the visions and works of the ultimate experimentalists—Russian scientists!

53. For a fascinating examination of these visions through the medium of Soviet cinema, see Emma Widdis, *Visions of a New Land: Soviet Film from the Revolution to the Second World War* (New Haven: Yale University Press, 2003).

54. A. Sviatogor, P. Ivanitskii, V. Zikeev, and E. Grozin, "Deklaratsiia Kreatoriia rossiiskikh i moskovskikh anarkhistov-biokosmistov," *Izvestiia*, January 4, 1922; 3; and *Biokosmist*, 1922, 1: 1.

55. Several examples of biocosmists' writings are available in German translation in Boris Groys and Michael Hagemeister, eds., *Die Neue Menschheit: Biopolitische Utopien in Russland zur Beginn des 20. Jahrhunderts*. Translated from the Russian by Dagmar Kassek (Frankfurt: Surhkamp Verlag, 2005).

56. See *Bessmertie. Organ severnoi gruppy biokosmistov-immortalistov* (Petrograd, 1922).

57. Aleksandr Sviatogor, "Nashi utverzhdeniia," *Biokosmist*, 1922, 1: 3–4.

58. On Fedorov and his *Philosophy of the Common Cause*, see Michael Hagemeister, *Nikolaj Fedorov. Studien zu Leben, Werk und Wirkung* (München: Sagner, 1989); George M. Young, *The*

Russian Cosmists: The Esoteric Futurism of Nikolai Fedorov and His Followers (New York: Oxford University Press, 2012); and *N. F. Fedorov: Pro and Contra* (St. Petersburg: RKhGI, 2004), vol. 1–2. For an English translation of Fedorov's selected writings, see Nikolai F. Fedorov, *What Was Man Created For?* (Hyperion Books, 1990).

59. See Michael Hagemeister, "Russian Cosmism in the 1920s and Today," in Bernice Glatzer Rosental, ed., *The Occult in Russian and Soviet Culture* (Ithaca, London: Cornell University Press, 1997), 185–202; I. V. Vishev, *Problema bessmertiia cheloveka v russkoi filosofii: Personalii i idei* (Cheliabinsk: Iuzhno-Ural'skii Universitet, 1999), vol. 2, especially, chapter 6, "Immortalizm i interplanetarizm," 118–35; and Asif A. Siddiqi, "Imagining the Cosmos: Utopians, Mystics, and the Popular Culture of Spaceflight in Revolutionary Russia," *Osiris*, 2008, 23: 260–88. These scholars, however, have completely neglected the "bio" component of the biocosmists' programs, ignoring the questions of what the root "bio" in biocosmism signified, where it came from, and why.

60. See Aleksandr Sviatogor, "'Doktrina ottsov' i anarkhizm-biokosmizm," *Biokosmist*, 1922, 3–4: 3–22, cit. on 18–19, fn. 1

61. See A. Sviatogor, "Biokosmicheskaia poetika," in Aleksandr Sviatogor and Pavel Ivanitskii, *Biokosmizm* (Moscow: Kreatorii biokosmistov, 1921), 3–11, here–4.

62. N. F. Fedorov, *Filosofiia obshchego dela. Stat'i, mysli i pis'ma Nikolaia Fedorovicha Fedorova, izdannye pod redaktsiei V. A. Kozhevnikova i N. P. Petersona*. Reprint edition. (Gregg International Publishers Limited England, 1970), Vol. 1, 289, italics are added. In fact, when Fedorov did mention "science" and "technology" as instruments that could be deployed in the quest for immortality and resurrection, he referred not to experimental biology and medicine, but to a Haeckelian monism, which for Fedorov was the latest advance in biological knowledge.

63. See Prof. N. P. Kravkov, "Dannye i perspektivy po ozhivleniiu tkanei umershikh," *Bessmertie*, 1922, 1: 2–3. The article had originally appeared just a few months earlier in a scientific publication; see N. P. Kravkov, "Dannye i perspektivy po ozhivleniiu tkanei umershikh," *Sbornik nauchnykh trudov v chest' 50-letiia nauchno-vrachebnoi deiatel'nosti glavnogo vracha Obukhovskoi bol'nitsy A. A. Nechaeva* (Petrograd: n. p., 1922), 5–7.

64. Aleksandr Iaroslavskii, "Professoru Kravkovu," *Bessmertie*, 1922, 1: 6.

65. See the advertisement for forthcoming publications on the back cover of a collection of writings by various members of the group, *Biokosmisty. Desiat' shtuk* (Petrograd: Komitet poezii biokosmistov-immortalistov, 1923). For more on Steinach and rejuvenation, see Chapter 5.

66. Aleksandr Iaroslavskii, *Poema anabioza* (Petrograd: Komitet poezii biokosmistov-immortalistov, 1922). For more on the poem and its author, see Chapter 3.

67. See P. Ivanitskii, "Proletarskaia etika," in Aleksandr Sviatogor and Pavel Ivanitskii, *Biokosmizm* (Moscow: Kreatorii biokosmistov, 1921), 12–14, here–13.

68. On the regime's antireligious stance, see Daniel Peris, *Storming the Heavens: the Soviet League of Militant Godless* (Ithaca: Cornell University Press, 1998); William B. Husband, *"Godless Communists." Atheism and Society in Soviet Russia, 1917–1932* (De Kalb: Northern Illinois University Press, 2000); and Paul Froese, *The Plot to Kill God: Findings from the Soviet Experiment in Secularization* (Berkeley: University of California Press, 2008).

69. On the history of this institution and its journal, see Michael David-Fox, *Revolution of the Mind: Higher Learning among the Bolsheviks, 1918–1929* (Ithaca, NY: Cornell University Press, 1998).

70. "Ot redaktsii," *Pod znamenem marksizma* (hereafter—*PZM*), 1922, 1–2: 3–4.

71. V. I. Lenin, "O znachenii voinstvuiushchego materializma," *PZM*, 1922, 3: 29.

72. On the expulsion, popularly known as "a philosophers' steamboat," see M. E. Glavatskii, *"Filosofskii parokhod." God 1922: Istoriograficheskie etiudy* (Ekaterinburg: Ural'skii universitet, 2002); for a large collection of previously inaccessible documents from the secret police archives related to the expulsion, see V. G. Makarov and V. S. Khristoforov, eds., *Vysylka vmesto rasstrela: Deportatsiia intelligentsii v dokumentakh VChK-OGPU, 1921–1923* (Moscow: Russkii put', 2005); for an English-language study, see Stuart Finkel, *On the Ideological Front* (New Haven: Yale University Press, 2007).

73. See, for instance, a brochure on "Immortality and Life," written by Vasilii Sokol'skii, a well-known theologian, which went through five printings in as many years; V. P. Sokol'skii, *Bessmertie i zhizn'* (Kazan': Kazanskii universitet, 1909), or a voluminous monograph on "Individualism and immortality," by the editor of a well-known journal, *Theosophical Life*, N. I. Ryshkovskii, *Individualizm i bessmertie* (Smolensk: n. p., 1910), or numerous works by Sergei Glagolev, a prolific professor of the Moscow Theological Academy—for instance, S. S. Glagolev, *Bessmertie proshedshego* (Sergiev Posad: n.p., 1903); and idem, *Estestvenno-nauchnye voprosy v ikh otnoshenii k khristianskomu miroponimaniiu* (Sergiev Posad: M. S. Elov, 1914).

74. P. I. Bakhmet'ev, "Sedalishche dushi," *Nauchnoe obozrenie*, 1902, 9: 1–6; 10: 21–39. For more on Bakhmet'ev and his work, see Chapter 3.

75. Georgii Bosse, *Telo khoziain dushi* (Moscow: Proletarii, 1925).

76. See, for example, V. M. Bekhterev, *Bezsmertie chelovecheskoi lichnosti kak nauchnaia problema* (Petrograd: Vestnik znaniia, 1918); and D. Pletnev, "Smert' kak estestvennyi iskhod zhizni," *Vrachebnoe delo*, 1920, 12–20: 523–26.

77. G. V. Shor, *O smerti cheloveka. Vvedenie v tanatologiiu* (Leningrad: Kubuch, 1925). For Metchnikoff's original formulation of thanatology, see E. Metchnikoff, "La mort du papillon du mûrier, un chapitre de thanatologie," *Annales de l'Institut Pasteur*, 1915, 29 (10): 477–96.

78. See, for instance, A. S. Dogel', *Starost' i smert'* (Petrograd: Mysl', 1922); A. V. Nagornyi, *Zhizn', starost' i smert'* (Odessa: GIZ Ukrainy, 1923); N. Belousov, *Starost' i smert'* (Khar'kov: Glavpolitizdat, 1923); S. P. Grigorovich, *Zhizn' i smert' v prirode* (Kharkov: Put' prosveshcheniia, 1923); A. V. Nemilov, *Chto takoe smert'? Smert' s tochki zreniia estestvennykh nauk* (Moscow-Petrograd: GIZ, 1923); idem, *Zhizn' i smert'* (Leningrad: GIZ, 1924); E. S. London and I. I. Kryzhanovskii, *Bor'ba za dolgovechnost'* (Petrograd: Put' k znaniiu, 1924); P. Iu. Shmidt, *Bor'ba so starost'iu* (Leningrad: GIZ, 1924); M. S. Mil'man, *Uchenie o roste, starosti i smerti* (Baku: n.p., 1926); I. I. Shmal'gauzen, *Problema smerti i bessmertiia* (Moscow-Leningrad: GIZ, 1926); K. Kh. Kekcheev, *O zhizni, starosti i smerti* (Moscow: Krest'ianskaia gazeta, 1926); M. A. Gremiatskii, *Smert' i ozhivlenie* (Moscow-Leningrad: GIZ, 1926); A. A. Bogomolets, *Zagadka smerti* (Moscow: Narkomzdrav, 1927); N. Vvedenskii, *Zhizn' organov vne tela* (Moscow-Leningrad: Moskovskii rabochii, 1928), and many others.

79. M. Gremiatskii, "Novosti nauchno-populiarnoi literatury o zhizni i smerti," *Knigonosha*, 1925, 20: 20–21.

80. P. Chek-vich, "Zhizn' posle smerti," *Novaia vecherniaia gazeta*, 1925, May 25, 1925, 6; "Chudesa nauki. Smert' i voskresenie na operatsionnom stole," *Vecherniaia Moskva* (hereafter—*VM*), January 24, 1925, 2; Nabliudatel', "Nauka vyshe boga," *ibid*, June 20, 1925, 2; S. B-de, "Nauka, pobezhdaiushchaia smert'," *ibid*, February 11, 1928, 2; "Mertvye ozhivaiut," *ibid*, August 15, 1928, 3.

81. Three recent monographs on the Bolshevik "militant godless" (Peris, *Storming the Heavens*; Husband, *Godless Communists*; and Froese, *The Plot to Kill God*) pay scant attention to the role of science and scientists (to say nothing of their explorations of issues of life and death) in the

formation of the atheistic worldview in Soviet Russia. Recently, Michael David-Fox has argued forcefully for the need to include science in any discussion of the "Soviet faith"; see Michael David-Fox, "Religion, Science, and Political Religion in the Soviet Context," *Modern Intellectual History*, 2011, 8 (2): 471–84.

82. B. Andreev, "O populiarizatsii estestvoznaniia," *Knigonosha*, 1923, 29: 12–13, here–13, italics are in the original.

83. See "Materializm ili idealizm. Chto takoe 'dusha'? Disput v Politekhnicheskom muzee," *VM*, October 18, 1924, 2.

84. Nowadays we are used to seeing special sections on science and medicine (or health) in the media, and it is difficult to imagine that this was not always so. But until the 1890s, science and medicine only very rarely appeared on the pages of daily newspapers or weekly magazines. The rapid and dramatic increase of science's public visibility during the decades around the turn of the twentieth century undoubtedly represents a revolution in the societal position and cultural importance of science and scientists in industrialized nations.

85. Literary scholars have regularly examined the recurrent motif of abolishing death in nineteenth- and twentieth-century Russian literature. Alas, they have paid almost no attention to the concurrent scientific developments that aimed—literally—at abolishing death and that provided underpinnings for much of this literature. See, for example, Irene Masing-Delic, *Abolishing Death: a Salvation Myth of Russian Twentieth-Century Literature* (Stanford, CA: Stanford University Press, 1992); Wolfgang Kasack, *Der Tod in der russischen Literatur: Aufsätze und Materialien aus dem Nachlass*, herausgegeben von Frank Göbler (München: Sagner, 2005); and Michael Hagemeister, "'Unser Körper muß unser Werk sein.' Beherrschung der Natur und Überwindung des Todes in russischen Projekten des frühen 20. Jahrhunderts," in Hagemeister and Groys, eds., *Die Neue Menschheit*, 19–67.

86. On the phenomenon of Russian "thick" journals, see Robert A. Maguire, *Red Virgin Soil* (Princeton: Princeton University Press, 1968), 36–66.

87. K. Sluchevskii, "Professor bessmertiia," *Russkii vestnik*, 1891, 4: 295–356. All of the subsequent quotations from the novella are from this edition, with page numbers given in brackets. The following year, the novella was reissued in book format, K. Sluchevskii, *Professor bessmertiia* (St. Petersburg: Obshchestvennaia pol'za, 1892).

88. Although I did not find a 1920s work of fiction bearing exactly this title, both "professor" and "immortality" were a staple of 1920s Russian science fiction.

89. For a voluminous study of this association, see Michael Ruse, *Monad to Man: The Concept of Progress in Evolutionary Biology* (Cambridge, MA: Harvard University Press, 1996).

90. Both Spencer's and Haeckel's works were available in numerous Russian translations and were highly popular among the Russian intelligentsia. On the popularity of Spencer in Russia, see Alexander Vucinich, *Social Thought in Tsarist Russia. The Quest for a General Science of Society* (Chicago: University of Chicago Press, 1976); and Shoshana Knopp, "Herbert Spencer in Čexov's 'Skučnaja istorija' and 'Duél': The Love of Science and the Science of Love," *Slavic and East European Journal*, 1985, 29 (3): 279–96. Unfortunately, there are no similar historical works examining the popularity of Haeckel in Russia. But, as witnessed by a special 1914 issue of the *Herald of Knowledge* dedicated to Haeckel's eightieth birthday, the contemporary Russian scientific community acknowledged and appreciated his influence, see *VZ*, 1914, 3.

91. The adjective *psychical* in modern English has a common meaning of something relating to ghosts or the spirits of the dead, but in this case it means the "activity of the psyche."

92. In addition to Darwin, Spencer, Haeckel, Lobachevskii, and Riemann, Sluchevskii also makes his character refer to Hermann Helmholtz's "psycho-physics" and Wilhelm Wundt's "experimental psychology" among the theoretical underpinnings of his own theory.

93. Hereafter I use the abbreviation "SF" to refer to all the permutations of the labels for this new genre, even though it is usually used to mean "science fiction"—the label that would come into wide use only in the 1940s. For the classic account of the literary history of SF, see Darko Suvin, *Metamorphoses of Science Fiction: On the Poetics and History of a Literary Genre* (New Haven: Yale University Press, 1979); for a comparative perspective on Western and Soviet SF, see Bernd Rullkötter, *Die wissenschaftliche Phantastik der Sowjetunion: eine vergleichende Untersuchung der spekulativen Literatur in Ost und West* (Bern: H. Lang, 1974); and John Griffiths, *Three Tomorrows: American, British and Soviet Science Fiction* (London: MacMillan, 1980).

94. For a summary of these debates based heavily on Anglophone SF, see Edward James and Farah Mendleson, eds., *The Cambridge Companion to Science Fiction* (New York: Cambridge University Press, 2003).

95. For a historical analysis of this process in the United States, see Ronald R. Kline, "Constructing 'Technology' as 'Applied Science': Public Rhetoric of Scientists and Engineers in the United States, 1880–1945," *Isis*, 1995, 86: 194–221.

96. For more on the development of the genre, see Chapter 6.

97. On the history of early Soviet science fiction, see R. Nudel'man, "Fantastika, rozhdennaia revoliutsiei," in *Fantastika*, 3 (Moscow: Molodaia gvardiia, 1966), 330–69; A. F. Britikov, *Russkii sovetskii nauchno-fantasticheskii roman* (Leningrad: Nauka, 1970); Darko Suvin, "The Utopian Tradition of Russian Science Fiction," *Modern Language Review*, 1971, 66 (1): 139–59; Patrick L. McGuire, *Red Stars: Political Aspects of Soviet Science Fiction* (Ann Arbor, MI: UMI Research Press, 1985); Vsevolod Revich, "Perekrestok utopii. U istokov sovetskoi fantastiki," in N. Berkova, ed., *Orion* (Moscow: Moskovskii rabochii, 1985), 309–48; Rafail Nudelman, "Soviet Science Fiction and the Ideology of Soviet Society," *Science Fiction Studies*, 1989, 16 (1): 38–66; B. Ionsher, "Utopia, fantastika, nauchnaia fantastika—k poniatiinoi sisteme v russkoi sovetskoi literature 20-ikh gg.," *Zeitschrift für Slawistik*, 1990, 35 (3): 360–65; Anindita Banerjee, "The genesis and evolution of science fiction in fin de siècle Russia, 1880–1921," PhD diss., University of California, Los Angeles, 2000; V. I. Okulov, *O zhurnal'noi fantastike pervoi poloviny XX veka* (Lipetsk: Krot, 2008); and Muireann Maguire, "Soviet Gothic-Fantastic: A Study of Gothic and Supernatural Themes in Early Soviet Literature," PhD diss., University of Cambridge, 2008. This list is far from exhaustive. All of these works, however, look mostly at the "fiction" part of SF, leaving the "science" part out of their analyses.

98. E. Zamaitin, *Gerbert Uells* (Petrograd: Epokha, 1922), 47.

99. See, for instance, S. Zubov, "Syvorotka bessmertiia," *Mir prikliuchenii*, 1923, 4: 12–28; Mark Volosov, "Omolozhenie," *KP*, 1925, 5:10–11; Gr. Iamskii, "Biotransformator," *Mir prikliuchenii*, 1926, 7: 53–66; N. Dolin, "Krov' mira," *ibid*, 1927, 6: 2–12; V. D. Nikol'skii, "Luchi zhizni," *ibid*, 1927, 10: 6–20; E. Fortunato, "Bor'ba so smert'iu," *ibid*, 1928, 8: 13–19.

CHAPTER 2

1. See, for example, Paul H. Stahl, *Histoire de la décapitation* (Paris: Presses universitaires de France, 1986); Geoffrey Abbott, *Severed Heads: British Beheadings through the Ages* (London: Andre Deutsch, 2003).

2. See Regina Jones, *Losing Our Heads: Beheadings in Literature and Culture* (New York: New York University Press, 2005).

3. A. Beliaev, "Golova professora Douelia," *Rabochaia gazeta*, June 16–21, 24–26, and July 2–6, 1925. The name of the main protagonist in Russian is *Douel'*. To avoid unnecessary confusion, I use the transliteration "Dowell" as it appeared in the first English publication of the story; see Alexander Beliaev, *Professor Dowell's Head*, translated by Antonina W. Bouis (New York, London: Macmillan Publishing, 1980).

4. On Beliaev's life and works, see B. V. Liapunov, *Aleksandr Beliaev* (Moscow: Sovetskii pisatel', 1967).

5. See A. R. Beliaev, "Golova professora Douelia," *Vsemirnyi sledopyt*, 1925, 3: 17–27; 4: 24–31. All further citations are from this edition.

6. A. Beliaev, *Golova professora Douelia* (Moscow-Leningrad: Zemlia i Fabrika, 1926).

7. Aleksandr Beliaev, *Golova professora Douelia* (Moscow-Leningrad: Sovetskii pisatel', 1938). *Professor Dowell's Head* became one of the most popular Soviet SF novels and, since its appearance in 1938, various publishers reissued it no fewer than 50 times! It opened Beliaev's eight-volume *Collected Writings* published in 1963–64, and was turned into a movie in 1984. In 1980, Macmillan issued an English translation of the novel with a short but enthusiastic introduction by Theodore Sturgeon; see Beliaev, *Professor Dowell's Head*.

8. Beliaev, "Golova professora Douelia," *Vsemirnyi sledopyt*, 1925, 3: 8.

9. In various Western publications his name has been spelled in several different ways: Brukhonenko, Bryuchonenko, Bruchonenko, Brykhonenko, Brychonenko, and even Brjukhenenko. I will use throughout the book the spelling of his name according to the Library of Congress transliteration rules: Briukhonenko. For a brief biography of Briukhonenko, see G. Sirotkina and V. S. Gutkin, *S. S. Briukhonenko* (Moscow: Meditsina, 1972). For an analysis of his research, see I. E. Konstantinov and V. V. Alexi-Meskishvili, "Sergei S. Brukhonenko: the development of the first heart-lung machine for total body perfusion," *Annals of Thoracic Surgery*, 2000, 69(3): 962–66; and W. Bottcher and V. V. Alexi-Meskishvili, "Sergej Sergejewitch Brychonenko (1890–1960): Pionier des kardiopulmonalen Bypasses in der Sowietunion," *Deutsche Gesellschaft für Kardiotechnik*, 2003, 2. http://www.dgfkt.de/content/hlm/Byrochoenko.pdf.

10. A version of the report is preserved among Briukhonenko's personal papers kept in the Archive of the Russian Academy of Sciences (hereafter—ARAN), *fond* (collection) 1534, *opis'* (inventory) 1, *delo* (file) 2, *listy* (lists) 1–2. Such references will be further given as ARAN, f. 1534, op. 1, d. 2, ll. 1–2.

11. See reports on the congress, "II Vsesoiuznyi s"ezd patologov," *Zhurnal dlia usovershenstvovaniia vrachei*, 1925, 10: 571–74; P. P. Dvizhkov, "Vtoroi Vserossiiskii s"ezd patologov," *Klinicheskaia meditsina*, 1926, 4: 77–79.

12. See S. S. Briukhonenko and S. I. Chechulin, "Opyty po izolirovaniiu golovy sobaki," *Trudy Vtorogo Vsesoiuznogo s"ezda fiziologov* (Leningrad: Glavnauka, 1926), 289–90.

13. Vl. O., "Opyty s ozhivlennoi golovoi sobaki," *VM*, November 9, 1926, 2; all the following citations are from this source.

14. "Dva chasa zhizni bez serdtsa i bez legkikh," *Krasnaia gazeta*, November 10, 1926, 2.

15. G. Grebnev, "Golova, otrezanaia ot tulovishcha, zhivet," *Komsomol'skaia pravda*, May 22, 1927, 4.

16. See "Iz komnatushki v laboratoriiu. Komakademiia pomozhet doktoru Briukhonenko," *Komsomol'skaia pravda*, May 24, 1927, 4.

17. S. S. Briukhonenko and V. V. Zverev, “O deistvii nekotorykh koaguliruiushchikh i stabiliziruiushchikh krov’ veshchestv in vivo,” in *Trudy Tret’ego Vsesoiuznogo s”ezda fiziologov* (Leningrad: Glavnauka, 1928), 62–63; S. S. Briukhonenko and V. A. Briukhonenko, “Koaguliatsiia i stabilizatsiia krovi i belkov,” in *ibid*, 63–64; S. S. Briukhonenko, “Primenenie ‘Bauer 205’ dlia perelivaniia stabilizirovannoi krovi,” in *ibid*, 64–65.

18. S. S. Briukhonenko, “Iskusstvennoe krovoobrashchenie tselogo organizma (sobaki) s vykliuchennym serdtsem (demonstratsiia),” in *Trudy Tret’ego Vsesoiuznogo s”ezda fiziologov* (Leningrad: Glavnauka, 1928), 213–14.

19. S. S. Briukhonenko and S. S. Chechulin, “Opyty na izolirovannoi golove sobaki,” in *Trudy Tret’ego Vsesoiuznogo s”ezda fiziologov* (Leningrad: Glavnauka, 1928), 214–15.

20. See *Trudy nauchnogo khimiko-farmatsevticheskogo instituta*, 1928, 20. Many, but not all, of Briukhonenko’s early publications have been reprinted posthumously in a volume of his collected works; see S. S. Briukhonenko, *Iskusstvennoe krovoobrashchenie* (Moscow: Nauka, 1964).

21. S. S. Briukhonenko and S. I. Chechulin, “Opyty po izolirovaniiu golovy sobaki,” in *Izuchenie novykh metodov iskusstvennogo krovoobrashcheniia i perelivaniia krovi* (Moscow: NTO VSNKh, 1928), 7–43; S. S. Briukhonenko, “Iskusstvennoe krovoobrashchenie tselogo organizma (sobaki) s vykliuchennym serdtsem,” *ibid*, 44–72; idem, “Apparat dlia iskusstvennogo krovoobrashcheniia (teplokrovnykh),” *ibid*, 73–80; O. A. Steppun and S. S. Briukhonenko, “Primenenie ‘Bayer 205’ dlia stabilizatsii perelivaemoi krovi,” *ibid*, 81–95.

22. See S. S. Briukhonenko, “Apparat dlia iskusstvennogo krovoobrashcheniia (teplokrovnykh),” *Zhurnal eksperimental’noi biologii i meditsiny*, 1928, 10(26): 296–306; S. S. Briukhonenko and S. I. Chechulin, “Perezhivanie tsentral’noi nervnoi sistemy golovnogo mozga v usloviiakh iskusstvennogo krovoobrashcheniia,” *ibid*, 307–21; S. Briukhonenko, “Iskusstvennoe krovoobrashchenie organizma,” *ibid*, 1929, 11(30): 21–27.

23. In 1930, Briukhonenko moved his laboratory from the Chemical-Pharmaceutical Institute to the Institute of Blood Transfusion.

24. See N. N. Terebinskii, *Materialy po izucheniiu otkrytogo dostupa k atrioventrikuliarnym klapanam serdtsa* (Moscow-Leningrad: Medgiz, 1940). For a historical account of Terebinskii’s work, see Vladimir V. Alexi-Meskishvili, Evgenij V. Potapov, Erik A. K. Beyer, and Roland Hetzer, “Nikolai Terebinski: A Pioneer of the Open Valve Operation,” *Annals of Thoracic Surgery*, 1998, 66: 1440–43.

25. Briukhonenko kept clippings of articles about his research from various newspapers and magazines. Several large files of more than one hundred pages in toto of such clippings are preserved among Briukhonenko’s personal papers; see ARAN, f. 1534, op. 3, d. 29; and op. 2, d. 15.

26. See “Na vsesoiuznom s”ezde fiziologov,” *VM*, May 31, 1928, 2; “Opyty s golovoi sobaki,” *ibid*, June 1, 1928, 4; “Novosti nauki i tekhniki: Tretii vsesoiuznyi s”ezd fiziologov,” *Khochu vse znat’*, July 1928, 13: 390–91; “III s”ezd fiziologov,” *Iskra*, 1928, 8: 28.

27. See K. L., “Iskusstvennoe krovoobrashchenie s vykliuchennym serdtsem,” *VZ*, 1928, 16: 830; M. P. Vinogradov, “Ozhivlenie otrezannoi golovy,” *ibid*, 1928, 19: 936–38; S. Briukhonenko and S. Chechulin, “Vozmozhna li zhizn’ golovy otrezannoi ot tulovishcha?” *Iskry nauki*, 1928, 5: 161–66; 6: 201–05; A. Kuliabko, “Opyty ozhivleniia serdtsa i golovy i ikh znachenie,” *ibid*, 1928, 7: 241–44; 8: 285–89; S. I. Chechulin, “Problema ozhivleniia golovnogo mozga,” *Nauchnoe slovo*, 1928, 9: 42–56; S. S. Briukhonenko, “Perezhivanie tsentral’noi nervnoi sistemy v iskusstvennykh usloviiakh,” *Prozhektor*, 1928, 41: 10–11; M. Slavina, “Perezhivanie tsentral’noi nervnoi sistemy,”

Gigiena i zdorov'e rabochei i krest'ianskoi sem'i, 1928, 19: 1; S. Chechulin, "O novom metode izoliatsii tsentral'noi nervnoi sistemy," *Iskra*, 1928, 12: 19–21; "Mozhet li zhit' izolirovannaia golova?" *Khochu vse znat'*, 1928, 19: 596–97; P. Iu. Shmidt, "Opyty ozhivleniia golovy," *Chelovek i priroda*, 1929, 1: 14–23.

28. S. Shirokov, "Zhizn' organov vne tela," *Zhenskii zhurnal*, 1930, 4: 49; I. Maksimov, "Ozhivlennaia golova," *Ekran*, 1929, 13: 7.

29. See front covers of *Iskry nauki*, 1928, no. 5; and *Gigiena i zdorov'e rabochei i krest'ianskoi sem'i*, 1928, no. 19.

30. See V. Pleskov, "Na grani zhivogo i mertvogo. Zhizn' golovy bez tulovishcha," *Krasnyi voskhod*, July 26, 1928, 4; "Blestiashchee dostizhenie sovetskoi nauki. Zastavili zhit' otrezannuiu golovu," *Komsomol'skaia pravda*, September 20, 1928, 4; "Nauka, pobezhdaiushchaia smert'," *VM*, September 20, 1928, 3; Il. Poltavskii, "Ulybka budushchego," *VM*, October 9, 1928, 2; and I. I. Slitsan, "Stranitsa iz knigi zhizni i smerti," *Nasha gazeta*, December 9, 1928, 6.

31. In the next few years, Briukhonenko in collaboration with his student V. D. Ianovskii conducted intensive studies of heparin and created an effective technology for the industrial production of heparin in the USSR. For an early history of blood transfusions in Soviet Russia, see Krementsov, *A Martian Stranded on Earth*.

32. Slitsan, "Stranitsa iz knigi zhizni i smerti," 6.

33. Poltavskii, "Ulybka budushchego," *VM*, October 9, 1928, 2.

34. For an engaging account of the long history of this association between the soul and the brain, see Carl Zimmer, *Soul Made Flesh* (New York: Free Press, 2005).

35. The first records of the brain's electrical activity (what later would be called electroencephalography—EEG) were obtained in the nineteenth century. On the early experiments with recording EEG, see Mary A. Brazier, *A History of Neurophysiology in the 19th Century* (New York: Raven Press, 1988); on the Russian contributions, see Chapter 12, 210–41.

36. "Blestiashchee dostizhenie sovetskoi nauki. Zastavili zhit' otrezannuiu golovu," *Komsomol'skaia pravda*, September 20, 1928, 4.

37. See "Zhizn' v iskusstvennykh usloviiakh," *VM*, September 12, 1928, 2.

38. See "Sovetskii den'," *Izvestiia*, September 13, 1928, 4.

39. See "Blestiashchee dostizhenie sovetskoi nauki. Zastavili zhit' otrezannuiu golovu," *Komsomol'skaia pravda*, September 20, 1928, 2; Il. Poltavskii, "Ulybka budushchego," *VM*, October 9, 1928, 2; I. I. Slitsan, "Stranitsa iz knigi zhizni i smerti," *Nasha gazeta*, December 9, 1928, 6. The new funding enabled Briukhonenko to expand his laboratory considerably: he hired nine assistants and technicians, bought equipment and chemicals (some of which were imported from Germany), and renovated the rooms; see financial documentation of Briukhonenko's laboratory in ARAN, f. 1534, op. 3, d. 66, ll. 1–11.

40. For technical reasons, one of the articles had to be divided into two parts. On the details of this publication, see Briukhonenko's correspondence with the journal's editor Eugene Gley in ARAN, f. 1534, op. 3, d. 65, ll. 4, 6, 9–14; for published articles, see S. Briukhonenko, "Appareil pour la circulation artificielle du sang des animaux à sang chaud," *Journal de Physiologie et de Pathologie Générale*, 1929, 27: 12–18; S. Briukhonenko and S. Chechuline, "Expériences avec la tête isolée du chien. I. Technique et conditions des expériences," *ibid*, 31–45; S. Briukhonenko and S. Chechuline, "Expériences avec la tête isolée du chien. II. Résultats des expériences," *ibid*, 64–79; S. Briukhonenko, "Circulation artificielle du sang dans l'organisme entier d'un chien avec cœur exclu," *ibid*, 257–72.

41. S. Briukhonenko, "La surie artificielle du système nerveux central," *Monde*, October 13, 1928, 9.

42. S. Briukhonenko, "Der lebende, Körperlose Hundekopf," *Die Umschau*, January 19, 1929, 33: 50.

43. J. Fernan-Perez, "La survivance du système nerveux central dans des conditions artificielles," *La Presse Médicale*, (February 6) 1929, 37: 183–84.

44. Joseph H. Kraus, "The living head," *Science and Invention*, 1929, (February), 922–23.

45. "Información Rusa—Colaboración Grafica," *El Día Medico* (Buenos Aires), September 2, 1929, 93–94, 100. I am grateful to Mario Rovere for his help in obtaining a copy of this publication.

46. The French original and a Russian translation of the letter are preserved among Briukhonenko's personal papers; see K. Mass to S. Briukhonenko, December 7, 1928, ARAN, f. 1534, op. 3, d. 29, l. 27 (Russian translation); d. 65, l. 22 (French original).

47. "Shaw will sich köpfen lassen, wenn . . . Ein Privatbrief des Dichters über ein neues, abschreckendes Tierexperiment," *Berliner Tageblatt*, März 17, 1929; "Shaw Feels 'Tempted' To Have Head Cut Off," *New York Times*, March 17, 1929, 5; a Russian translation of the letter appeared in May in *Evening Moscow*; see "O golove Bernarda Shou," *VM*, May 25, 1929, 3. The Russian translation was apparently prepared by Briukhonenko himself or someone from his circle: a handwritten text of the Russian translation from the German publication is preserved among Briukhonenko's papers; see ARAN, f. 1534, op. 3, d. 29, ll. 45–46.

48. This and all following quotations are from the *New York Times* publication cited above.

49. For historical explorations of this image, see a special issue of *Studies in History and Philosophy of Biological and Biomedical Sciences*, 2004, 35, especially the editor's introduction, Cathy Gere, "The brain in a vat," *Studies in History and Philosophy of Biological and Biomedical Sciences*, 2004, 35: 219–25; see also, Fernando Vidal, "Ectobrains in the Movies," in William Tronzo, ed., *The Fragment: An Incomplete History* (Los Angeles: Getty Research Institute, 2009), 193–211.

50. J. D. Bernal, *The World, the Flesh and the Devil: An Enquiry into the Future of the Three Enemies of the Rational Soul* (London: K. Paul, Trench, Trubner & co., ltd.; New York: E. P. Dutton & co., 1929), 34.

51. On Bernal and Haldane's relations, see Gary Werskey, *The Visible College* (New York: Holt, 1978).

52. Indeed, some fifteen years later, in 1943, Haldane provided an introduction to a Soviet documentary film about Briukhonenko's work, which was widely distributed in the United States and Britain. A copy of this film is available at http://archive.org/details/Experime1940 .

53. See, for instance, Alexander Kazantsev, "No science without fiction," *The UNESCO Courier*, November, 1984; http://www.questia.com/library/1G1-3499586/no-science-without-fiction#articleDetails.

54. See A. G. Lapchinskii, "S. S. Briukhonenko—osnovopolozhnik iskusstvennogo krovoobrashcheniia," in Briukhonenko, *Iskusstvennoe krovoobrashchenie*, 6–12, here–9.

55. See P. M. Galetti and G. A. Brecher, *Heart-lung bypass. Principles and techniques of extracorporeal circulation* (New York: Grune & Stratton, 1962). For historical accounts of the techniques of perfusion and extracorporeal circulation, see L. E. Curtis, "An early history of extracorporeal circulation," *Journal of Cardiovascular Surgery* (Torino), 1966, 7 (3): 240–47; K. G. Lee, "The history of extracorporeal circulation," in *Essays on the History of Anaesthesia* 1996, 98–104; Wolfgang Boettcher, Frank Merkle, and Heinz-Hermann Weitkemper, "History of Extracorporeal

Circulation," *Journal of Extra-Corporeal Technology*, 2003, 35: 172–91; Richard Jenson, "History of Perfusion," *ibid*, 270–74.

56. See J. J. C. Legallois, *Expériences sur le Principe de la Vie* (Paris: Chez D'Hautel, 1812). An English translation appeared the next year; see J. J. C. LeGallois, *Experiments on the principle of life* (Philadelphia: M. Thomas, 1813). For an account of post-revolutionary French debates involving severed heads, see Ludmila Jordanova, "Medical Mediations: Mind, Body and the Guillotine," *History Workshop Journal*, 1989, 28: 39–52.

57. See Ch. Brown-Séquard, "Recherches expérimentales sur les propriétés physiologique et les usages du sang rouge et du sang noir et de leurs principaux éléments gazeux, l'oxygène et l'acide carbonique," *Journal de la Physiologie l'Homme et des Animaux*, 1858, 1: 95–122, 353–67, 729–35.

58. For historical accounts of research on isolated organs, see Erich Bauereisen, "Carl Ludwig as the Founder of Modern Physiology," *Physiologist*, 1962, 4: 293–99; L. E. Curtis, "An early history of extracorporeal circulation," *Journal of Cardiovascular Surgery*, 1966, 7: 240–47; W. B. Fye, "H. Newell Martin and the isolated heart preparation: the link between the frog and open heart surgery," *Circulation*, 1986, 73: 857–64; Heinz-Gerd Zimmer, "The Isolated Perfused Heart and Its Pioneers," *News in Physiological Sciences*, 1998, 13: 203–10.

59. A. A. Kuliabko, "Opyty nad izolirovannym serdtsem ptits," *Izvestiia Imperatorskoi akademii nauk*, 1901, 15(4): 471–72.

60. A. A. Kuliabko, "Opyty ozhivleniia serdtsa," *Izvestiia Imperatorskoi akademii nauk*, 1902, 16 (4): 175–83.

61. A. A. Kuliabko, "Dal'neishie opyty ozhivleniia serdtsa. Ozhivlenie chelovecheskogo serdtsa," *Izvestiia Imperatorskoi akademii nauk*, 1902, 17(5): 189–210.

62. A. A. Kuliabko, "Primenenie iskusstvennoi tsirkuliatsii na otrezannoi ryb'ei golove," *Zapiski Imperatorskoi akademii nauk*, 1907, 20(7): 1–25.

63. See N. Kravkov, "O razlichnykh fazakh deistviia iadov na izolirovannoe serdtse," *Russkii vrach*, 1911, 9(41): 1565–71.

64. See F. A. Andreev, *O narushenii funktsii serdtsa pod vliianiem gal'vanicheskogo toka. Eksperimental'noe issledovanie* (Moscow: Moscow University, 1910); idem, "Smert' ot elektricheskogo toka," *Prakticheskii vrach*, 1912, 8: 123–27, 9: 146–50; idem, "Opyty vosstanovleniia deiatel'nosti serdtsa, dykhaniia i funktsii tsentral'noi nervnoi sistemy," *Voprosy nauchnoi meditsiny*, 1913, 2: 137–72.

65. See S. S. Briukhonenko and S. I. Chechulin, "Opyty po izolirovaniiu golovy sobaki," *Trudy Vtorogo Vsesoiuznogo s"zeda fiziologov* (Leningrad: Glavnauka, 1926), 289.

66. S. S. Briukhonenko, "Apparat dlia iskusstvennogo krovoobrashcheniia (teplokrovnykh)," in *Izuchenie novykh metodov iskusstvennogo krovoobrashcheniia i perelivaniia krovi* (Moscow: NTO VSNKh, 1928), 73. This long lineage notwithstanding, the worldwide (and particularly Soviet) excitement over Briukhonenko's works stemmed from the fact that he was the first to perform his experiments on the heads of dogs—mammals, which are much closer to humans on the evolutionary scale than cold-blooded fish or frogs utilized in the earlier research—and that in his experiments severed heads "survived" for long periods of time.

67. See Maurice Renard, *Le docteur Lerne, sous-Dieu* (Paris: Editions G. Crès, 1919); on Renard's works see, Arthur B. Evans, "The fantastic fiction of Maurice Renard," *Science Fiction Studies*, 1994, vol. 21 (3); at http://www.depauw.edu/sfs/backissues/64/evans.htm. On the contemporary French fascination with isolated heads and brains, see J. Tresch, "In a solitary place:

Raymond Roussel's brain and the French cult of unreason," *Studies in History and Philosophy of Biological and Biomedical Sciences*, 2004, 35: 307–32.

68. See, for example, Moris Renan [Maurice Renard], *Doktor Lern. Polubog* (St. Petersburg: M. G. Kornfeld, 1912); and idem, *Novyi zver' (Doktor Lern)* (Moscow-Petrograd: Puchina, 1923).

69. See Carl Grunert, "Mr. Vivacius Style," in idem, *Der Marsspion und andere Novellen. Nachdruck der 1908 erschienenen Erstausgabe im Neusatz* (Auflage: Nachdr. d. Erstausg, 1908), reprint (Norderstedt: Books on Demand GmbH, 2005), 107–18. On Grunert's life and works, see Detlef von Münch, *Carl Grunert (1865–1918), der Pionier der deutschen Kurzgeschichten Science Fiction* (Dortmund: Synergen, 2006).

70. See Evg. Brandis, "[Predislovie]," *Iskatel'*, 1978, 1: 152; for a recent Russian republication of the story, see Karl Grunnert, "Golova mistera Staila," *ibid*, 1978, 1: 151–59.

71. See, for instance, Karl Grunert, *Gibel' zemli. Shpion s Marsa* (St. Petersburg: Khronos, 1911).

72. Unfortunately, Beliaev's personal papers perished during the Nazi occupation of the city of Pushkin where he had resided during his last years. A collection of his papers kept in the St. Petersburg Central State Archive of Literature and Arts (hereafter TsGALI SPb, f. 215, op. 1, dd. 1–9) consists of only nine thin folders with almost no original documents, and thus it is impossible to study drafts, research notes, or any other materials relevant to the creation of "Professor Dowell's Head."

73. A. Beliaev, "O moikh rabotakh," *Detskaia literatura*, 1939, 5: 23–24.

74. See *Rabochaia gazeta*, June 16, 1925, 5.

75. "Posleslovie," *Rabochaia gazeta*, July 6, 1925, 5; following quotations are from this source.

76. See A. Carrel, "On the permanent life of tissues outside of the organism," *Journal of Experimental Medicine*, 1912, 15: 516–28. On Carrel and his work on tissue cultures, see Jan A. Witkowski, "Alexis Carrel and the mysticism of tissue culture," *Medical History*, 1979, 23: 279–96; idem, "Dr. Carrel's Immortal Cells," *Medical History*, 1980, 24: 129–40. Interestingly, in the late 1930s Carrel, in collaboration with the famous pilot and inventor Charles Lindbergh, worked on creating an artificial heart. See a popular account of this work in David M. Friedman, *The Immortalists* (New York: Harper Collins Publishers, 2007).

77. See "Ot redaktsii," *Vsemirnyi sledopyt*, 1925, 3: 16–17, here–17.

78. *Ibid*. Italics are in the original.

79. See A. Beliaev, "Fantastika i nauka," in idem, *Bor'ba v efire* (Moscow-Leningrad: Molodaia Gvardiia, 1928), 309–23.

80. A. Beliaev, "Posleslovie k romanu 'Chelovek-amphibiia,'" *Vokrug sveta*, 1928, 13: 200–02.

81. See [Al. Luchinskaia], "Taina Professora," *KP*, 1925, 45: 12–14; 46: 12–14; 47: 3–6. The story's publication omitted the name of its author altogether, giving only the name of its "translator," which strongly suggests that "translation" was fictitious.

82. This resurrection was recorded in many contemporary diaries and memoirs, see, for instance, N. P. Okunev, *Dnevnik moskvicha* (Moscow: Voenizdat, 1997); vol. 1, 1917–1920; vol. 2, *1920–1924*.

83. See Mikhail Bulgakov, "Sorok sorokov," *Nakanune*, April 15, 1923.

84. See N. Kravkow [Kravkov], "Über funktionelle Eigenschaften der Blutgefässe isolierter (normaler und pathologischer) Organe von Tieren und Menschen," *Zeitschrift für die gesamte experimentelle Medizin*, 1922, 27(3–4): 127–73; and N. P. Kravkov, "Dannye i perspektivy po ozhivleniiu taknei umershikh," *Sbornik nauchnykh trudov v chest' 50-letiia nauchno-vrachebnoi deiatel'nosti glavnogo vracha Obukhovskoi bol'nitsy A. A. Nechaeva* (Petrograd: n.p., 1922), 5–7; for

the scientific community's reaction to Kravkov's research, see N. K. Kol'tsov, "Issledovaniia N. P. Kravkova o funktsional'nykh osobennostiakh izolirovannykh organov u zhivotnykh i cheloveka," *Uspekhi eksperimental'noi biologii*, 1922, 1(3–4): 375–81.

85. See, for instance, "Doklady prof. N. P. Kravkova," *Pravda*, March 2, 1924, 4; and N. Kol'tsov, "Blestiashchie dostizheniia russkoi nauki (N. P. Kravkov i ego raboty)," *Pravda*, June 8, 1924, 6.

86. See Nina Tumarkin, *Lenin Lives! The Lenin Cult in Soviet Russia* (Cambridge, MA: Harvard University Press, 1997), revised edition.

87. See "Smert' professora Kravkova," *Pravda*, April 25, 1924, 4; "Smert' prof. Kravkova," *Izvestiia*, April 25, 1924, 6; "Vecher pamiati N. P. Kravkova," *Izvestiia*, May 15, 1924, 7.

88. See "Problema zhizni i smerti," *Izvestiia*, March 14, 1925, 4.

89. "Ot redaktsii," *Vsemirnyi sledopyt*, 1925, 3: 17.

90. See S. S. Briukhonenko, "Nekotorye dannye o deistvii limonnokislogo natra i peptona pri sypnom tife," *Moskovskii meditsinskii zhurnal*, 1921, 1: 7–10.

91. See Briukhonenko's unpublished manuscript on "the history of the development of the idea of artificial circulation and its uses in biology and medicine," S. S. Briukhonenko, "Istoriia razvitiia idei iskusstvennogo krovoobrashcheniia i ee primenenie v biologii i meditsine," ARAN, f. 1534, op. 3, d. 15, l. 10.

92. Compare, for instance, M. Boutillette, C. Coveney, S. Kun, L. J. Menides, "The influence of science fiction films on the development of biomedical instrumentation," in *Proceedings of the IEEE 25th Annual Northeast Bioengineering Conference, April 8–9, 1999*, 143–44; Allen Thiher, *Fiction Rivals Science: The French Novel from Balzac to Proust* (Columbia: University of Missouri Press, Missouri, 2001); and idem, *Fiction Refracts Science: Modernist Writers from Proust to Borges* (Columbia: University of Missouri Press, 2005).

93. See, for example, Ludmila Jordanova, ed., *Languages of Nature: Critical Essays on Science and Literature* (New Brunswick, NJ: Rutgers University Press, 1986); John H. Cartwright and Brian Baker, *Literature and Science: Social Impact and Interaction* (Santa Barbara, CA: ABC-CLIO, 2005); Philip Coleman, ed., *On Literature and Science: Essays, Reflections, Provocations* (Dublin, Ireland; Portland, OR: Four Courts Press, 2007); Alison Sinclair, "Social imaginaries: the literature of eugenics," *Studies in History and Philosophy of Biological and Biomedical Sciences*, 2008, 39: 240–46, and many others. For more on the interrelations between science and fiction, see Chapter 6.

94. Grebnev, "Golova, otrezanaia ot tulovishcha, zhivet," 4.

95. See W. R. Probert and D. G. Melrose, "An early Russian heart-lung machine," *British Medical Journal*, 1960, 1: 1047–48.

96. "Shaw Feels 'Tempted' To Have Head Cut Off," *New York Times*, March 17, 1929, 5.

97. See A. V. Lavrov and Dzhon Mal'mstad, eds. *Andrei Belyi i Ivanov-Razumnik. Perepiska* (St. Petersburg: Atheneum-Feniks, 1998).

98. *Ibid*, 408. Italics are in the original.

99. *Ibid*, 409.

100. *Ibid*. Italics are in the original.

101. On the Haldane-Russell exchange and its impact on British scientists and writers, see Adams, "Last Judgment," 457–91; and Charles T. Rubin, "Daedalus and Icarus Revisited," *The New Atlantis: the Journal of Technology and Society*, 2005, 8: 73–91. On a more general issue of interrelations between biology and values in Britain, see Roger Smith, "Biology and Values in Interwar Britain: C. S. Sherrington, Julian Huxley and the Vision of Progress," *Past and Present*,

2003, 178: 210–42; for a more general discussion of the situation in the post-World War I Britain, see Christopher Lawrence and Anna- K. Mayer, eds., *Regenerating England: Science, Medicine and Culture in Inter-War Britain* (Amsterdam: Editions Rodopi, 2000).

102. Julian Huxley, "The tissue-culture king," *The Yale Review*, 1926, 15: 487–504, here–504. Italics are in the original.

103. As witnessed by the widely publicized "Joint statement upon the relations of science and religion" issued in June 1923 and the subsequent infamous "Scopes Trial" of 1925, in the United States a similar debate centered on religious, not ethical, concerns spurred by biological science.

104. For examples of contemporary reaction to Bulgakov's novella, see V. Pravdukhin, "Nedra. Kniga shestaia," *KN*, 1925, 3: 287–89; A. Voronskii, "Pisatel', kniga, chitatel'," *ibid*, 1927, 1: 237–38; N. Korotkov, "Nedra. Kniga shestaia," *Rabochii zhurnal*, 1925, 3: 156–57; A. Pridorogin, "Nedra. Kniga shestaia," *Knigonosha*, 1925, 6: 18; M. Lirov, "Nedra. Kniga shestaia,"*Pechat' i revoliutsiia*, 1925, 5–6: 517–19; I. M. Nusinov, "Put' M. Bulgakova," *ibid*, 1929, 4: 40–53.

105. K. Loks, "A. Beliaev. Golova Professora Douelia," *Pechat' i revoliutsiia*, 1926, 8: 200–01. Similarly, another reviewer praised the gripping story but did not comment upon its ethical dimensions; see Bor. Ol., "A. Beliaev. Golova professora Douelia," *Sovetskii iug*, September 24, 1926, 4.

106. This was the title of an article on Briukhonenko's works published in *Evening Moscow*; see "Nauka, pobezhdaiushchaia smert'," *VM*, September 20, 1928, 3.

107. A. Nizhnii, "Sobaki i liudi," *VM*, March 22, 1929, 2.

108. For a vivid depiction of antivivisectionists' stance toward experiments with dogs in the United States and the biomedical community's strategies of dealing with this kind of critique, see Susan E. Lederer, "Political Animals: The Shaping of Biomedical Research Literature in Twentieth-Century America," *Isis*, 1992, 83: 61–79.

109. V. I. Lenin, "Lozhka degtia v bochke meda," in V. I. Lenin, *Polnoe sobranie sochinenii*, (Moscow: Politizdat, 1970), 5th ed., vol. 45: 206–07.

110. See A. Irkutov, "Bessmertie," *Bor'ba mirov*, 1924, 3: 30–37.

111. See, for instance, reminiscences of Iurii Dolgushin, another Soviet writer, who met Briukhonenko and even participated in his experiments. Briukhonenko became a prototype of one of the characters in Dolgushin's novel, *The Generator of Miracles* (1939). See Iurii Dolgushin, "Ot avtora," in idem, *Generator chudes* (Moscow: Trudrezervizdat, 1959), 5–10.

112. This episode is recorded in the diaries of Bulgakov's wife. See Elena and Mikhail Bulgakovy, *Dnevnik Mastera i Margarity* (Moscow: Vagrius, 2003), 174–75, 180, 184–85. I am grateful to Galina Savina for calling my attention to this source.

CHAPTER 3

1. A. Beliaev, "Ni zhizn', ni smert'," *Vsemirnyi sledopyt*, 1926, 5: 3–15; 6: 3–14. All of the following quotations from the novella are from this source.

2. For brief historical accounts of Bakhmet'ev's research on anabiosis, see S. N. Matsko, "P. I. Bakhmet'ev i problema anabioza," *Biulleten' MOIP*, 1973, 78 (6): 137–140; L. A. Frolova, "O rabotakh P. I. Bakhmet'eva po anabiozu nasekomykh i mlekopitaiushchikh," *Istoriko-biologicheskie issledovaniia*, 1978, 6: 167–73; and Nikolina M. Sretenova, "A Contribution of Bulgarian Science to European Science at the End of the 19th Century—Porfirij Bakhmetiev's Theory of Anabiosis and the Idea of Self-organization," in Brigitte Hoppe, ed., *Biology Integrating Scientific Fundamentals* (Munich: Institute für Geschichte der Naturwissenschaften, 1997), 126–48.

3. For the only, but heavily fictionalized, book-length biography of Bakhmet'ev, see A. G. Chulkov and V. I. Azanov, *Zaveshchanie Bakhmet'eva* (Saratov: Privolzhskoe knizhnoe izdatel'stvo, 1980).

4. "U professora Bakhmet'eva," *Vol'skaia zhizn'*, June 13, 1913, 3.

5. Some information on Kleiner (1849–1916) and his views of physics could be found in Jos Uffink, "Insuperable difficulties: Einstein's statistical road to molecular physics," *Studies in History and Philosophy of Modern Physics*, 2006: 37(1): 36–70.

6. P. I. Bakhmet'ev, "Magnitnyi moment zheleznykh provolochnykh puchkov," *Zhurnal Russkogo phiziko-khimicheskogo obshchestva* (hereafter—ZhRFKhO), 1883, 15: 142–46; idem, "Zametka o zvuchashchikh trubakh," *ibid*, 1883, 15: 166–68; idem, "Nekotorye iavleniia ostatochnogo magnetizma," *ibid*, 1883, 15: 173–200.

7. P. I. Bakhmet'ev, "Teplovye iavleniia magnetizma," *ZhRFKhO*, 1884, 16: 81–135; idem, "Magnetizm zheleznykh provolok, zakliuchennykh ne vsei svoei dlinoi v namagnichennuiu spiral'," *ibid*, 1884, 16: 213–22; idem, "Teplota namagnichivaniia kol'tseobraznogo elektromagnita," *ibid*, 1884, 16: 257–62;idem, "Vliianie lineinogo szhatiia na magnitnost' zheleznykh, stal'nykh i nikelevykh sterzhnei," *ibid*, 1884, 16: 427–52; idem, "Zavisimost' mezhdu diamagnitnost'iu i teplotoi plavlenia tel," *ibid*, 1884, 16: 519–24.

8. Another version states that Bakhmet'ev himself gave the passport to one of his friends to bring to Russia in order to send it to his father, who was supposed to arrange its exchange for a new one. See Vl. Belgorodskii, "Professor P. I. Bakhmet'ev," *Russkoe slovo*, October 15, 1913, 6.

9. See Chulkov and Azanov, *Zaveshchanie Bakhmet'eva*, 56–57.

10. See P. I. Bakhmet'ev, "K teorii vliianiia mekhanicheskogo i termicheskogo izmeneniia razmerov tela na ego magnitnost'," *ZhRFKhO*, 1886, 18: 31–47; idem, "Termoelektricheskie izsledovaniia. Predvaritel'noe soobshchenie," *ibid*, 1886, 18: 47–51; idem, "Termoelektricheskie izsledovaniia," *ibid*, 1889, 21: 264–87; idem, "Zavisimost' paramagnitnoi i diamagnitnoi sposobnosti tel ot atomnogo vesa," *ibid*, 1889, 21 (4): 39–43; idem, "Chto predstavliaiut soboi deformatsionnye toki 'Brauna'?" *Vestnik opytnoi fiziki i elementarnoi matematiki* (hereafter—VOFEM), 1889, 64: 68–73; idem, "Luchi elektricheskoi sily," *ibid*, 1889, 68: 153–57.

11. For a description of this device, see P. I. Bakhmet'ev, "Novyi telefotograf," *Elektrichestvo*, 1885, 1: 2–7.

12. *Vysshee Uchilishche*, an analogue of the German *Hochschule*.

13. For the text of Kleiner's letter, see Nikolina Sretenova, *Universitet't i fizitsite: Nachalo* (Sofia: Kheron Press, 2000) (in Bulgarian), 147–48. This book contains a wealth of materials on Bakhmet'ev's activities at Sofia University and his role in the establishment of the university's Physics Institute.

14. For an extensive list of secondary literature (mostly in Bulgarian) on Bakhmet'ev's work in the mathematics and physics department of Sofia University, see the website of the Bulgarian Academy of Sciences' History of Physics Museum: http://www.issp.bas.bg/museum/m11-page1.html. I would like to express my deepest gratitude to the museum's curator, Ganka Kamisheva, for her help in obtaining copies of several rare publications, including her own works, on the subject. For a nearly complete list of Bakhmet'ev's publications (including 223 items in Russian, Bulgarian, and German), see *Trudy Saratovskoi uchenoi arkhivnoi komissii*, 1913, 30: 277–89.

15. See G. Kamisheva and A. Vavrek, "Milestones in the development of physics in the Sofia University (1889–1945)," *Bulgarian Journal of Physics*, 2000, 27(4): 55–8; idem, "Contents of the courses in physics in the Sofia University (1889–1945)," *ibid*, 2000, 27(4): 59–62.

16. P. I. Bakhmet'ev, "Prichina vliianiia namagnichivaniia na termoelektricheskie svoistva zheleza i nikelia," *ZhRFKhO*, 1891, 23: 301–23; idem, "Vliianie lineinogo szhatiia provolok na ikh termoelektricheskie svoistva," *ibid*, 1891, 23: 430–35; idem, "Zemnye elektricheskie toki," *ibid*, 1894, 26 (1): 31–76; (3): 159–99.

17. See P. I. Bakhmet'ev and Ia. Vezharov, "Termometricheskie izsledovaniia amal'gam," *ZhRFKhO*, 1893, 25(7): 238–55; P. I. Bakhmet'ev and P. Penchev, "Kalorimetricheskie izsledovaniia colloidal'nogo serebra," *ibid*, 1893, 25(5): 138–55; P. I. Bakhmet'ev and N. Penchev, "Elektricheskie toki prosachivaniia," *ibid*, 1894, 26 (5): 225–48; P. I. Bakhmet'ev and G. Stamboliev, "Elektricheskie toki, poluchaemye pri nagrevanii odnorodnykh metalicheskikh provolok," *ibid*, 1895, 27 (1): 1–24; P. I. Bakhmet'ev and G. Stamboliev, "Elektricheskie toki, poluchaemye vsledstvie okhlazhdeniia rasplavlennykh metalov," *ibid*, 1895, 27 (4): 65–102; P. I. Bakhmet'ev and P. Vaskov, "Vliianie okruzhaiushchei sredy na uprugoe posledeistvie metalicheskikh provolok," *ibid*, 1896, 28: 217–20.

18. Bakhmet'ev relayed the story of his newfound passion for butterflies to a colleague; see Vl. A. Petrov, "Biologicheskie issledovaniia P. I. Bakhmet'eva," *Estestvoznanie i geografiia*, 1903, 3: 57–75.

19. On Bakhmet'ev's entomological studies in Bulgaria, see Ivan Buresh, *Istoriia na entomologichnoto prouchvane na Bulgaria* (Sofia, 1924), 46–54 (in Bulgarian); see also a short but detailed account of Bakhmet'ev's life in Bulgaria written by the same author, Bakhmet'ev's close friend and director of the Sofia Zoo, Ivan Buresh, "Bibliografski ocherk v"rkhu nauchno-pisatelskata tvorba na fizika-biolog professor Porfirii Bakhmet'ev," *Godishnik na B"lgarskiia Bibliografski Institut, (1945–1946)*, 1948, 1: 68–83 (in Bulgarian). I am grateful to Ganka Kamisheva for providing me with a copy of this rare publication.

20. The monograph also contained a preliminary synopsis of Bulgarian flora and the results of meteorological observations during the prior ten years. See P. I. Bakhmet'ev, "Babochki Bolgarii," *Horae Societie entomologicae Rossicae*, 1902, 35: 355–466. This work also came out as a separate volume under the auspices of the Imperial Academy of Sciences in St. Petersburg, see P. I. Bakhmet'ev, *Babochki Bolgarii* (St. Petersburg: Imperatorskaia Akademiia Nauk, 1902).

21. See, for instance, P. Bachmetjew, "Ueber die Dimensionen der bulgarischen Schmetterlinge im Vergleich zu den westeuropäischen. I. Familie *Satyridae*," *Societas entomologica*, 1899, 14(6): 43–45; P. I. Bakhmet'ev, "Izmenchivost' dliny kryl'ev u *Aporia crataegi* L. v Rossii i ee zavisimost' ot meteorologicheskikh elementov," *Zapiski Imperatorskoi akademii nauk* (Fiziko-matematicheskoe otdelenie), 1910, 25(7): 1–47.

22. The following description of Bakhmet'ev's techniques is based on his articles, P. Bakhmet'ev, "Temperatura nasekomykh," *Nauchnoe obozrenie*, 1898, 9: 1602–11; and P. Bachmetjew, "Die Temperatur der Insekten," *Krancher's Entomologische Jahrbuch*, 1898, 8: 121–31.

23. In all of his experiments, Bakhmet'ev used the Celsius temperature scale.

24. The following description of Bakhmet'ev's experiments is taken from his articles, P. Bakhmet'ev, "Sobstvennaia temperatura pchel i voobshche nasekomykh," *Russkii pchelovodnyi listok*, 1899, 3: 84–90; 4: 114–19; idem, "Ueber die Temperatur der Insekten nach den Beobachtung in Bulgarien," *Zeitschrift für wissenschaftliche Zoologie*, 1899, 66: 521–604: idem, "Der kritische Punkt und die normale Erstarrung-Temperatur der Insektensafte," *Societas entomologica*, 1899, 14 (1): 1–2.

25. See P. Bakhmet'ev, "Pereokhlazhdenie zhidkostei," *ZhRFKhO*, 1900, 32 (otdel 1) (8): 218–41; idem, "Unterkaltungs-Erscheinungen bei schwimmenden Para-Nitrotoluol-Kugelchen," *Zapiski Imperatorskoi akademii nauk*, (Fiziko-matematicheskii otdel), 1900, 10 (7): 1–63.

26. See P. Bachmetjew, "Die Abhangigkeit des kritischen Punktes bei Insekter von doren Abkuhlungs-Geschwindigkeit," *Zeitschrift für wissenschaftliche Zoologie*, 1900, 67: 529–50; idem, "Das vitale Temperaturminimum bei Insekten abhangig von der Zeit," *Societas entomologica*, 1900, 15(6): 41–43; (7): 49–52; idem, "Der Variation des kritischen Punktes bei verschiedenen Exemplaren einer und derselben Insekten-Art," *ibid*, 1900, 15(1): 1–2.

27. See P. Bachmetjew, "Verkhu temperaturata na nasekomite v Bolgarii," *Ministerski Sbornik* (Sofia) 1900, 11–12: 82–159; idem, "Lahmung bei Lepidoptera infolge erhohter Temperatur ihres Korpers," *Societas entomologica*, 1900, 15(12): 89–91; (13): 97–101; (14): 105–10.

28. P. Bakhmet'ev, "Vital'nyi temperaturnyi minimum zhivotnykh s peremennoi temperaturoi krovi," *Arkhiv biologicheskikh nauk*, 1900, 8 (3): 239–60.

29. See W. Preyer, "Über die Anabiose," *Biologisches Zentralblatt*, 1891, 11: 1–5. On the early history of research on anabiosis, see D. Keilin, "The Problem of Anabiosis or Latent Life: History and Current Concept," *Proceedings of the Royal Society of London. Series B, Biological Sciences*, 1959, 150: 149–91. However, this detailed overview does not even mention Bakhmet'ev's research on the subject. See also, P. Iu. Shmidt, *Anabioz* (Moscow-Leningrad: AN SSSR, 1955); Audrey U. Smit, *Biological Effects of Freezing and Suprecooling* (London: Edward Arnold, 1961); E. V. Maistrakh, *Gipotermiia i anabioz* (Moscow-Leningrad: Nauka, 1964); A. M. Goldovskii, *Anabioz i ego prakticheskoe znachenie* (Leningrad: Nauka, 1986); and Veselin Denkov, *Na grani zhizni* (Moscow: Znanie, 1988).

30. P. Bakhmet'ev, "Anabioz," *Nauchnoe obozrenie*, 1900, 1: 17–28.

31. See P. Bakhmet'ev, "Retsept dozhit' do XXI veka," *Estestvoznanie i geografiia*, 1901, 8: 1–13.

32. *Ibid*, 10.

33. He borrowed the idea from the work of a prominent French anesthesiologist, Rafael Dubois, "Autonarcose carbonico-acétonémique, ou sommeil hivernal de la Marmotte," *Comptes Rendus l'Académie des Sciences* (Paris), 1895, 120: 458–60.

34. Bakhmet'ev, "Retsept dozhit' do XXI veka," 13.

35. P. Bachmetjew, *Experimentelle entomologische Studien*, 1. *Temperaturverhaltnisse bei Insekten* (Leipzig: Wilhelm Engelmann, 1901).

36. See, for instance, Jeoffrey Smith, "The temperature of insects," *The Zoologist* (fourth series), 1902, 6: 287–93.

37. W. M. W[heeler], "Temperature of Insects," *American Naturalist*, 1902, 36: 401–05. All the following quotations are from this source.

38. Petrov, "Biologicheskie issledovaniia P. I. Bakhmet'eva," 75.

39. See the protocol of the society's general meeting on March 1, 1904, in *Trudy Saratovskogo obshchestva estestvoispytatelei*, 1916, 7 (3): 34.

40. Bakhmet'ev recounted these conversations in his article, P. I. Bakhmet'ev, "Itog moikh issledovanii ob anabioze nasekomykh i plan ego issledovaniia u teplokrovnykh zhivotnykh," *Izvestiia Imperatorskoi akademii nauk*, 1902, 17 (4): 161–66.

41. See "Elizabeth Thompson Science Fund," *Science*, 1902, 15, (372): 276; and Charles S. Minot, "The Elizabeth Thompson Science Fund," *ibid*, 1902, 16 (392): 35–36. For a general description of the fund and its activities, see the 25th-anniversary report of its board of trustees, *The Elizabeth Thompson Science Fund, 1886–1911* (n.p., c1911). Surprisingly, there is no historical study of this very interesting foundation, which supported a number of prominent researchers worldwide. For a rather brief analysis, see Hal F. Starnes, *Patterns in the funding of science: a case study of the Elizabeth Thompson Science Fund, 1886 to 1911*, B. A. Honors Thesis, Harvard University, 1975.

42. See P. Bakhmet'ev, "Zaveshchanie milliardera," *Estestvoznanie i geografiia*, 1904, 10: 1–24; 11: 1–31; 12: 1–47. For more details, see Chapter 4.

43. See Charles S. Minot, "The Elizabeth Thompson Science Fund," *Science*, 1905, 21(537): 596–97.

44. The letter written by D. Gerilovski has been found and published by the Bulgarian historian Nikolina Sretenova, *Universitet't i fizitsite*, 144–46.

45. For details of this affair, see P. A. Tribunskii, "P. N. Miliukov v Bolgarii, 1897–1899," *Novaia i noveishaia istoriia*, 2002, 4: 151–68.

46. For details, see Sretenova, *Universitet't i fizitsite*, 11–83.

47. For details, see G. Kamisheva, M. Borisov, and A. Vavrek, "Niakoi dokumenti okolo konflikta na Porfirii Ivanovich Bakhmet'ev s"s Sofiiskii universitet," *Fiziko-matematichesko spisanie*, 1991, 1–2: 3–15, (in Bulgarian). I am grateful to Ganka Kamisheva for providing me with a copy of this publication.

48. For details, see Sretenova, *Universitet't i fizitsite*, 11–83, 141–49.

49. P. Bakhmet'ev, "Biologicheskie analogii u kristallov," *Estestvoznanie i geografiia*, 1906: 1–16.

50. Bakhmet'ev had published his research on earth electric currents, which he presented as his dissertation five years earlier; see Porfirij Bachmetjew, "Der gegenwärtige Stand der Frage über elektrische Erdströme," *Zapiski Imperatorskoi akademii nauk*, 1902, 12(3): 1–53.

51. The documents related to Bakhmet'ev's submission of his dissertation have been preserved in the Zurich University Archives kept in the State archive of the Zurich canton: StAZ U 110e.11. I am profoundly grateful to the archivist Dr. Karin Huser for her help in locating the documents and providing me with copies.

52. P. Bachmetjew, *Experimentelle entomologische Studien vom physikalisch-chemischen Standpunkt aus. Band II. Einfluss der äusseren faktoren auf Insekten* (Sofia: Staatsdrukerei, 1907).

53. Since Bakhmet'ev published the volume in Sofia at his own expense, it had limited distribution and did not result in the same flood of reviews in all biological journals that had greeted the first volume of his *Studies*. For reviews of the second volume, see F. A. D., "Experimental entomology," *Nature*, 1908, 78: 241–42; and Frank E. Lutz, "The effect of the environment upon animals," *American Naturalist*, 1909, 43 (505): 55–57.

54. P. Bachmetjew, *Experimentelle entomologische Studien*, VII.

55. P. Bachmetjew, "Die Flugellange bei *Aporia crataegi* in Rumanien, 1905," *Bulletin de la Societe des sciences de Bucarest*, 1909, 17 (5–6): 299–305; idem, "Die Beziehung zwischen der Flugellange von *Aporia crataegi*, welche in Rumanien 1905 gefangen wurden, und den meteorologischen Elementen," *ibid*, 306–16; idem, "Die Varabilitat der Flugellange von *Aporia crataegi* in Sophia (Bulgarien) als Resultat siebenjahriger nacheinander folgender Untersuchungen," *Zeitschrift für wissenschaftliche Insektenbiologie*, Berlin, 1909, 5 (2): 110–13; (5): 141–47; (6): 186–96; idem, "Analytisch-statistische Untersuchungen der Flugelhaken bei Bienen und die daraus hervorgehenden Consequenzen," *Zeitschrift für wissenschaftliche Zoologie*, 1909, 94 (1): 1–80; P. I. Bakhmet'ev, "Izmenchivost' dliny kryl'ev u *Aporia crataegi* L. v Rossii i ee zavisimost' ot meteorologicheskikh elementov," *Zapiski Imperatorskoi akademii nauk* (Fiziko-matematicheskoe otdelenie), 1910, 25(7): 1–47.

56. P. Bakhmet'ev, "Popytka ustanovit' periodicheskuiu sistemu paleoarkticheskikh babochek (Opyt prognoza novykh vidov nasekomykh)," *Trudy Saratovskogo obshchestva estestvoispytatelei*, 1903–1904, 4(3): 126–94.

57. "Izvlecheniia iz protokolov," *Vremennik obshchestva sodeistviia uspekham opytnykh nauk i ikh prakticheskikh primenenii imeni Kh. S. Ledentsova* (hereafter—*Vremennik*), 1912, 1: 32. For a brief history of this society and its founder, see E. D. Panov, "Nauka-trud-liubov'-dovol'stvo: Khristofor Semenovich Ledentsov," *Vestnik RAN*, 2004, 74(1): 63–70.

58. The following description is based on P. Bakhmet'ev, "Kak ia nashel anabioz u mlekopitaiushchikh," *Priroda*, 1912, 5: 606–22; and P. Bakhmet'ev, "Rezul'taty, poluchennye pri issledovanii letuchikh myshei, podverzhennykh deistviiu nizkikh temperatur," *Vremennik*, 1913, 1: 51–53. See also recollections of two participants of this experiment, P. Petkov and I. Buresh: P. Petkov, "Dvaiset godini ot smertta na znamenitiia uchen prof. Bakhmet'ev," *Literaturen glas*, 1933, 209 (November 19); and Iv. Buresh, "Na granitsata mezhdu zhivota i smertta. Izsledvaniia na Sofiiskii professor Parfirii Bakhmet'ev verkhu problemite na anabiozata," *Priroda* (Sofia), 1952, 1 (1): 54; cited in Sretenova, 186–87.

59. It lived in the lab for several more weeks, until it died from starvation.

60. "Moskovskie vesti," *Russkie vedomosti*, February 19, 1912, 5. See also, "Noveishie issledovaniia prof. P. I. Bakhmet'eva po voprosu ob anabioze," *Estestvoznanie i geografiia*, 1912, 2: 90; N. O., "Anabioz u mlekopitaiushchikh," *VZ*, 1912, 7: 214–15; P. I. Bakhmet'ev, "Rezul'taty, poluchennye pri issledovanii letuchikh myshei, podverzhennykh deistviiu nizkikh temperatur," *Vremennik*, 1913, 1: 51–53.

61. See the society statute in "Ustav obshchestva," *Vremennik*, 1910, 1: 3–9. See also S. A. Fedorov, "Pamiati Khristofora Semenovicha Ledentsova," *Vremennik*, 1911, 1: 35–43.

62. See P. I. Bakhmet'ev, "Teoreticheskie i prakticheskie sledstviia iz moikh issledovanii anabioza u zhivotnykh," *Priroda*, 1912, 12: 1427–42; idem, "Anabioz i ego znachenie v sel'skom khoziaistve," *Sel'skoe khoziastvo i lesovodstvo*, 1912, 240: 345–54.

63. See "Izvlechenie is protokolov," *Vremennik*, 1913, 1: 14–15.

64. On the organization of the society, see M. S. Zernov, *Obshchestvo Moskovskogo nauchnogo instituta v pamiat' 19 fevralia 1861 goda* (Moscow: Sviridov, 1913); and A. F. Petrovskii, "Moskovskii Nauchnyi Institut v pamiat' 19 fevralia 1861 goda," in *Trudy 13-go s"ezda russkikh estestvoispytatelei i vrachei v g. Tiflise 16–24 iiunia 1913* (Tiflis: n.p., 1915), 124–30. For the society's statute, see *Ustav Obshchestva Moskovskogo nauchnogo instituta (V pamiat' 19 fevralia 1861 g.)* (Moscow: b. p., 1912). Alas, there is no detailed historical study of this extremely interesting institution and its accomplishments.

65. See "V nauchnom institute," *Russkie vedomosti*, March 27, 1913, 5–6; M. Z[arochentsev], "Biologicheskii institut i laboratoriia nizkikh temperatur v Moskve," *Kholodil'noe delo*, 1913, 3: 294–95; R. S., "Beseda s prof. P. I. Bakhmet'evym," *ibid*, 1913, 4: 401–02; "V nauchnom institute," *ibid*, 1913, 4: 402.

66. See advertisements for the courses in *Russkie vedomosti*, April 2, 1913, 1; and April 11, 1913, 4, which listed Bakhmet'ev as one of the key lecturers.

67. See V. I. Sklabinskii, "Novye sposoby promyshlennogo khraneniia ryby," *Astrakhanskii listok*, August 11, 1913, 3–4; P. Bakhmet'ev, "Deistvie nizkikh temperatur na organizmy," *Kholodil'noe delo*, 1913, 5: 439–45.

68. See "K voprosu o konservirovanii ikry," *Astrakhanskii listok*, July 6, 1913, 4; "O konservirovanii ikry," *ibid*, September 4, 1913, 6.

69. See "Moskovskie vesti," *Russkie vedomosti*, March 29, 1913, 3; A. Bespalov, "Opyty zamorazhivaniia zhivoi ryby prof. P. I. Bakhmet'evym na Moskovsko-Kazanskom kholodnom sklade," *Kholodil'noe delo*, 1913, 5: 498–99.

70. On the history of these particular educational institutions, known as "people's universities" in Russia, see L. V. Dubrovina, *Narodnye universitety* (Moscow: Znanie, 1963).

71. See V. S[klabinski], "Prof. P. I. Bakhmet'ev," *Astrakhanskii listok*, April 5, 1913, 4; idem, "Porfirii Ivanovich Bakhmet'ev," *ibid*, April 7, 1913, 3; idem, "Ne zhizn' i ne smert'," *ibid*, April 10, 1913, 4.

72. See "Moskovskie vesti," *Russkie vedomosti*, May 31, 1913, 4.

73. R. S., "Dar nauke," *Kholodil'noe delo*, 1913, 6: 694–95.

74. "K voprosu o predstoiashchikh opytakh P. I. Bakhmet'eva," *Kholodil'noe delo*, 1913, 6: 695–96.

75. See "Doklad prof. P. I. Bakhmet'eva," *Saratovskii listok*, June 2, 1913, 5; "Mezhdu zhizn'iu i smert'iu. Doklad prof. P. I. Bakhmet'eva ob anabioze," *ibid*, June 6, 1913, 3; V. Abutkov, "K priezdu professora P. I. Bakhmet'eva," *Vol'skaia zhizn'*, June 18, 1913, 2; "Na lektsii professora P. I. Bakhmet'eva," *ibid*, June 23, 1913, 3; "Lektsii professora P. I. Bakhmet'eva," *Astrakhanskii listok*, August 9, 1913, 3; "K lektsii professora P. I. Bakhmet'eva," *Iuzhnyi krai* (Khar'kov), August 27, 1913, 3; "Na grani mezhdu zhizn'iu i smert'iu," *ibid*, August 28, 1913, 5; "Zakon smerti," *ibid*, August 9, 1913, 6; A., "Lektsiia professora P. I. Bakhmet'eva," *Kievskaia mysl'*, September 16, 1913, 3; "Lektsiia Prof. Bakhmet'eva," *Poslednie novosti* (Kiev), September 15, 1913, 4; "Lektsiia P. I. Bakhmet'eva," *Kievlianin*, September 17, 1913, 4; Is. Sh., "Na grani zhizni i smerti (Lektsiia prof. P. I. Bakhmet'eva)," *Odesskie novosti*, September 8, 1913, 4–5; Kalina, "U sovremennogo Fausta," *Russkoe slovo* (Moscow), March 14, 1913, 6; "Laboratoriia imeni P. I. Bakhmet'eva," *ibid*, October 29, 1913, 7.

76. Mishel', "Anabioz," *Saratovskii listok*, June 7, 1913, 4.

77. V. Untovskii, "Zigzagi," *Iuzhnyi krai*, October 20, 1913, 6.

78. See O. Runova, "Pis'mo v redaktsiiu," *Saratovskii listok*, June 6, 1913, 3.

79. See "Lektsiia professora P. I. Bakhmet'eva," *Astrakhanskii listok*, August 14, 1913, 4.

80. See V. I. Sklabinskii, "Novye sposoby promyshlennogo khraneniia ryby," *Astrakhanskii listok*, August 11, 1913, 3–4.

81. See "Vyzdorovlenie professora P. I. Bakhmet'eva," *Vol'skaia zhizn'*, September 6, 1913, 3.

82. A., "Lektsiia professora P. I. Bakhmet'eva," *Kievskaia mysl'*, September 15, 1913, 3; "Lektsiia professora P. I. Bakhmet'eva," *Poslednie novosti* (Kiev), September 15, 1913, 4.

83. A. Bezpalov, "Prof. P. I. Bakhmet'ev," *Kholodil'noe delo*, 1913, 10: 1064–66.

84. See Sergei Iablonovskii, "Pamiati Prof. P. I. Bakhmet'eva," *Russkoe slovo*, October 15, 1913, 2; Nik. Kol'tsov, "P. I. Bakhmet'ev," *Russkie vedomosti*, October 15, 1913, 6; "Prof. P. I. Bakhmet'ev," *Iuzhnyi krai*, October 16, 1913, 3; A. Achkasov, "P. I. Bakhmet'ev," *Kievskaia mysl'*, October 15, 1913, 3; "Konchina prof. P. I. Bakhmet'eva," *Vol'skaia zhizn'*, October 18, 1913, 2; "U groba Prof. P. I. Bakhmet'eva," *Saratovskii listok*, October 18, 1913, 3; P. L-v, "Pamiati Prof. P. I. Bakhmet'eva," *Astrakhanskii listok*, October 18, 1913, 5; and many others.

85. See, for instance, Nik. Kulagin, "Pamiati P. I. Bakhmet'eva," *Priroda*, 1913: 1127–30; N. K. Kol'tsov, "Pamiati P. I. Bakhmet'eva," *Estestvoznanie i geografiia*, 1913, 9: 88–94; N. Kuznetsov, "Pamiati P. I. Bakhmet'eva," *Russkoe entomologicheskoe obozrenie*, 1913, 13 (3–4): 568–70; M. M. Zavadovskii, "Porfirii Ivanovich Bakhmet'ev," *VZ*, 1914, 5: 323–27.

86. See the plans to continue Bakhmet'ev's research discussed by the Society of Moscow Scientific Institute reported in newspapers: "Laboratoriia imeni P. I. Bakhmet'eva," *Russkoe slovo*, October 29, 1913, 7; "Prodolzhenie rabot P. I. Bakhmet'eva," *Russkie vedomosti*, October 29, 1913, 5.

87. See minutes of the meeting of the Physics and Mathematics Division on November 13, 1913, in *Protokoly zasedanii fiziko-matematicheskogo otdeleniia Imperatorskoi akademii nauk*, 1914, 15: #705.

88. See *Protokoly zasedanii fiziko-matematicheskogo otdeleniia Imperatorskoi akademii nauk*, 1914, 1 (January 8): #42; 3 (February 5): #113.

89. It also rejected a similar offer received a few weeks later from another enthusiast. See *Protokoly zasedanii fiziko-matematicheskogo otdeleniia Imperatorskoi akademii nauk*, 1914, 3 (February 5): # 117.

90. See, for instance, S. W. Britton, "Effects of lowering the temperature of homoitermic animals," *Quarterly Journal of Experimental Physiology*, 1923, 13: 55–68; for historical overviews of Western research, see P. M. Gehenio and B. J. Luyet, "The mechanism of injury and death by low temperature," *Biodynamica*, 1940, 3 (60): 33–99; and Keilin, "The Problem of Anabiosis or Latent Life," 149–91.

91. Aside from a few eulogies and short entries in Soviet encyclopedias, there is practically no secondary literature on the life of this remarkable biologist, even though a large collection of his personal papers is preserved in the St. Petersburg Branch of the Archive of the Russian Academy of Sciences. For a brief biographical sketch, see Margarita Shmidt, *Shmidty: Biogenealogicheskie ocherki* (St. Petersbug: SRP "Pavel" VOG, 2008), 8–55.

92. See, for instance, P. Shmidt, "Ueber das Leuchten der Zuckmücken," *Zoologische Jahrbucher*, 1895, 8: 52–66; idem, "Beiträge zur Kenntnis der niederen Myriapoden," *Zeitschrift für wissenschaftliche Zoologie*, 1895, 59: 436–510; idem, "Beitrag zur Kenntnis des Laufspinnen (*Aranea Citigradae* Thor.) Russlands," *Zoologische Jahrbucher*, 1895, 8: 439–84;idem, "K morfologii nizshikh *Myriapoda*," *Trudy Imperatorskogo St.-Peterburgskogo obshchestva estestvoispytatelei*, 1895, 26 (1): 18–22; idem, "Materialy k poznaniiu fauny Semirechenskoi oblasti," *Zapiski Zapadno-Sibirskogo otdeleniia Russkogo geograficheskogo obshchestva*, 1896, 20: 1–32; idem, "K poznaniiu roda *Aeolosoma*," *Trudy Imperatorskogo St.-Peterburgskogo obshchestva estestvoispytatelei*, 1896, 27 (1/5): 161–63.

93. P. Shmidt, *Ryby vostochnykh morei Rossiiskoi imperii* (St. Petersburg: Izdatel'stvo Imperatorskogo Russkogo Geograficheskogo Obshchestva, 1904), and idem, *Morskie promysly Sakhalina* (St. Petersburg, G. P. Pozharov, 1905).

94. See P. Iu. Shmidt, *Lektsii po zoologii, chitannye na St.-Peterburgskikh sel'skokhoziaistvennykh kursakh v 1907–1908 g* (St. Petersburg: Bogdanov, c1909).

95. P. Iu. Shmidt, *Raboty zoologicheskogo otdela na Kamchatke v 1908–1909 gg. Obshchii ocherk rabot* (Moscow: Riabushinskie, 1916).

96. See, for instance, P. Iu. Shmidt, *Bor'ba so starost'iu: Ocherk* (Petrograd: P. P. Soikin, 1915); and idem, *Privivka zhizni. Biologicheksii ocherk* (Petrograd: P. P. Soikin, 1916).

97. See Konrad Keller, *Zhizn' moria. Zhivotnyi i rastitel'nyi mir moria, ego zhizn' i vsaimootnosheniia*, translated by P. Iu. Shmidt (St. Petersburg: A. F. Devrien, 1896–97), 2 vols. This was a translation of the famous work by the Swiss zoologist Conrad Keller, *Das Leben des Meeres* (Leipzig: C. H. Tauchnitz, 1895).

98. See P. Iu. Shmidt, *Tsarstvo zhivotnykh v kartinakh. 250 izobrazhenii dlia nagliadnogo obucheniia s ob"iasnitel'nym tekstom* (St.-Petersburg: A. F. Devrien, 1896); and idem, *Tsarstvo rastenii i mineralov. 270 izobrazhenii dlia nagliadnogo obucheniia s ob"iasnitel'nym tekstom* (St.-Petersburg: A. F. Devrien, 1896).

99. See P. Iu. Shmidt, *Strana utrennego spokoistviia. Koreia i ee obitateli* (St. Petersburg: O. N. Popova, 1903); idem, *Iaponiia i ee obitateli* (St. Petersburg: Vestnik obrazoavniia, 1904); idem, *Ostrov izgnaniia (Sakhalin)* (St. Petersburg: O. N. Popova, 1905).

100. See P. Iu. Shmidt, *Biologicheskie etiudy* (St.-Petersburg: Nauchnoe obozrenie, 1895); idem, *Kletka i ee zhizn'* (St.-Petersburg: Nauka i iskusstvo, 1912); idem, *Iz oblasti zoopsikhologii. Strakh*

i ugroza u zhivotnykh (St.-Petersburg-Moscow: A. S. Panafidina, 1913); idem, *Na granitse zhizni i smerti* (Petrograd-Moscow: A. S. Panafidina, 1916).

101. P. Iu. Shmidt, *Osnovy zhizni. Populiarnoe vvedenie v biologiiu* (Petrograd: E. I. Blek, 1916).

102. See P. Bakhmet'ev, "Temperatura nasekomykh," *Nauchnoe obozrenie*, 1898, 9: 1602–11; and P. Shmidt, "V strane zhivykh iskopaemykh," *ibid*, 9: 1536–47.

103. P. Iu. Shmidt, "Anabiosis," *Russkoe bogatstvo*, 1913, 12: 96–107.

104. P. Iu. Shmidt, *Na granitse zhizni i smerti* (Petrograd-Moscow: A. S. Panafidina, 1916).

105. Shmidt, *Na granitse zhizni i smerti*, 54.

106. In the late nineteenth and early twentieth centuries, while Bakhmet'ev was investigating the anabiosis induced by freezing, many other researchers were studying "latent life" induced by drying. In 1894, Alfred Giard, an eminent French zoologist, even coined a special name for this form of "latent life": anhydrobiosis—life without water. See A. Giard, "L'anhydrobiose ou ralentissement des phénomènes vitaux sous l'influence de la déshydratation progressive," *Comptes Rendus de la Société de Biologie* (Paris), 1894, 46: 497–500.

107. P. Iu. Shmidt and T. V. Shchepkina, "K voprosu ob anabioze. Opyty nad dozhdevymi cherviami," *Izvestiia Petrogradskoi biologicheskoi laboratorii*, 1917, 16: 98–103; a preliminary note on these experiments also appeared in French; see P. Schmidt [Shmidt] et T. Stchepkina [Shchepkina], "Sur l'anabiose des vers de terre," *Comptes Rendus des Seances. Société de Biologie* (Paris), 1917, (January 17): 366–68. Perhaps Shmidt's involvement with the Agricultural College directed his attention to the survival of animals and plants under the conditions of extreme dryness, which were very relevant to Russian agriculture because of repeated droughts.

108. See Peter Schmidt [Shmidt], "Anabiosis of the Earth Worm," *Journal of Experimental Zoology*, 1918, 27(1): 59–72; the same article in a slightly revised form also appeared in Russian; see P. Iu. Shmidt, "Anabioz dozhdevykh chervei," *Trudy Petrogradskogo obshchestva estestvoispytatelei*, 1920, 50(1): 11–28.

109. Schmidt [Shmidt], "Anabiosis of the Earthworm," 69.

110. *Ibid*, 70.

111. For a detailed analysis of the material conditions of urban life in the aftermath of the Bolshevik Revolution, see A. A. Il'iukhov, *Zhizn' v epokhu peremen. Material'noe polozhenie gorodskikh zhitelei v gody revoliutsii i grazhdanskoi voiny* (Moscow: ROSSPEN, 2007).

112. For a detailed analysis of the situation in Petrograd during the Revolution and the Civil War, see V. A. Shishkin, ed., *Petrograd na perelome epokh: gorod i ego zhiteli v gody revoliutsii i grazhdanskoi voiny* (St. Petersburg: Dmitrii Bulanin, 2000).

113. For an analysis of black market economy during the War Communism, see A. Iu. Davydov, *Meshochniki: Nelegal'noe snabzhenie rossiiskogo naseleniia i vlast', 1917–1921* (Moscow: Nauka, 2002).

114. See, for instance, P. Iu. Shmidt, *Omolazhivanie* (Petrograd: P. P. Soikin, 1923); idem, *Sila zhizni* (Petrograd: Seiatel', 1923); idem, "Psikhicheskaia zhizn' chelovekoobraznykh obez'ian," *Zvezda*, 1924, 6: 214–26; idem, *Zanimatel'naia zoologiia* (Leningrad: Nachatki znanii, 1924), and many others.

115. See P. Iu. Shmidt, "Zimniaia spiachka zhivotnykh," *V masterskoi prirody*, 1919, 1: 11–18.

116. P. Iu. Shmidt, *Zagadki zhizni* (Peterburg: GIZ, 1920). The title of the book—*Riddles of Life*—was a clear allusion to Ernst Haeckel's *Die Welträtsel*, which in Russian was translated as "Mirovye zagadki" (Riddles of the Universe).

117. P. Iu. Shmidt, *Osnovy zhizni. Populiarnoe vvedenie v biologiiu* (Petrograd: GIZ, 1920); idem, *Osnovy zhizni* (Moscow-Petrograd: GIZ, 1924), 3d edition.

118. P. Iu. Shmidt, "Opyty nad vysushivaniem zhivotnykh," *Trudy I Vserossiiskogo s"ezda zoologov, anatomov i gistologov* (Petrograd: n.p., 1922), 165–66.

119. P. Iu. Shmidt, *Anabioz (Iavlenie ozhivaniia)* (Peterburg-Moscow: L. D. Frenkel', 1923).

120. Shmidt, *Anabioz*, 170.

121. See, for instance, "Anabioz v bor'be s vrediteliami," *Nauka i tekhnika*, 1923, 16: 9; V. Timofeev, "Iavleniia skrytoi zhizni (anabioz)," *ibid*, 1924, 4: 6–7; M. P. Vinogradov, "Anabioz ili iavlenie ozhivaniia," *Krasnyi zhurnal*, 1924, 12: 749–57; V. Ryzhkov, *Mezhdu zhizn'iu i smert'iu* (Khar'kov: Iunyi leninets, 1925).

122. "Anabioz," in P. Iu. Shmidt, ed. *Malaia biologicheskaia entsiklopediia* (Moscow-Leningrad: L. D. Frenkel', 1924), 8.

123. P. Iu. Shmidt, "Anabioz," *Chelovek i priroda*, 1924, 3: 197–208; "Problema zhizni i smerti," *Izvestiia*, 14 March 1925, 4.

124. P. Iu. Shmidt, "Smert' i bessmertie s biologicheskoi tochki zreniia," *Novyi mir*, 1925, 5: 130–41.

125. P. Iu. Shmidt, "Na grani zhizni i smerti," *Iskry nauki*, 1926, 3: 18–21; 6: 10–15; B. D. Morozov, "Mezhdu zhizn'iu i smert'iu (anabioz)," *Iskra*, 1927, 3: 14–18; P. Iu. Shmidt, *Zhizn', smert' i bessmertie* (Moscow: Gudok, 1928).

126. At least, my search through 1920s Soviet biological and medical periodicals has yielded only ten articles that in any way touched upon anabiosis.

127. See D. F. Sinitsyn, "Problema anabioza u mlekopitaiushchikh," *Zapiski Belorusskogo gosudarstvennogo instituta sel'skogo khoziaistva*, 1923, 1: 21–37.

128. See G. N. Pavlov, "Vliianie nizkoi temperatury na spermatozoidy," *Russkii fiziologicheskii zhurnal*, 1927, 10 (3–4): 291–99.

129. See V. V. Efimov, "Vymerzanie i pereokhlazhdenie prosteishikh," *Russkii protistologicheskii zhurnal*, 1922, 1: 153–68; N. N. Plavil'shchikov, "Zametki ob anabioze u kolovratok (*Rotatoria*). 1 Anabioz u *Philodina roseola*," *Russkii gidrobiologicheskii zhurnal*, 1925, 4 (3–6): 98–102; M. P. Vinogradov, "Anabioz pri vysushivanii. Opyty s dozhdevymi cherviami," *Zapiski Leningradskogo sel'sko-khoziastvennogo instituta*, 1927, 4: 689–702; S. A. Zernov, "O zimovke vodnykh organizmov vo l'du i merzloi zemle, po materialam N. V. Boldyrevoi, P. P. Sharminoi i Iu. D. Shmelevoi. O 'pagone'–novyi termin," *Russkii gidrobiologicheskii zhurnal*, 1928, 7(1–2): 1–8; N. L. Sakharov, "K izucheniiu kholodostoikosti nasekomykh," *Zhurnal opytnoi agronomii Iugo-Vostoka* (Saratov), 1928, 6(2): 85–104; the same article also appeared in English: N. Sacharov, "Studies in cold resistance in insects," *Ecology*, 1930, 11(3): 505–17.

130. For a brief overview of Shmidt's research in the 1920s, see Margarita Shmidt, *Shmidty*.

131. On the Institute of Biophysics and its research, see the director's report of 1928 in GARF, f. A-482, op. 1, d. 635, ll. 354–59 rev.

132. For a brief biography of Iaroslavskii, see V. L. Genis, *Nevernye slugi rezhima: Pervye sovetskie nevozvrashchentsy* (Moscow: n. p., 2009), vol. 1: 379–439.

133. Aleksandr Iaroslavskii, *Poema anabioza* (Petrograd: Komitet poezii biokosmistov-iimortalistov, 1922). All the following quotations are from this source.

134. On the "new anarchism" and its break with the "old anarchism" or, as biocosmists themselves termed it, the "doctrines of fathers," see Aleksandr Sviatogor, "'Doktrina ottsov' i anarkhizm-biokosmizm," *Biokosmist*, 1922, 3–4: 3–21.

135. On the goals of biocosmists, see "Deklarativnaia rezoliutsiia," *Biokosmist*, 1922, 1: 1; and Aleksandr Sviatogor, "Nashi utverzhdeniia," *Biokosmist*, 1922, 1: 3–5. The latter article articulated the "twelve commandments" of biocosmism.

136. A few years later, in 1927, Iaroslavskii also published an SF novel, tellingly entitled the *Argonauts of the Universe*, that would further advance this belief and would also employ anabiosis. See A. Iaroslavskii, *Argonavty Vselennoi* (Moscow, 1927).

137. One of the earliest uses of anabiosis as a kind of time machine is an SF novel, *The Coming World*, written by Iakov Okunev, and serialized in the *Worker's Gazette* in the fall of 1922; see Ia. Okunev, "Griadushii mir," *Rabochaia gazeta*, September 3 to November 29, 1922. The next year the novel was issued in a book format.

138. In the late twentieth century, "cryogenic prisons" would become a staple of SF novels and particularly movies, including *Knight Rider 2000* (1991), *Demolition Man* (1993), or *Minority Report* (2002).

139. See John Hunter, "Lectures on the principles of Surgery," in James F. Palmer, ed., *The complete works of John Hunter* (Philadelphia: Haswell, Barrington, and Haswell, 1841), vol. 1, Chapter VIII, "On the heat of animal," 70–88, here–76.

140. Shmidt, *Anabiosis*, 1923, 118.

141. See Aleksei Tolstoi, "Fabrika molodosti," in idem, *Sobranie sochinenii v 15-ti tomakh* (Moscow: Nedra, 1930), 13: 114–80.

142. Indeed, two years earlier, in a short story entitled "Blue cities," Tolstoy had already used this idea in passing. The protagonist of the story envisions himself in the year 2024 and explains to his audience: "A half a century ago, when I had been dying as a very old man, the government included me in the 'list of youth.' One could get on that list only for great services to the people. I had been subjected to a 'full rejuvenation,' according to the newest system: I had been frozen in a chamber filled with nitrogen and exposed to the powerful magnetic currents that changed the very molecular structure of the body. Then [my] entire internal secretion had been refreshed by the transplantations of monkey glands." See A. N. Tolstoi, "Golubye goroda," *KN*, 1925, 4: 3–32, here–6.

143. A character by the same name but with the title of "professor" also figures in another short story that placed the questions of life and death at its center: See Aleksei Tolstoy, "Podkidnye duraki," *Novyi mir*, 1928, 11: 59–71; for a brief literary analysis of this story, see Irene Masing-Delic, *Abolishing Death: a Salvation Myth of Russian Twentieth-Century Literature* (Stanford, CA: Stanford University Press, 1992), 195–96.

144. For an analysis of the concurrent "rejuvenation craze" and various methods of rejuvenation pursued in Russia during the 1920s, see Chapter 5.

145. Bor. Pil'niak and N. M. Fedorovskii, "Delo smerti," *Novyi mir*, 1928, 2: 133–40.

146. On Pil'niak, see T. R. N. Edwards, *Three Russian Writers and the Irrational: Zamyatin, Pilnyak, and Bulgakov* (New York: Cambridge University Press, 1982); Gary L. Browning, *Boris Pilnyak: Scythian at a Typewriter* (Ann Arbor, MI: Ardis Publishers, 1985); and *Boris Pil'niak: opyt segodniashnego prochteniia* (Moscow: Nasledie, 1995).

147. It is difficult to attribute to each author a particular part in creating the story. Stylistically, it shows Pil'niak's hand.

148. For a short, fictionalized biography of Fedorovskii, see I. V. Paramonov, *Chelovek redkoi sud'by* (Moscow: Politizdat, 1973); for an analysis of Fedorovskii's scientific activities, see I. V. Paramonov and N. P. Korobochkin, *Nikolai Mikhailovich Fedorovskii (1886–1956)* (Moscow: Nauka, 1979); for a bibliography of his publications, see *Nikolai Mikhailovich Fedorovskii: Materialy k bibliografii uchenykh SSSR* (Moscow: Nauka, 1986).

149. For more details on Kol'tsov and his institute, see Chapters 5 and 6.

150. On the history of Russian eugenics, see Mark B. Adams, "Eugenics in Russia," in *The Well-born Science: Eugenics in Germany, France, Brazil, and Russia*, edited by M. B. Adams (New York: Oxford University Press, 1989), 153–229; and Nikolai Krementsov, "From 'Beastly Philosophy' to Medical Genetics: Eugenics in Russia and the Soviet Union," *Annals of Science*, 2011, 68 (1): 61–92.

151. For details of the debate over which method should be employed to preserve Lenin's corpse, see Iu. M. Lopukhin, *Pravda i mify o bolezni, smerti i bal'zamirovanii Lenina* (Moscow: Meditsina, 1997). For an English-language account, see Ilya Zbarsky and Samuel Hutchinson, *Lenin's Embalmers* (London: Harvill Press, 1998).

152. "Grey, dear friend, is all theory and green the golden tree of life."

CHAPTER 4

1. See P. Bakhmet'ev, "Zaveshchanie milliardera," *Estestvoznanie i geografiia*, 1904, 10: 1–24; 11: 1–31; 12: 1–47.

2. For historical accounts of this institution, see the five-volume *Centennial History of the Carnegie Institution of Washington* (New York: Cambridge University Press, 2004–6), vol. 1–5.

3. See, for instance, a response to Bakhmet'ev's vision by an eminent professor of zoology at Iur'ev University, K. Sent-Iler, "Po povodu internatsional'nykh kongressov," *Spravochnyi listok biologa*, 1907, 10–11: 145–50; or the vision of an "Institute of Experimental Sciences" advanced by a well-known physicist, Boris Veinberg, *Liudi zhizni, dumaite o griadushchikh pokoleniiakh* (Moscow: I. D. Sytin, 1907).

4. *Svobodnaia assotsiatsiia dlia razvitiia i rasprostraneniia polozhitel'nykh nauk* (Petrograd: n. p., 1917), 5.

5. *Ibid*, 6.

6. "Rech' Akademika I. P. Pavlova," *ibid*, 25–29, here–27.

7. For details, see Krementsov, "Big Revolution, Little Revolution"; for a general analysis of the innovative nature of institution-building under the Bolsheviks, see M. S. Bastrakova, *Stanovlenie sovetskoi sistemy organizatsii nauki (1917–1922)* (Moscow: Nauka, 1973); and Loren R. Graham, "The Formation of Soviet Research Institutes: A Combination of Revolutionary Innovation and International Borrowing," *Social Studies of Science*, 1975, 5: 303–29.

8. Ernest H. Starling, "The Croonian Lectures on 'The chemical correlations of the functions of the body,'" *Lancet*, 1905, 2: 339–41, 423–25, 501–03, 579–83; here–339, italics are added.

9. Louis Berman, *The Glands Regulating Personality; A Study of the Glands of Internal Secretion in Relation to the Types of Human Nature* (Macmillan, 1922), 263, italics are added. On Berman and his role in the development of endocrinology in the United States, see Christer Nordlund, "Endocrinology and expectations in 1930s America: Louis Berman's ideas on new creations in human beings," *British Journal of the History of Science*, 2007, 40(1): 83–104. For a more general analysis of the visionary elements in early endocrinology in the United States, see J. E. Rechter, "'The glands of destiny': a history of popular, medical and scientific views of the sex hormones in 1920s America," PhD diss., University of California, Berkeley, 1997.

10. For a detailed analysis of the institutional development of endocrinology in Russia as compared to other countries, see Krementsov, "Hormones and the Bolsheviks."

11. For a comprehensive overview of the intellectual development of research on internal secretions, see Victor Cornelius Medvei, *The history of clinical endocrinology: a comprehensive account of endocrinology from earliest times to the present day* (Pearl River, NY: Parthenon, 1993).

12. For a short essay on the history of Russian research on internal secretions up to the 1920s, see D. M. Rossiiskii, "Ocherk istorii razvitiia endokrinologii v Rossii," *Mediko-biologicheskii zhurnal*, 1926, 1: 64–72. The same author also compiled a complete bibliography of Russian publications on endocrinology from 1860 to 1926, which includes more than two thousand entries; see D. M. Rossiiskii, *Sistematicheskii ukazatel' russkoi literatury po endokrinologii i organoterapii ot ee vozniknoveniia do 1926 g.* (Moscow-Leningrad: Gosizdat, 1926); idem, "Sistematicheskii ukazatel' russkoi literatury po endokrinologii i organoterapii za 1926 god," *Mediko-biologicheskii zhurnal*, 1928, 2 (Supplement): 1–19.

13. See L. Galuzinskii, *O znachenii nadpochechnykh zhelez v zhivotnom organizme* (St. Petersburg: P. A. Kulish, 1862).

14. On Brown-Séquard and the rise of organotherapy, see Merriley Borell, "Brown-Séquard's Organotherapy and its Appearance in America at the End of the 19th century," *Bulletin of the History of Medicine*, 1976, 50: 309–20; idem, "Organotherapy, British Physiology, and Discovery of the Internal Secretions," *Journal of the History of Biology*, 1976, 9: 235–68; idem, "Organotherapy and the Emergence of Reproductive Endocrinology," *ibid*, 1985, 18: 1–30.

15. For the only English-language account of the Russian pharmaceutical industry, see Mary Schafferer Conroy, *In Health and Sickness: Pharmacy, Pharmacists, and the Pharmaceutical Industry in Late Imperial, Early Soviet Russia* (New York: Columbia University Press, 1994). Unfortunately, this voluminous tome contains very little information on Pel's and Ferrain's firms, and even less on organotherapy.

16. See A. V. Pel', *Spermin. Novoe stimuliruiushchee sredstvo* (St. Petersburg: Ettinger, 1890); A. von Poehl, J. Tarchanoff, Alf. von Poehl, and P. Wachs, *Rational Organotherapy with Reference to Urosemiology* (London: J. & A. Churchill, 1906).

17. For a contemporary account of Pel's laboratory, see A. V. Pel', "Obzor 25-letnei deiatel'nosti Khimicheskoi laboratorii prof. doktora khimii Pelia," *Zhurnal meditsinskoi khimii i farmatsii*, 1898, 19–20: 245–62.

18. For a brief historical account of Pel's involvement with organotherapy, see Ingrid Kastner, "Alexander Poehl (1850–1908) und die Organotherapie," in Regine Pfrepper, Ingrid Kastner, and Dietriechvon Engelhardt, eds., *Von Samuel Gottleib Reise durch Russland bis zum Niedergang Apothekerfamilie Poehl* (Shaker Verlag, 2001), 251–71.

19. For a contemporary overview of endocrinology in the United States, see Benjamin Harrow, *Glands in Health and Disease* (New York: E. P. Dutton & Co., 1922).

20. See *Ukazatel' 1-oi oblastnoi vystavki kozovodstva* (Moscow, 1914), 22–23.

21. Tobolkin's 1925 Curriculum Vitae preserved in the archives is very opaque on details of his pre-revolutionary life and career. See GARF, f. A-482, op. 41, d. 3347, ll. 2–6. I am profoundly grateful to Tobolkin's niece, Ol'ga Georgievna Amerikova, for clarifying certain details of her uncle's life.

22. For the history of this institute, see *Istoriia Kazanskogo gosudarstvennogo veterinarnogo instituta imeni N. E. Baumana. Ocherki* (Kazan': Tatarskoe knizhnoe izdatel'stvo, 1973).

23. In 1893, the Kazan' institute had only 212 students in toto; see T. I. Mineeva, *Istoriia veterinarii* (St. Petersburg: Lan', 2005), 228.

24. See A. E. Ivanov, *Vysshaia shkola Rossii v kontse XIX-nachale XX veka* (Moscow: n. p., 1991), 91–93; and Mineeva, *Istoriia veterinarii*, 230.

25. For a brief survey of the history of Russian veterinary science and education, see Mineeva, *Istoriia veterinarii*.

26. See *Vsia Moskva* (Moscow: A. S. Suvorin, 1899), 271, 722; *Vsia Moskva* (Moscow: A. S. Suvorin, 1900), 321–22, 849.

27. See I. V. Korneev, "Zoologicheskaia epopeia, (1890–1910)," on the website of the Leningrad Zoo at http: //www.lenzoo.ru, accessed on January 15, 2006.

28. See *Vsia Moskva* (Moscow: A. S. Suvorin, 1910), 435.

29. See *Rossiiskoe kozovodstvo*, 1911, 6: 14.

30. See GARF, f. A-482, op. 25, d. 18, l. 2.

31. See M. S. Karpov, "Pervaia vystavka moskovskogo otdela Rossiiskogo obshchestva kozovodstva," *Rossiiskoe kozovodstvo*, 1911, 5: 4–7.

32. In Russia this condition was named after the German physician Carl Adolph Basedow (1799–1854), who in 1840 produced a classic description of its symptoms. In English-speaking countries, it was known as Graves' disease, after the Irish physician Robert James Graves (1796–1853).

33. See *Rossiiskoe sel'sko-khoziastvennoe zhivotnovodstvo i rossiiskoe kozovodstvo*, 1912, 5: 17–18.

34. See, for instance, D. Burghart, "Beitrage zur Organotherpie," *Deutsche meditzinische Wochenschrift*, 1899, 25: 610–11, 627–29; Otto Lanz, "Weitere Mitteilung über serotherapeutische Behandlung des Morbus Basedowii," *Muenchener meditzinische Wochenschrift*, 1903, 50 (4): 146–49.

35. See "Moloko ot koz s udalennymi shchitovidnymi zhelezami kak lechebnoe sredstvo pri Basedovoi bolezni," in *Ukazatel' 1-oi oblastnoi vystavki kozovodstva* (Moscow, 1914), 29–30.

36. See N. Babushkin, "K voprosu o moloke destrumirovannykh koz i izmeneniiakh v obmene veshchestv u operirovannykh," in *Ukazatel' 1-oi oblastnoi vystavki kozovodstva* (Moscow, 1914), 24–27.

37. For a brief biography of Shervinskii, see A. M. Tsiklik, *V. D. Shervinskii* (Moscow: Meditsina, 1978).

38. See his memoirs preserved in the Museum of the History of Medicine, the First Moscow Medical University (hereafter—MIM), OF 605/699: NVF 737/295.

39. See V. D. Shervinskii, *O zhirovoi embolii* (Moscow: A. Gattsuk, 1879).

40. On the professionalization and professional activities of Russian physicians of that era, see Nancy M. Frieden, *Russian Physicians in an Era of Reform and Revolution, 1865–1905* (Princeton, NJ: Princeton University Press, 1981).

41. See V. D. Shervinskii, "O vnutrennei sekretsii i ee klinicheskom znachenii," *Meditisnskoe obozrenie*, 1910, 73 (11): 1035–74.

42. See V. D. Shervinskii, "K voprosu ob izmeneniiakh v organizme tireodektomirovannoi kozy," *Trudy Moskovskogo terapevticheskogo obshchestva*, 1913: 224.

43. V. D. Shervinskii, "O lechenii bazedovoi bolezni," *Voprosy nauchnoi meditsiny*, 1913, 5–6: 465–96.

44. On the early history of Narkomzdrav, see Susan G. Solomon and John F. Hutchinson, eds., *Health and Society in Revolutionary Russia* (Bloomington: Indiana University Press, 1990). Although there are numerous articles commemorating Semashko's role in creating the Soviet system of medical services and public health, there is no book-length analytical biography of this remarkable man. For a brief, Soviet-era, hagiographic biography, see B. D. Petrov and B. M. Potulov, *N. A. Semashko* (Moscow: Meditsina, 1974).

45. On the hostility of the majority of Russian physicians to the Bolshevik regime, see Peter F. Krug, "Russian Public Physicians and Revolution: The Pirogov Society, 1917–1920." PhD diss. University of Wisconsin-Madison, 1979.

46. GARF, f. A-482, op. 1, d. 98, ll. 45.

47. See Bonch-Bruevich's collection in NIOR RGB, f. 369. I am grateful to Marina Sorokina for calling my attention to this collection.

48. Bauman would attain a "saint's" status as a martyr of the first Russian revolution of 1905–06, and after the Revolution the Kazan' Veterinary Institute would be given his name. For his heavily fictionalized biography, see M. A. Novoselov, *Nikolai Ernestovich Bauman* (Moscow: Molodaia gvardiia, 1955).

49. See GARF, f. 102, op. 3, d. 383, l. 1; d. 880, l. 4; op. OO-1905, d. 104, part 6, ll. 8, 13–13 rev. I m grateful to Marina Sorokina for calling my attention to this collection.

50. See GARF, f. 102, op. OO-1907, d. 9, part 35, l. 185.

51. See his memoirs, V. D. Bonch-Bruevich, *V. I. Lenin v Petrograde i v Moskve* (Moscow: Politizdat, 1955).

52. For an English-language biography of Krupskaia, see Robert H. McNeal, *Bride of the Revolution: Krupskaya and Lenin* (Ann Arbor: University of Michigan Press, 1972). For a short biography of Velichkina, see M. Ia. Iarovinskii, *V. M. Bonch-Bruevich (Velichkina), 1868–1918* (Moscow: Meditsina, 1990).

53. NIOR RGB, f. 369, karton 212, d. 14, l. 1.

54. GARF, f. A-482, op. 1, d. 98, l. 48.

55. GARF, f. A-482, op. 1, d. 98, l. 59.

56. See GARF, f. A-482, op. 25, d 18, l. 2; op. 1, d. 98, ll. 75.

57. GARF, f. A-482, op. 25, d 18, ll. 4–7 rev.

58. GARF, f. A-482, op. 1, d. 82, ll. 205, 207–207-rev.

59. GARF, f. A-482, op. 1, d. 82, l. 193.

60. For the early history of this institution, see the special volume issued for its fifth anniversary, L. A. Tarasevich and V. A. Liubarskii, eds., *Gosudarstvennyi institut narodnogo zdravookhraneniia imeni Pastera ("GINZ"), 1919–1924* (Moscow: GINZ, 1924). Hereafter references to this collection will be given as *GINZ*, 1924.

61. See GARF, f. A-482, op. 1, d. 82, ll. 116–123-rev.

62. GARF, f. A-482, op. 1, d. 82, ll. 223, 227.

63. GARF, f. A-482, op. 1, d. 82, ll. 252-rev-253-rev.

64. GARF, f. A-482, op. 1, d. 82, l. 249.

65. GARF, f. A-482, op. 1, d. 116, ll. 24-rev.-25, 130.

66. In 1922, Pletnev would become a professor of therapy at the Kremlin clinic; see *Kremlevskaia meditsina (ot istokov do nashikh dnei)* (Moscow: Izvestiia, 1997), 89.

67. GARF, f. A-482, op. 1, d. 82, l. 303; and op. 25, d 18, ll. 16–16 rev.

68. GARF, A-482, op. 1, d. 82, ll. 316–25. All the following quotations are from this source.

69. These huge sums reflect not the actual value of the preparation and the projected facility for its manufacture, but the enormous inflation of the Russian currency induced by the Bolshevik Revolution.

70. GARF, f. A-482, op. 25, d. 91, ll. 6–7. All the following quotations are from this source.

71. GARF, f. A-482, op. 1, d. 82, l. 303.

72. For a short biography of Bakh, see L. A. Bakh and I. A. Oparin, *Aleksei Nikolaevich Bakh* (Moscow: Izd. AN SSSR, 1957).

73. GARF, f. A-482, op. 25, d. 41, l. 47.

74. For the actual description of Bakh's technique and its use at the farm, see A. N. Bakh and S. Zubkova, "Kolichestvennoe opredelenie katalazy, proteazy, perooksidazy i esterazy v kaple krovi," in *GINZ*, 1924, 223–225; A. N. Bakh and E. P. Kheraskova, "Pokazatel' katalazy u tireodektomirovannykh koz," in *GINZ*, 1924, 227.

75. See GARF, f. A-482, op. 25, d. 41, l. 46; and op. 1, d. 82, l. 365, 370.

76. See GARF, f. A-482, op. 1, d. 121, l. 72.

77. GARF, f. A-482, op. 25, d. 41, ll. 55–56.

78. "Cheka" was the popular abbreviation for the name of the "extraordinary commission to combat counter-revolution, sabotage, and speculation," which was the main instrument of the Red Terror during the Civil War and immediately after.

79. Indeed, according to internal party reports, nearly one-third of all visits to the doctor by top-level party functionaries were related to TB. See the Russian State Archive of Socio-Political History (hereafter—RGASPI), f. 17, op. 84, d. 704, l. 27.

80. See Shervinskii's collection in MIM.

81. See GARF, f. A-482, op. 25, d. 103, l. 9-rev.

82. GARF, f. A-482, op. 25, d 18, ll. 36–36 rev.

83. GARF, f. A-482, op. 25, d.104, l. 59.

84. GARF, f. A-482, op. 25, d. 18, l. 38.

85. GARF, f. A-482, op. 25, d. 104, ll. 55–60.

86. On the tactics and strategies of survival under the Bolshevik regime, see T. M. Smirnova, *"Byvshie liudi" Sovetskoi Rossii: Strategii vyzhivaniia i puti integratsii, 1917–1936 gg.* (Moscow: Mir Istorii, 2003).

87. GARF, f. A-482, op. 1, d. 82, l. 253.

88. See *GINZ*, 1924, 25–27.

89. GARF, f. A-482, op. 25, d. 18, ll. 39–42.

90. GARF, f. A-482, op. 25, d. 104, l. 151; d. 177, l. 5.

91. See *GINZ*, 1924, 21–23.

92. GARF, f. A-482, op. 25, d. 177, l. 158-rev.

93. GARF, f. A-482, op. 25, d. 236, l. 7-rev.

94. See B. I. Il'in and L. V. Redina, "Khimicheskii sostav moloka tireodektomirovannykh koz," in *GINZ*, 1924, 152–54; O. P. Molchanova, "K voprosu o vliianii gipofiza na rost zhivotnykh," *ibid*, 192–93.

95. See Nabliudatel', "Laboratoriia v pivnoi," *VM*, September 30, 1925, 2.

96. GARF, f. A-482, op. 25, d. 279, ll. 32–33.

97. GARF, f. A-482, op. 25, d. 279, ll. 72–74.

98. GARF, f. A-482, op. 25, d. 310, ll. 5–8. All the following quotations are from this document.

99. See N. G. Zalkind, *Moskovskaia shkola antropologii* (Moscow: MGU, 1975).

100. On the history of Soviet social hygiene, see Susan G. Solomon, *Soviet social hygiene, 1917–1936*, unpublished manuscript, 1997. I am grateful to Susan G. Solomon for the opportunity to read this work.

101. The report was published in the first issue of *Vestnik endokrinologii*, 1925, 1 (1): V–VII.

102. See "Moskovskii organoterapevticheskii institut," *Izvestiia*, April 11, 1924, 7.

103. GARF, f. A-2307, op. 17, d. 115, l. 284.

104. GARF, f. A-482, op. 28, d. 3, l. 20.

105. GARF, f. A-482, op. 28, d. 3, l. 19.

106. See the protocol of this meeting in *Vestnik endokrinologii*, 1925, 1 (2): 169–70.

107. V. D. Shervinskii, "Rech' pri otkrytii Rossiiskogo endokrinologicheskogo obshchestva," *Vestnik endokrinologii*, 1925, 1 (1): 67–76. All the following quotations are from this source.

108. See "V endokrinologicheskom obshchestve," *Izvestiia*, December 21, 1924, 5; "V endokrinologicheskom obshchestve," *Izvestiia*, March 25, 1925, 6; "V endokrinologicheskom obshchestve," *Izvestiia*, February 21, 1926, 5; "V endokrinologicheskom obshchestve," *Izvestiia*, May 28, 1926, 4.

109. GARF, f. A-482, op. 25, d. 310, ll. 100, 109, 116–17.

110. GARF, f. A-482, op. 25, d. 320, ll. 1–4.

111. GARF, f. A-482, op. 25, d. 311, ll. 76–78.

112. See Milton Hoefle, "The early history of Parke-Davis and Company," *Bulletin for the History of Chemistry*, 2000, 25 (1): 28–34.

113. The Moscow firm Karl Ferrein and the Petrograd firm Aleksandr Pel' produced a variety of adrenal extracts. Pel' even advertised a preparation called "adrenal-chloride." There are some indications that during World War I Ferrein began research on the manufacture of adrenalin, but it remains unclear whether either firm actually produced pure adrenalin in any quantity.

114. See the unsigned and untitled editorial in the first issue of *Herald of Endocrinology*, *Vestnik endokrinologii*, 1925, 1 (1): III–IV.

115. See "Institut eksperimental'noi endokrinologii," *Izvestiia*, October 8, 1925, 3.

116. See *Protokoly zasedanii Rossiiskogo endokrinologicheskogo obshchestva* (Moscow, 1926–1929), Vols. 1–6. For the protocols of the Leningrad Branch's sessions, see *Vestnik endokrinologii*, 1927, 2 (2): 185–90.

117. For more details on the station's establishment, see Chapter 5.

118. See "Insulin budet," *VM*, June 16, 1927, 2.

119. See "Dostizheniia sovetskoi meditsiny," *Izvestiia*, January 20, 1928, 4; "Vypushchen v prodazhu insulin," *Pravda*, July 31, 1928, 8.

120. "Pobeda sovetskikh laboratorii. Sovetskii insulin," *VM*, January 11, 1928, 2.

121. See "Khodataistva o nagrazhdenii zvaniem zasluzhennykh deiatelei nauki," *Izvestiia*, October 21, 1928, 2; "V Sovnarkome RSFSR," *ibid*, December 22, 1928, 2; "Sovnarkom prisvoil zvanie zasluzhennogo deiatelia nauki direktoru gosudarstvennogo instituta eksperimental'noi endokrinologii V. D. Shervinskomu," *Pravda*, December 22, 1928, 4.

122. See V. D. Shervinskii and G. P. Sakharov, eds., *Osnovy endokrinologii* (Leningrad: Prakticheskaia meditsina, 1929).

123. See *Klinicheskaia endokrinologiia. Osnovy meditsinskoi endokrinologii dlia vrachei i studentov* (Leningrad: Kubuch, 1930).

124. See GARF, f. A-482, op. 28, d. 211, l. 2 rev.

125. Of course, the story of Tobolkin-Shervinskii's partnership does not cover the entire institutional development of endocrinology in the Soviet Union. The multiplicity of governmental agencies supporting science in the 1920s allowed other researchers to find patrons outside Narkomzdrav. Based on the former research laboratory of the Ferrein pharmaceutical firm in Moscow, a small endocrinology department emerged within the Chemical-Pharmaceutical Institute founded by the Supreme Council for National Economy (VSNKh) in 1921. Another group created an endocrinological laboratory at the Timiriazev Biological Institute established the same year under the patronage of Narkompros. Yet another department appeared in 1925 within

Moscow University's Clinical Institute for Experimental Therapy, also funded by Narkompros. Moreover, since each republic of the Soviet Union had its own system of governmental agencies in charge of science and medicine, the development of endocrinology reached well beyond the traditional centers of Moscow and Leningrad. During the 1920s, several important endocrinological institutions emerged in the Ukraine. For a contemporary overview of the development of endocrinology in the Ukraine, see M. S. Shindler and E. Ia. Reznitskaia, *Ukrainskaia endokrinologiia i organoterapiia za 15 let revoliutsii* (Khar'kov: NKZ USSR, 1936). For a short historical account of organotherapy and endocrinology in the Ukraine, see N. P. Arzhanov and E. G. Arzhanova, "Ot organoterapii do endocrinologii," *Provizor*, 1999, no. 7, at http: //www.provisor.com.ua/archive/1999/N7/arganov.htm, accessed February 17, 2006.

126. See, for instance, "Izlechenie sakharnoi bolezni," *Izvestiia*, April 4, 1924, 4; "Depeshi chelovescheskogo organizma," *ibid*, June 12, 1924, 7; P. Iu. Shmidt, "Fabrika gormonov," *VM*, May 8, 1926, 2; Il. Poltavskii, "Standarty bodrosti," *ibid*, May 16, 1927, 2; M. Ia. Breitman, "Nastoiashchee i budushchee v uchenii ob organakh vnutrennei sekretsii," *VZ*, 1927, 6: 333–36; S. Fainshtein, "Lechenie organopreparatami," *Iskry nauki*, 1927, 8: 313–15; N. Bakunin, "Zhivaia voda dvadtsatogo veka," *ibid*, 1927, 10: 337–38; B. Zavadovskii, *Ocherki vnutrennei sekretsii* (Leningrad: Priboi, 1928); and many others.

127. See P. V. Bochkarev, *Za zhivoi vodoi* (Moscow: Narkomzdrav, 1928).

128. See, for instance, A. Irkutov, "Bessmertie," *Bor'ba mirov*, 1924, 3: 30–37; and N. Dolin, "Krov' mira," *Mir prikliuchenii*, 1927, 6: 2–12.

129. See Aleksandr Beliaev, "Chelovek, poteriavshii svoe litso," *Vokrug sveta*, 1929, 19: 1–5; 20: 6–9; 21: 18–22; 22: 18–22; 23: 19–23; 24: 19–23; 25: 21–23.

130. His last name *Sorokin* is a clear allusion to Serge Voronoff, a French surgeon of Russian extraction, who achieved extraordinary notoriety in the 1920s by transplanting monkey glands to humans. *Soroka* means magpie in Russian and *vorona* means crow. Magpies and crows are closely related species and the close links between magpie and crow are played out in a number of Russian colloquialisms. On Voronoff, his works, and his popularity in Russia, see Chapter 5.

131. Dr. Hascombe in Julian Huxley's 1926 story "The tissue-culture king" did exactly the opposite, transforming normal individuals into the obese, dwarfs, giants, and other "circus freaks" to win the support of his captors' king.

132. Aldous Huxley's *Brave New World* (1932) provides a similar illustration to the contemporary infatuation with "hormonal control."

CHAPTER 5

1. The first edition of the novella appeared in Paris in 1968; see Mikhail Bulgakov, *Sobach'e serdtse* (Paris: IMCA-Press, 1968). The first English translation came out the same year; see Mikhail Bulgakov, *The Dog's Heart*, transl. by Michael Glenny (New York: Harcourt, Brace & Wolff, 1968). The majority of the later English translations, however, use the title "The Heart of a Dog," which is somewhat inaccurate: translated back to Russian it would read "*Serdtse sobaki*," not "*Sobach'e serdtse*." Furthermore, "The Heart of a Dog" does not correspond to Bulgakov's thinking about a title for this novella. Bulgakov had considered several variants, and originally wanted to title the novella, "The Dog's Life" (*sobach'ia zhizn'*), which in Russian has very specific connotations: "terrible life" and "living like a stray dog." Apparently, this title was too provocative for a novella that depicted life in Bolshevik Moscow and thus Bulgakov decided to use a

more neutral title: "Sobach'e serdtse" instead of "Sobach'ia zhizn'." See the foremost authority on Bulgakov, his biographer and commentator Marietta Chudakova's notes to the first publication of *The Dog's Heart* in Russia, in M. Chudakova, "Posleslovie," *Znamia*, 1987, 6: 135–41. For the text of the first publication of the novella in the Soviet Union, see Mikhail Bulgakov, "Sobach'e serdtse," *Znamia*, 1987, 6: 73–134; all of the following citations come from this edition with page numbers given in brackets.

2. The "verdict" is cited in Chudakova, *Zhizneopisanie Mikhaila Bulgakova*, 326.

3. See, for instance, Peter Doyle, "Bulgakov's Satirical View of Revolution in *Rokovye Iaitsa* and *Sobach'e Serdtse*," *Canadian Slavonic Papers*, 1978, 20 (4): 467–82.

4. Various scholars have observed in passing that the contemporary boom of interest in rejuvenation likely inspired the plot of Bulgakov's novella. See, for instance, Chudakova, "Posleslovie," *Znamia*, 1987, 6: 136; and Ellendea Proffer, *Bulgakov: Life and Work* (Ann Arbor: Ardis, 1984), 122. Yet aside from these superficial remarks, literary historians have failed to connect Bulgakov's novella to the contemporary experimental revolution in the life sciences. See also Diana Burgin, "Bulgakov's Early Tragedy of the Scientist-Creator: An Interpretation of *Dog's Heart*," *Slavic and East European Journal*, 1978, 22 (4): 494–508; Susanne Fusso, "Failures of Transformation in Bulgakov's Sobach'e serdce," *ibid*, 1989, 33 (3): 386–99; Mikhail Zolotonosov, "Masturbanizatsiia: Erogennye zony sovetskoi kuil'tury 1920–1930-kh godov," *Literaturnoe obozreniie*, 1991, 11: 93–99; Ronald D. LeBlanc, "Feeding a Poor Dog a Bone: The Quest for Nourishment in Bulgakov's Sobach'e Serdtse," *Russian Review*, 1993, 52(1): 58–78; E. A. Iablokov, *Khudozhestvennyi mir Mikhaila Bulgakova* (Moscow: Iazyki slavianskoi kul'tury, 2001), 53–57; Zsuzsa Hetenyi, "Rokovye serdtsa 20-kh. O povesti Bulgakova 'Sobach'e serdtse,'" *Studia Slavica* (Hungary), 2004, 49 (3–4): 365–79. The only attempt at reading the novella in the context of contemporary biomedical research is Yvonne Howell's article, "Eugenics, Rejuvenation, and Bulgakov's Journey into the Heart of Dogness," *Slavic Review*, 2006, 65(3): 544–62.

5. For instance, the choice of dog as the subject of Preobrazhenskii's experiment was a clear allusion to the Nobelist Ivan Pavlov, whose seminal work on conditional reflexes in dogs, *Twenty-year Experience in Studying the Highest Nervous Activity*, was issued with great fanfare by the state publishing house in late 1923. Furthermore, Pavlov was certainly one of the actual prototypes for Professor Preobrazhenskii. For instance, Preobrazhenskii's famous diatribe against "reading Soviet newspapers" most likely came from a lengthy article published in early 1924 by Nikolai Bukharin, a member of the Politburo, in the popular literary journal *Red Virgin Soil*. Entitled "On world revolution, our country, culture and other issues (A response to academician I. Pavlov,)" the article cited one of Pavlov's recent lectures, which, among other criticisms of the new regime, lambasted Bolshevik newspapers. See N. Bukharin, "O mirovoi revoliutsii, nashei strane, kul'ture i prochem (Otvet akademiku I. Pavlovu)," *KN*, 1924, 1: 170–88; 2: 104–99. (For Pavlov's attitude toward the Soviet press and Bukharin's mocking assessment, see no. 1: 176–78). Bulgakov read the journal on a regular basis and obviously was familiar with Bukharin's "response."

6. For dated but still useful overviews of the history of this quest, see Eric J. Trimmer, *Rejuvenation. The History of Idea* (London: Robert Hale, 1967); and Joseph T. Freeman, *Aging, Its History and Literature* (New York: Human Sciences Press, 1979).

7. On Brown-Séquard's life and works, see J. M. D. Olmstead, *Charles-Eduard Brown-Séquard: A Nineteen-Century Neurologist and Endocrinologist* (Baltimore: Johns Hopkins University Press,

1946); and Michael J. Aminoff, *Brown-Séquard: A Visionary of Science* (New York: Raven Press, 1993).

8. On Claude Bernard and his concept of internal secretions, see Frederic L. Holmes, *Claude Bernard and Animal Chemistry: The Emergence of a Scientist* (Cambridge: Harvard University Press, 1974); and Alan G. Wasserstein, "Death and the internal milieu: Claude Bernard and the origins of experimental medicine," *Perspectives in Biology and Medicine*, 1996, 39 (3): 313–26.

9. On Brown-Séquard and the rise of organotherapy, see Borell, "Brown-Séquard's Organotherapy and its Appearance in America at the End of the 19th Century"; idem, "Organotherapy, British Physiology, and Discovery of the Internal Secretions"; idem, "Organotherapy and the Emergence of Reproductive Endocrinology"; on public perception of organotherapy in the United States, see Bert Hansen, "New Images of a New Medicine: Visual Evidence for the Widespread of Therapeutic Discoveries in America after 1885," *Bulletin of the History of Medicine*, 1999, 73: 629–78; and idem, *Picturing Medical Progress from Pasteur to Polio* (Newark, NJ: Rutgers University Press, 2009).

10. See Elie Metchnikoff, "Études biologiques sur la vieillesse," *Annales de l'Institut Pasteur*, 1901, 15: 865–79; 1902, 16: 912–17; idem, *Études sur la nature humaine: Essai de philosophie optimiste* (Paris: Masson & C-ie, 1903).

11. Unfortunately, there is no book-length examination of Metchnikoff's ideas of senescence. For a brief discussion, see A. L. Cochrane, "Elie Metschnikoff and his theory of an 'instinct de la mort'"; Mark Aveline, "Elie Metschnikoff and his theory of an 'instinct de la vie'"; and Tomas Hugh, "Early thoughts on death, disease, and sex," *International Journal of Epidemiology*, 2003, 32 (1): 32–36.

12. For overviews of the history of gerontology in the US context, see W. Andrew Achenbaum, *Crossing Frontiers: Gerontology Emerges as a Science* (New York, Cambridge: Cambridge University Press, 1995); and Stephen Katz, *Disciplining Old Age: The Formation of Gerontological Knowledge* (Charlottesville: University Press of Virginia, 1996).

13. For a contemporary summary of the work on unicellular organisms, see H. S. Jennings, *Life and Death; Heredity and Evolution in Unicellular Organisms* (Boston: Gorham Press, 1920); for a historical assessment of this work, see A. J. Lustig, "Sex, Death, and Evolution in Proto- and Metazoa, 1876–1913," *Journal of the History of Biology*, 2000, 33: 221–46.

14. For a thorough contemporary overview of this concept, see Eugene Korschelt, *Lebensdauer, Altern und Tod* (Jena: Fischer, 1922). For a similar survey in English, see T. Brailsford Robertson, *The Chemical Basis of Growth and Senescence* (Philadelphia and London: J. B. Lippincott company, 1923).

15. One such scientist was Alexander Bogdanov. For details, see Krementsov, *A Martian Stranded on Earth*.

16. On Steinach's works, see Chandak Sengoopta, *The Most Secret Quintessence of Life: Sex, Glands, and Hormones, 1850–1950* (Chicago: University of Chicago Press, 2006); on Voronoff's works, see David Hamilton, *The Monkey Gland Affair* (London: Chatto & Windus, 1986). Unfortunately, neither book even mentions the situation in Russia. On the 1920s rejuvenation craze in various Western countries, see also B. P. Setchell, "The Testis and Tissue Transplantation: Historical Aspects," *Journal of Reproductive Immunology*, 1990, 18(1): 181–88; Erica R. Freeman, David A. Bloom, and Edward J. McGuire, "A Brief History of Testosterone," *Journal of Urology*, 2001, 165(2): 371–73; Diana Wyndham, "Versemaking and Lovemaking—W. B. Yeats' 'Strange

Second Puberty': Norman Haire and the Steinach Rejuvenation Operation," *Journal of the History of the Behavioral Sciences*, 2003, 39 (1): 25–50; Arnold Kahn, "Recovering Lost Youth: The Controversial and Colorful Beginnings of Hormone Replacement Therapy in Aging," *Journals of Gerontology: Series A; Biological sciences and medical sciences*, 2005, 60 (2): 142–47; E. M. Cuperschmid and T. P. R. de Campos, "Dr. Voronoff's Curious Glandular Xenoimplants," *Historia, Ciencias, Saude—Manguinhos* (Rio de Janeiro), 2007, 14 (3): 1–24; and Pope Brock, *Charlatan: America's Most Dangerous Huckster, the Man Who Pursued Him, and the Age of Flimflam* (New York: Crown Publishing. 2008).

17. The article was also issued as a brochure; Eugen Steinach, *Verjüngung durch experimentelle Neubelebung der alternden Pubertätsdrüse* (Berlin: Verlag von Julius Springer, 1920).

18. Serge Voronoff, *Vivre. Études des moyens de relever l'énergie vitale et de prolonger la vie* (Paris: B. Grasset, 1920).

19. On Kol'tsov's life and activities, see V. Polynin, *Prorok v svoem otechestve* (Moscow: Sovetskaia Rossiia, 1969), and B. L. Astaurov and P. F. Rokitskii, *Nikolai Konstantinovich Kol'tsov* (Moscow: Nauka, 1975).

20. For a description of the Big Practicum and its influence, see M. M. Zavadovskii, *Stranitsy zhizni* (Moscow: Izd. MGU, 1991), 37–56.

21. For a description of the colloquium's reports and discussions during its first two years of operation, see M. M. Zavadovskii, "Biologicheskii kollokvii N. K. Kol'tsova," *Nauchnye biulleteni*, 1914, 1: 153–72.

22. See the protocols of the colloquium's sessions kept among Zavadovskii's personal papers in ARAN, f. 1657, op. 1. d. 152, ll. 1–82.

23. See, for instance, Alexander Serebrovskii's report "on the determination of sex" delivered at the colloquium; A. S. Serebrovskii, "K voprosu ob opredelenii pola," *Nauchnye biulleteni*, 1914, 1: 168–69.

24. I. G. Kogan, "Proizvol'noe prevrashchenie samtsov v 'samok,' " *Nauchnye biulleteni*, 1914, 1: 1–20.

25. On Mikhail Zavadovskii's life and works, see his memoirs Zavadovskii, *Stranitsy zhizni*; and memoirs of his daughter, M. M. Zavadovskaia-Sachenko, "Pamiati moego otsa," *Repressirovannaia nauka* (St. Petersburg: Nauka, 1994), 2: 286–97.

26. S. Perel'muter, "Vtorichnye polovye priznaki," *Vestnik shaniavtsev*, 1918, 3: 64–65.

27. See Mikh. Zavadovskii, "Askaniia Nova," *KN*, 1922, 2(6): 286–98. On the history of the preserve, see E. V. Boreiko, *Askania Nova: Tiazhkie versty istorii, 1826–1993* (Kiev: Kievskii ekologo-kul'turnyi tsentr, 2001).

28. See N. Frishman, "Konferentsiia GINZa," *Vrachebnoe delo*, 1920–21, 1–6: V. Of course, Kol'tsov was not the only Russian scientist engaged into propaganda of Steinach's work. In December 1920, Gavriil Voskresenskii, a well-known urologist, delivered a series of lectures at various meetings of Moscow physicians on the "Struggle against old age in the experiments of Steinach and his school," assessing "scientific and practical import of these investigations" (see Gavr. Dm. Voskresenskii, "Bor'ba so strarost'iu v opytakh prof. Steinach'a i ego shkoly; nauchnoe i prakticheskoe znachenie etikh issledovanii," *Referativnyi meditsinskii zhurnal*, 1921, 8–9: 486–88). In January 1921, Voskresenskii also gave a talk on "What we could get from Steinach's newest works on rejuvenation" to a scientific conference of the Red Army's doctors held at the House of Sanitary Education (see "Khronika," *Pravda*, January 18, 1921, 4).

29. See *Referativnyi meditsinskii zhurnal*, 1920, 1(1–2): 8–11.

30. See *Referativnyi meditsinskii zhurnal*, 1921, 1(8–9): 481–86.

31. The younger brother of Mikhail Zavadovskii, Boris, prepared the Moscow edition, while Boris Slovtsov, an eminent physiologist at the Institute of Experimental Medicine, arranged for the Petrograd one. See P. Kamerer, *Omolazhivanie i dolgovechnost'* (Petrograd-Berlin: Z. I. Grzhebin, 1922); P. Kamerer, *Omolozhenie i prodlenie lichnoi zhizni (opyty na rasteniiakh, zhivotnykh i cheloveke v obshchedostupnom izlozhenii)* (Moscow: GIZ, 1922). For the original, see Paul Kammerer, *Über Verjüungun und Verlängerung des persönlichen Lebens. Die Versuche an Pflanze, Tier und Mensch gemeinverständlich dargestellt* (Stutgart, Berlin: DVA, 1921). On Vivarium and its remarkable cohort of researchers, see Wolfgang L. Reiter, "Zerstort und vergessen: Die Biologische Versuchsanstalt und ihre Wissenschaftler/innen," *Osterreichischezeitschrift für Geschichtswissenschaften*, 1999, 10 (4): 585–614.

32. See N. K. Kol'tsov, "Opyty Shteinakha i Voronova po omolozheniiu organizma," *Uspekhi eksperimental'noi biologii*, 1922, 1 (1): 94–119. He also issued his 1920 lecture on Steinach's operation as a brochure; see idem, *Omolozhenie organizma po metodu Shteinakha* (Petrograd: Vremia, 1922).

33. S. S. Girgolav, "Operatsiia 'omolozheniia' po Steinakh'u pri samoproizvol'nykh gangrenakh," *Sbornik nauchnykh trudov v chest' piatidesiatiletiia nauchno-vrachebnoi deiatel'nosti glavnogo vracha Obukhovskoi bol'nitsy professora A. A. Nechaeva* (Petrograd: n.p., 1922), 265–69.

34. See, for instance, S. Abramov, *Problema omolozheniia po Shteinakhu* (Berlin: Russkoe universal'noe izdatel'stvo, 1921); A. Krontovskii, [Retsenziia], *Vrachebnoe delo*, 1921, 25–26: 365; idem, "Ob izmenenii seksual'nosti i omolozhenii po Steinach'u," *Kievskii meditsinskii zhurnal*, 1922, 1: 48–58; "Protokol 11-go zassedaniia Turkestanskogo medistsinskogo obshchestva 22 fevralia 1921," *Turkestanskii meditsinskii zhurnal*, 1924, 3(1–8): 179.

35. See A. V. Palladin, *Omolozhenie* (Khar'kov: Put' prosveshcheniia, 1923); P. Iu. Shmidt, *Omolazhivanie* (Petrograd: Soikin, 1923); V. A. Shtark, *Omolozhenie. Popytka resheniia problemy vechnoi molodosti* (Baku: n.p., 1923); D. Gober, *Starost' i omolozhenie* (Rostov-na-Donu: Vserokompom, 1923); B. M. Zavadovskii, *Problema starosti i omolozheniia v svete ucheniia o vnutrennei sekretsii* (Moscow: Krasnaia nov', 1923); E. S. London, and I. I. Kryzhanovskii, *Bor'ba za dolgovechnost'* (Petrograd: Put' k znaniiu, 1924); V. E. Krasheninnikov, *Biologicheskie osnovy starosti, ee mekhanizm i omolozhenie* (Moscow: n.p., 1924); P. Iu. Shmidt, *Bor'ba so starost'iu* (Leningrad: GIZ, 1924); M. A. Gremiatskii, *Chto takoe omolozhenie* (Moscow-Leningrad: GIZ, 1925); and many others.

36. N. K. Kol'tsov, ed., *Omolozhenie* (Moscow-Petrograd: n.p., 1923); idem, ed., *Omolozhenie* (Moscow-Petrograd: GIZ, 1924).

37. N. K. Kol'tsov, "Novaia literatura po omolozheniiu," *Uspekhi eksperimental'noi biologii*, 1923, 2(1–2): 140–44; idem, "Noveishaia amerikanskaia literatura v oblasti operativnovo omolozheniia cheloveka," in idem, ed. *Omolozhenie* (Moscow-Petrograd: GIZ, 1924), 124–47.

38. S. Voronov, *Omolozhenie peresadkoi polovykh zhelez* (Leningrad: Prakticheskaia meditsina, 1924); idem, *Peresadka polovykh zhelez* (Khar'kov: Nauchnaia mysl', 1924); idem, *Omolozhenie. Sorok tri privivki ot obez'iany k cheloveku* (Leningrad-Moscow: Kniga, 1924); idem, *Vivre* (Moscow: Sabashnikov, 1924); idem, *Prakticheskoe primenenie peresadki organov v zhivotnovodstve* (Leningrad: Akademicheskoe izd-vo, 1925).

39. See P. Shmidt [Peter Schmidt], *Teoriia i praktika omolozheniia* (operatsii Shteinakha) (Petrograd: Prakticheskaia meditsina, 1923). This was a translation of Peter Schmidt, *Theorie und Praxis der Steinach'schen Operation* (Wien, Leipzig, München: Rikola Verlag, 1922).

40. On Zamkov and his association with IEB, see Eric Naiman, "Injecting Communism: A. A. Zamkov, Soviet Endocrinology and the Stalinist Body," at http: //www.virginia.edu/crees/ Naiman, accessed on September 12, 2003; and idem, *Sex in Public: The Incarnation of Early Soviet Ideology* (Princeton: Princeton University Press, 1997).

41. See I. G. Kogan, "Peresadka polovykh zhelez u zhivotnykh i cheloveka," *Biulleten' MOIP* (otdel eksperimental'noi biologii), 1924, 1: 179–95.

42. *Omolozhenie v Rossii* (Leningrad: Meditsina, 1924).

43. A. V. Nemilov, "Fiziologicheskie osnovy 'omolozheniia,'" in *Omolozhenie v Rossii*, 5–36.

44. V. A. Gorash, "Prakticheskoe primenennie operatsii 'omolozheniia' i ee obosnovanie," in *Omolozhenie v Rossii*, 37–97.

45. L. N. Voskresenskii, "Opyty nabliudenia nad 'omolozheniem' liudei i krupnykh sel'skokhoziastvennykh zhivotnykh," in *Omolozhenie v Rossii*, 98–126.

46. See S. Eiber, "Nabliudeniia nad operatsiei 'omolozheniia' po Shteinakhu," in *Omolozhenie v Rossii*, 127–31; S. V. Ligin, "Svodka rezul'tatov 13 sluchaev operatsii Shteinakha," in *ibid*, 131–34; L. M. Vasilevskii, "Perspektivy 'omolozheniia,'" in *ibid*, 134–39.

47. I. M. Shmyrov, "Literatura po voprosu 'omolozheniia,'" in *Omolozhenie v Rossii*, 139–44.

48. See, for example, F. I. Rodionov, "Shteinakhovskaia operatsiia u bol'noi s arterioskleroticheskim psikhozom," *Zhurnal psikhologii, nevrologii i psikhiatrii*, 1924, 4: 144–51; A. Ia. Damskii, "K probleme omolozheniia," *Nauchnye izvestiia gosudarstvennogo smolenskogo universiteta*, 1923, 1: 119–26; M. A. Shereshevskii, "Bor'ba so starost'iu i uchenie Shtenakha o pubertatnoi zheleze," *Trudy gosudarstvennogo meditsinskogo instituta v Moskve*, 1924, 1 (2): 101–14; V. N. Sukhov, "Ob operatsii omolozheniia po Steinach'u," *Leningradskii meditsinskii zhurnal*, 1925, 1: 24–28; and B. B. Dmitrieff[Dmitriev], "Zur Frage der Operation nach Stainach und der Hodentransplantation," *Zentralblatt für Chirurgie*, 1925, 21: 130–34.

49. See M. M. Zavadovskii, *Vozmozhna li bor'ba so starost'iu* (Moscow: GIZ, 1924); and idem, "Operatsiia po Shteinakhu na dikom kozle-ture (*Capra cylindricornis*, Blyth)," *Trudy laboratorii eksperimental'noi biologii Moskovskogo zooparka*, 1926, 1: 235–38. On the history of the Moscow Zoo during the first post-revolutionary decade, see O. P. Belozerov, "Ot zoosada k zooparky: Moskovskii zoologicheskii sad v pervoe posleoktiabr'skoe desiatiletie," *Istoriko-biologicheskie issledovaniia*, 2009, 1(1): 85–94.

50. In addition to the articles cited above, see F. I. Rodionov, "Operatsiia Shteinakha pri starcheskom marazme," *Vrachebnaia gazeta*, 1923, 23: 499; M. B. Fabrikant and A. V. Taft, "K voprosu ob omolozhenii (predvaritel'noe soobshchenie)," *Vrachebnoe delo*, 1923, 6–8: 162–69; M. P. Fridman, "K kazuistike peresadok zhelez vnutrennei sekretsii," *ibid*, 1923, 6–8: 169–70; N. M. Shustrov, S. G. Karpova, and I. V. Tikhomirov, "Opyty nad perezhivaniem i peresadkoi polovykh zhelez," *ibid*, 1924, 20–23: 1133–35. N. M. Shustrov, "Transplantatsiia polovykh zhelez," *ibid*, 1924, 20–23: 1210–14; D. K. Kustria, "Sluchai poiavleniia zubov u starogo kota, omolozhennogo posredstvom transplantatsii semennikov," *Russkii fiziologicheskii zhurnal*, 1924, 7(1–6): 361–62.

51. See report on the congress in *Turkestanskii meditsinskii zhurnal*, 1924, 3 (9): 223–26.

52. See M. M. Fedorovich and A. F. Kaizer, "K voprosu ob omolozhenii," *Turkestanskii meditsinskii zhurnal*, 1925, 4 (9): 544–49; "2-i nauchnyi s"ezd vrachei Srednei Azii, Tashkent, 26–30

aprelia 1925," *ibid*, 1925, 4(9): 561; "Vtoroi nauchnyi s"ezd vrachei Srednei Azii v Tashkente," *Novaia khirurgiia*, 1925, 3: 420–21; B. I. Neiman, "1-i Vseukrainskii s"ezd khirurgov," *Russkaia klinika*, 1926, 6(32): 884–90; N. Kabanov, "Vseukrainskii s"ezd terapevtov," *ibid*, 1926, 6(32): 891–93; "Oblastnoi s"ezd khirurgov levoberezhnoi Ukrainy," *Novyi khirurgicheskii arkhiv*, 1925, 8 (1): 152–56; "Kratkii otchet o zasedaniiakh khirurgicheskoi sektsii 3-go nauchnogo kongressa vrachei Gruzii v Tiflise 24–28 maia 1925," *ibid*, 1925, 8(3): 444–46.

53. See *Trudy VII Vsesoiuznogo s"ezda terapevtov, 25–31 maia 1925* (Leningrad, 1926), 523–30.

54. See, for example, B. B. Dmitriev, "Pis'mo iz Parizha," *Novyi khirurgicheskii arkhiv*, 1926, 10(4): 456–67.

55. See, for instance, "Omolozhenie 110-letnego starika," *Novaia vecherniaia gazeta*, February 24, 1926, 3.

56. See, for instance, an offer in Kol'tsov's archive, ARAN, f. 450, op. 4, d. 25, ll. 11–11 rev.

57. See "Pitomnik obez'ian," *Pravda*, October 1, 1926, 3.

58. See "Obez'iany v Sukhumi," *VM*, June 11, 1927, 3; for a contemporary account of the establishing of the monkey farm and its goals, see P. V. Bochkarev, *Obez'iany v Sukhumi: Ocherk istorii i sovremennogo sostoianiia Sukhumskogo Gosudarstvennogo nauhno-issledovatel'skogo pitomnika obez'ian* (Sukhumi: Abgiz, 1932).

59. See, for instance, I. I. Shirokogorov, "Kritika teorii Shteinakha," *Zhurnal teoreticheskoi i prakticheskoi meditsiny*, 1924, 1(1–2): 1–17.

60. See "Ne omolozhenie, a osvezhenie," *VM*, March 4, 1924, 2.

61. Thus, N. Dzhaparidze, a Georgian surgeon, reported that out of nine cases of Steinach's operation he had performed at his clinic, seven produced no effect and two improved the conditions of his patients "insignificantly." See "Kratkii otchet o zasedaniiakh khirurgicheskoi sektsii 3-go nauchnogo kongressa vrachei Gruzii v Tiflise 24–28 maia 1925," *Novyi khirurgicheskii arkhiv*, 1925, 8(3): 444–46.

62. See M. K. "Dva moskovskikh pravednika," *Vavilonskaia bashnia*, 1922, 1 (June): 15.

63. "Omolazhivanie organizmov," *Izvestiia*, October 29, 1920, 2. It is worth noting that the reportage appeared on the same page that carried information from the theaters of operation in the southern, western, and eastern fronts of the Civil War.

64. N. Kol'tsov, "Opyty Shteinakha po omolozheniiu organizma," *Priroda*, 1921, 1–3: 3–26.

65. B. Zavadovskii, "Obzor inostrannoi i russkoi literatury po voprosu ob omolozhenii," *Pechat' i revoliutsiia*, 1921, 2: 78–89. Like his older brother Mikhail, Boris had also studied with Kol'tsov. In 1918, he had graduated from Moscow University and two years later he became a professor of biology at the recently founded Sverdlov Communist University. On Boris Zavadovskii's life and works, see M. V. Kasatkin, "Akademik Boris Mikhailovich Zavadovskii. Vekhi zhizni i tvorchestva," *VIET*, 2004, 4: 144–55.

66. B. Zavadovskii, "Problema starosti i omolozheniia v svete noveishikh rabot Shteinakha, Voronova i drugikh," *KN*, 1921, 3: 146–77. Two years later he expanded the survey to a 130-page book; see idem, *Problema starosti i omolozheniia v svete ucheniia o vnutrennei sekretsii* (Moscow: Krasnaia nov', 1923).

67. See "Opyty professora Voronova," *Izvestiia*, October 31, 1922, 5; "O tak-nazyvaemom omolozhenii," *ibid*, March 4, 1923, 4; "Kino. Beseda s professorom Kol'tsovym o fil'me Omolozhenie," *ibid*, October 1, 1923, 5; "Omolozhenie," *ibid*, October 9, 1923, 4; "V dome uchenykh," *ibid*, November 13, 1923, 6; "Omolozhenie ludei i zhivotnykh," *ibid*, November 20, 1923, 5; "Opyt omolozheniia," *ibid*, December 1, 1923, 4; "Novosti nauki i tekhniki. Dostizheniia russkoi nauki,"

ibid, December 16, 1924, 6; N. K. Kol'tsov, "Omolozhenie organizma," *Pravda*, November 14, 1923, 2–3.

68. See "Blestiashchie opyty omolozheniia," *Poslednie novosti*, July 3, 1922, 1; S. V. "Omolazhivanie liudei," *ibid*, July 10, 1922, 1; A. Dvoretskii, "Bor'ba so starost'iu i omolozhenie organizma," *Rabochaia gazeta*, October 4, 1922, 2; Gr. Gri-v, "Bor'ba za dolgoletie i molodost'," *Gudok*, March 15, 1923, 4; B. Zavadovskii, "E. Petterer et Voronoff," *KN*, 1922, 3: 290–92; "Blestiashchie opyty omolozheniia," *Poslednie novosti*, July 3, 1922, 1; V. Muralevich, "Starost' i omolozhenie," *Rabochii zhurnal*, 1924, 2: 124–29; and many others.

69. See N. N. Anichkov, "O iavleniiakh starosti u cheloveka i zhivotnykh," *Chelovek i priroda*, 1920, 1: 61–63; V. Isaev, "Problema omolozheniia," *ibid*, 1922, 4: 37–64; A. V. Nemilov, "Noveishie opyty omolazhivaniia liudei," *ibid*, 1923, 4–5: 25–30; A. V. Nemilov, "'Omolozhenie' sel'skokhoziastvennykh zhivotnykh," *ibid*, 1925, 1: 23–36; V. Ryzhkov, "Bor'ba so starost'iu," *Znanie* (Khar'kov), 1923, 4: 27–30; "Omolozhenie," *V masterskoi prirody*, 1923, 7: 64; A. V. Nemilov, "O peresadke semennykh zhelez u mlekopitaiushchikh i cheloveka," *Priroda*, 1923, 7–12: 78–84; G. I. Azimov, "Bor'ba so starost'iu i smert'iu," *Iskra*, 1923, 2: 21–27; L. M. Vasilevskii, "Novye opyty omolozheniia v SSSR," *VZ*, 1925, 21–22: 1275–78, and many others.

70. V. Isaev, "Problema omolozheniia (opyty Shteinakha)," *Chelovek i priroda*, 1922, 4: 37–64.

71. L. V. "O peresadke i omolozhivanii," *Nauka i tekhnika*, 1923, 1: 9–11; 2: 10–12.

72. See G. Bosse, "Chto chitat' o starosti, smerti i bor'be so starcheskim odriakhleniem," *Knigonosha*, 1923, 10: 2–3.

73. P. Iul'ev, "Omolazhivanie," *Gigiena i zdorov'e rabochei sem'i*, 1923, 1: 1–3.

74. N. Pyzhov, "Problema omolozheniia," *Krasnaia niva*, 1923, 48: 26; "Omolazhivanie," *KP*, 1923, 6: 12; L. M. Vasilevskii, "Omolozhenie," *Zhurnal dlia zhenshchin*, 1924, 7: 24–25; V. Muralevich, "Starost' i omolozhenie," *Rabochii zhurnal*, 1924, 1: 124–29; Al. Terskii, "Bor'ba so starost'iu," *Ogonek*, 1924, 16.

75. P. M. "Ob omolazhivanii," *Peterburg*, 1921, 1: 26–27; B. Zavadovskii, "Problema starosti i omolozheniia v svete noveishikh rabot Shteinakha, Voronova i dr.," *KN*, 1921, 3: 146–76; L. M. Vasilevskii, "Novoe ob operatsii omolozheniia," *Zvezda*, 1924, 5: 188–99; Iu. M. Kharin, "Bor'ba so starost'iu," *Molodaia gvardiia*, 1924, 7: 224–39; P. Iu. Shmidt, "Smert' i bessmertie s biologicheskoi tochki zreniia," *Novyi mir*, 1925, 5: 130–41; idem, "Noveishie uspekhi russkoi biologii," *ibid*, 1925, 10: 122–30; O. I. Bronshtein, "Novye zavoevaniia meditsiny," *ibid*, 1925, 7: 118–27.

76. See "Disput ob omolozhenii," *VM*, February 29, 1924, 2.

77. For the text of his lecture, see N. K. Kol'tsov, "Chudesnye dostizheniia nauki," *Narodnyi uchitel'*, 1925, 4: 61–75.

78. See "Teoriia prof. Shteinakha v kino," *Poslednie novosti*, August 21, 1922, 3; "Kino. Beseda s professorom Kol'tsovym o fil'me 'Omolozhenie,'" *Izvestiia*, October 1, 1923, 5.

79. *Pravda*, June 4, 1924, 7.

80. See "Kuda ustraivat' ekskursii," *Krasnyi medrabotnik*, December 15, 1924, 4.

81. See "Kino, Kul'tob"edinenie," *Pravda*, September 7, 1924, 5.

82. See L. N. Voskresenskii, "Opyty nabliudeniia nad 'omolozheniem' liudei i krupnykh sel'skokhoziastvennykh zhivotnykh," *Omolozhenie v Rossii*, 98–126; "Ne omolozhenie, a osvezhenie," *VM*, March 4, 1924, 2. Alas, I was not able to find a copy of this film in the archives.

83. "Kogo nado omolazhivat'," *VM*, September 3, 1925, 3; "Kogo nuzhno omolozhivat'," *VM*, October 29, 1925, 3.

84. See Mikh. Bulgakov, "Seriia nol' shest' no. 0660243," *VM*, January 5, 1924, 2–3.

85. See "Disput ob omolozhenii," *VM,* February 29, 1924, 2; "Ne omolozhenie, a osvezhenie," *VM,* March 4, 1924, 2; "Omolozhenie liudei (beseda s prof. Zavadovskim)," *VM,* March 26, 1924, 2; "D-r S. Voronov ob omolozhenii liudei," *VM,* April 2, 1924, 2; "Samoomolozhenie d-ra Voronova," *VM,* May 9, 1924, 2.

86. "Omolozhenie Lloid-Dzhorzha," *VM,* September 10, 1924, 2; "1500 omolodivshikhsia," *VM,* September 15, 1924, 2.

87. See Bubus, "Obez'iany i chelovechestvo," *VM,* February 23, 1925, 3; "Omolozhenie zhivotnykh. Novaia stat'ia dokhoda dlia predpriimchivykh kapitalistov," *VM,* March 31, 1925, 2; "Posle omolozheniia. O konchine Rotshilda," *VM,* April 14, 1925, 3; "Bor'ba protiv omolozheniia v Indii," *VM,* June 12, 1925, 3; "Nauka vyshe boga," *VM,* June 20, 1925, 2; "Kogo nado omolazhivat'," *VM,* September 3, 1925, 3; "Pokhod protiv omolozheniia," *VM,* September 5, 1925, 3; L. M. Vasilevskii, "Noveishie opyty omolozheniia v Moskve," *VM,* September 11, 1925, 2; "Kogo nuzhno omolozhivat'," *VM,* October 29, 1925, 3; "D-r Voronov sobiraetsia v Moskvu," *VM,* November 11, 1925, 2; "Pervye itogi rabot prof. Voronova," *VM,* December 3, 1925, 3.

88. A. Nemilov, "Nado byt' ostorozhnym," *VM*, February 4, 1926, 2. All the following citations are from this source.

89. K. Viktorov, "Chto daiut operatsii Shteinakha po omolozheniiu," *VM*, February 23, 1926, 2. For a brief account of Viktorov's life and work, see D. S. Vorontsov, "K 30-letiiu nauchnoi i pedagogicheskoi deiatel'nosti prof. K. R. Viktorova," *Uchenye zapiski Kazanskogo veterianrnogo instituta*, 1934, 44: 3–12.

90. "Vozmozhno li vozvrashchenie molodosti," *VM,* March 13, 1926, 2.

91. "Pis'mo prof. Shteinakha o polemike v Vechernei Moskve," *VM*, April 3, 1926, 2.

92. A large file of correspondence with various individuals over the issues of rejuvenation is preserved among Kol'tsov's personal papers in ARAN, f. 450, op. 4, d. 25.

93. See Chandak Sengoopta, "The Modern Ovary: Constructions, Meanings, Uses," *History of Science*, 2000, 38: 425–88.

94. V. E. Krasheninnikov, *Biologicheskie osnovy starosti, ee mekhanizm i omolozhenie* (Moscow: n.p., 1924); and S. Rodionov, "Shteinakhovskaia operatsiia u bol'noi s arteriosklerotiheskim psikhozom," *Zhurnal psikhologii, nevrologii i psikhiatrii*, 1924, 4: 144–51.

95. Compare this with the situation in the United States vividly portrayed in a recent book, Pope Brock, *Charlatan: America's Most Dangerous Huckster, the Man Who Pursued Him, and the Age of Flimflam* (New York: Crown Publishing, 2008).

96. "Batumskie Shteinakhi," *Gigiena i zdorov'e rabochei sem'i*, 1924, 3: 16.

97. See also an inquiry about the cost of operation addressed to Kol'tsov in ARAN, f. 450, op. 4, d. 25, l. 13.

98. L. M. Vasilevskii, "Noveishie opyty omolozheniia v Moskve," *VM*, November 9, 1925, 2.

99. On Frolenko's life and revolutionary activities, see V. Figner, *Chlen ispolnitel'nogo komiteta partii "Narodnaia volia" M. F. Frolenko* (Moscow: Ob-vo politkatorzhan, 1928).

100. See his collected works, M. P. Frolenko, *Sobranie sochinenii* (Moscow: Ob-vo politkatorzhan, 1930), vols. 1–2. On the history of the society, see Ia. Leont'ev and M. Iunge, eds., *Vsesoiuznoe obshchestvo politkatorzhan i ssyl'noposelentsev. Obrazovanie, razvitie, likvidatsiia, 1921–1935* (Moscow: Zven'ia, 2004).

101. On Morozov's life and works, see *Nikolai Aleksandrovich Morozov—uchenyi-entsiklopedist* (Moscow: Nauka, 1982); and V. A. Tvardovskaia, *N. A. Morozov v russkom osvoboditel'nom dvizhenii* (Moscow: Nauka, 1983).

102. A part of this correspondence survived among Morozov's personal papers held at the Archive of the Russian Academy of Sciences; see ARAN, f. 453, op. 4, d. 2239, ll. 1–7.

103. For the edits, see ARAN, f. 453, op. 6, d. 36.

104. See Sergei Voronov, *Sorok tri privivki ot obez'iany cheloveku* (Leningrad: Kniga, 1924).

105. See ARAN, f. 453, op. 4, d. 2239, ll. 1–7.

106. ARAN, f. 450, op. 4, d. 25, l. 47.

107. See, for example, Lynne Attwood, *The New Soviet Man and Woman: Sex-Role Socialization in the USSR* (Bloomington: Indiana University Press, 1990); Wendy Z. Goldman, *Women, the State and Revolution: Soviet Family Policy and Social Life, 1917–1936* (Cambridge: Cambridge University Press, 1993); Naiman, *Sex in Public*; Dan Healey, *Homosexual Desire in Revolutionary Russia: The Regulation of Sexual and Gender Dissent* (Chicago: University of Chicago Press, 2001); Gregory Carleton, *Sexual Revolution in Bolshevik Russia* (Pittsburg: University of Pittsburg Press, 2005); Frances Lee Bernstein, *The Dictatorship of Sex: Lifestyle Advice for the Soviet Masses* (DeKalb: Northen Illinois University Press, 2007); Dan Healey, *Bolshevik Sexual Forensics: Diagnosing Disorder in Clinic and Courtroom, 1917–1939* (DeKalb, IL: Northen Illinois University Press, 2009); and many others.

108. For details, see Angus McLaren, *Impotence: A Cultural History* (Chicago: University of Chicago Press, 2007); on the relations of 1920s rejuvenation craze to issues of impotence in the West, see 181–208.

109. Eugen Steinach and Robert Lichtenstern, "Umstimmung der Homosexualitat durch Austausch der Pubertatsdrusen," *Munchen medische Woschenschrift*, 1918, 65: 145–48; and Eugen Steinach, "Histologische Beschaffenheit der Keimdruse bei homosexuellen Mannen," *Archiv für Entwicklungsmechanik der Organismen*, 1920, 46: 29–37. For a historical analysis of this work, see Chandak Sengoopta, "Glandular Politics: Experimental Biology, Clinical Medicine, and Homosexual Emancipation in Fin-de-Siècle Central Europe," *Isis*, 1998, 89 (3): 445–73.

110. On the long history of this debate, see C. Baker Jorgensen, *John Hunter, A. A. Bertold, and the Origins of Endocrinology* (Odense: Odense University Press, 1971).

111. See M. M. Zavadovskii, *Pol i razvitie ego priznakov. K analizu formoobrazovaniia u zhivotnykh* (Moscow: GIZ, 1922).

112. M. M. Zavadovskii, *Pol zhivotnykh i ego prevrashchenie (mekhanika razvitiia pola)* (Moscow-Petrograd: GIZ, 1923).

113. For an account of the use of Zavadovskii's data for sex education, see Bernstein, *The Dictatorship of Sex*, 41–73.

114. In the West, 1920s sex-change operations are largely associated with the name of Magnus Hirschfeld. For details, see Charlotte Wolff, *Magnus Hirschfeld: A Portrait of a Pioneer in Sexology* (Quartet Books, 1987); and Joanne Meyerowitz, *How Sex Changed: A History of Transsexuality in the United States* (Cambridge: Harvard University Press, 2004).

115. See V. L. Pogodin to GINZ, January 15–23, 1921, ARAN, f. 450, op. 4, d. 25, ll. 2–3.

116. The letter is undated and its addressee is identified only by the title "professor." Yet its contents strongly suggest that it was addressed to Zavadovskii and not before he published his books on the transformation of sex in chickens. See ARAN, f. 450, op. 4, d. 25, l. 5.

117. See, for instance, ARAN, f. 450, op. 4, d. 25, l. 64.

118. The letter was actually addressed to Semashko, who evidently forwarded it to Kol'tsov; see ARAN, f. 450, op. 4, d. 25, ll. 6–7 rev.

119. ARAN, f. 450, op. 4, d. 25, ll. 75–76.

120. See S. Braude, "Nedorazvitye liudi," *VM*, December 30, 1926, 2.

121. See Ia. I. Kirov, "K voprosu o geterotransplantatsii pri gomoseksualizme," *Vrachebnoe delo*, 1928, 20: 1587–1590. For a detailed analysis of the treatment of homosexuals by the Bolshevik regime, see Healey, *Homosexual Desire in Revolutionary Russia*.

122. A. Agienko, "O taktike," *Biokosmist*, 1922, 1 (March): 5; see also, P. Ivanitskii, "Proletarskaia etika," in A. Sviatogor and P. Ivanitskii, *Biokosmizm: Materialy, no. 1* (Moscow: Kreatorii biokosmistov, 1921): 12–14.

123. See "Biokosmisty," *Vavilonskaia Bashnia*, 1922, 3 (August): 2.

124. See "Mezhdunarodnaia propaganda," *Bessmertie*, 1922, 1 (November): 11–12.

125. See, for example, Semen Tikhonov, "Lavochka bessmertiia (Venskoi burzhuazii)," in *Biokosmisty. Desiat' Shtuk* (Petrograd: Komitet poezii biokosmistov-immortalistov, 1923): 14–15.

126. I. Lezhnev, "Dve privivki: Lenin i Shteinakh," *Rossiia*, 1923, 8: 13–17.

127. See the famous volume entitled "Oswald Spengler and the Dawn of Europe," with essays written by prominent Russian philosophers, including Fedor Steppun, Nikolai Berdyaev, and Sergei Frank, *Osval'd Shpengler i zakat Evropy* (Moscow: Bereg, 1922); and a special section in the second 1922 issue of *Red Virgin Soil* with the "Marxist" assessment of Spengler's book, *Kransnaia nov'*, 1922, 2: 196–241. For brief discussions of the impact of Spengler's work in Russia, see Carol Avins, *Border Crossings: The West and the Russian Identity in Soviet Literature, 1917–1934* (Berkley: University of California Press, 1983), 35–38; and Finkel, *On the Ideological Front*, 126–35.

128. Mark Krinitskii was the pen name of Mikhail Vladimirovich Samygin (1874–1952), a close friend of Valerii Briusov's and a prolific popular novelist of the 1910s. In 1915–1918, he published a fifteen-volume (!) set of his "Collected Works."

129. Mark Krinitskii, "Eliksir bessmertiia," *Ogoniek*, 1923, 10: 1–4; all of the following citations are from this source.

130. See, for instance, Mark Volosov, "Omolozhenie," *KP*, 1925, 5: 10–11; N. Miur, "Eliksir bessmertiia," *ibid*, 1926, 52: 2–3.

131. Here, Bulgakov clearly (and, perhaps, deliberately) blends together two contemporary scientists, the brothers Zavadovskii. The older one, Mikhail, conducted sex change experiments on chickens and was the scientific head of the Moscow Zoo where he indeed exhibited several specimens illustrating his research. The younger one, Boris, was a professor of biology at the Sverdlov Communist University and published extensively on rejuvenation, aging, and endocrinology, not only in scientific journals but also in a variety of literary magazines, which Bulgakov read on a regular basis.

132. Characteristically, in 1934, preparing a new version of the play for performance in Paris, Bulgakov removed the whole episode.

133. For a historical examination of the fate of "former people" in Soviet Russia, see Smirnova, *"Byvshie liudi" Sovetskoi Rossii*.

134. One of the earliest fictional renderings of the Steinach-Voronoff rejuvenation techniques in English was Bertram Gayton's novel, *The Gland Stealers* (Philadelphia & London: J. B. Lippincott Co., 1922).

135. See, for instance, Dzheims Hourd, "Izobretenie professora Gedda," *KP*, 1924, 19: 12–13; Robert Sharl', "Bessmertie," *ibid*, 1925, 41: 4–5; K. Fezandie, "Taina vechnoi molodosti," *Mir prikliuchenii*, 1925, 2:33–50; A. Verril', "Tragediia professora Gendersona," *V masterskoi prirody*, 1928, 5: 151–55, 6: 183–86.

136. See Noel' Rozhe, *Chelovek budushchego* (Moscow: Puchina, 1924). For the original, see Noëlle Roger, *Le nouvel Adam* (Paris: A. Michel, 1924).

137. Noel' Rozhe, *Griadushchii Adam* (Leningrad: GIZ, 1926). The same year the novel appeared in English; see Noëlle Roger, *The New Adam* (London: Stanley Paul & Co. Ltd., 1926).

138. Marsel' Rolan, *Osman-Omolazhivatel'* (Moscow-Leningrad: GIZ, 1926). A different Russian translation came out in 1929 in Riga under the title "Rejuvenated," see Marsel' Rolan, *Omolozhennyi (taina vracha omolazhevatelia)* (Riga: B. Shereshevskii, 1929). For the original French edition, see Marcel Roland, *Osmant le rajeunisseur* (Paris: A. Michel, 1925). On the contemporary French fascination with rejuvenation, see Brett A. Berliner, "Mephistopheles and Monkeys: Rejuvenation, Race, and Sexuality in Popular Culture in Interwar France," *Journal of the History of Sexuality*, 2004, 13(3): 306–25.

139. Portrayed by some critics as "a northern Jules Verne," Vsevold Valiusinskii (1899–1935) died at the age of thirty-six in a hunting accident. There is very little information available on his life and writings; see V. Borovoi, "Severnyi Zhiul' Vern," *Sovetskaia Onega*, November 6, 1975, 4. For a brief discussion of his novel, see Muireann Maguire, "Soviet Gothic-Fantastic," 96–101.

140. Vsevolod Valiusinskii, *Piat' bessmertnykh* (Khar'kov: Proletarii, 1928); all of the subsequent quotations are from this source.

CHAPTER 6

1. See numerous reports by Western visitors on the development of Soviet science in the 1920s and early 1930s—for instance, J. G. Crowther, *Science in Soviet Russia* (London: Williams & Norgate Ltd., 1930), or J. S. Huxley, *A Scientist among the Soviets* (London: Chatto and Windus, 1934).

2. This situation could perhaps be best illustrated by widely publicized experiments on the hybridization of apes and humans conducted in the mid-1920s by the Russian biologist Il'ia Ivanov. For details of this episode, see Kirill Rossiianov, "Beyond Species: Il'ya Ivanov and His Experiments on Cross-Breeding Humans with Anthropoid Apes," *Science in Context*, 2002, 15 (2): 277–316. Perhaps the best illustration of an entirely different situation in the United States is a "Joint statement upon the relations of science and religion" (initiated by the Nobel-winning physicist Robert A. Millikan and cosigned by thirty-five well-known theologians and church officials, eminent scientists, and "men of affairs") that appeared in the country's leading scientific journal *Science* on June 1, 1923. Its stated purpose was "to assist in correcting two erroneous impressions which seem to be current among certain groups of persons. The first is that religion today stands for medieval theology; the second, that science is materialistic and irreligious" (630–31). It is quite telling that Millikan also included this statement as an appendix into his exploration of the role of science in the post-World War I future of humanity in his *Science and Life* (Boston: Pilgrim Press, 1924), 86. On the general issue of "constructing scientific literature in a hostile public environment" in the United States, see Susan E. Lederer, "Political Animals," 61–79. For a historical analysis of the role of religion in shaping biomedical experimentation with, and clinical practices of, blood transfusion and organ transplants in the United States, see Susan E. Lederer, *Flesh and Blood: Organ Transplantation and Blood Transfusion in Twentieth-Century America* (New York: Oxford University Press, 2008), 185–207.

3. See, for instance, Thiher, *Fiction Rivals Science*; and idem, *Fiction Refracts Science*.

4. For an illuminating sketch of the long history of interactions between science and literature, see Andrea Battistini, "The Antagonistic Affair between Literature and Science," in Spinozzi and Hurwitz, eds., *Discourses and Narrations in Biosciences*, 61–72.

5. Although written with entirely different goals in mind, Fritz Ringer's 1969 classic provides a perfect example and a fine-grained analysis of such transformation in Germany; see Fritz Ringer, *The Decline of the German Mandarins: The German Academic Community, 1890–1933* (Cambridge, MA: Harvard University Press, 1969).

6. For a general overview of science's place in Russian culture before the Revolution, see Alexander Vucinich, *Science in Russian Culture, A History to 1860* (Stanford, CA: Stanford University Press, 1963) and idem, *Science in Russian Culture, 1861–1917* (Stanford, CA: Stanford University Press, 1970).

7. The centrality of literature in nineteenth-century Russian culture made literary fiction a subject of countless studies. For a sample of various approaches to this subject, and especially to the interrelation of Russian literature and another dominant component of culture—religion—see six volumes published under the general title *Christianity and Russian Literature, Khristianstvo i russkaia literatura* (St. Petersburg: Nauka, 1994–2010), vol. 1–6.

8. See, for instance, a recent volume devoted to the pedagogical potential of Dostoevskii's writings, *"Pedagogiia" F. M. Dostoevskogo. Sbornik statei: Pedagogicheskii potentsial russkoi literatury* (Kolomna: KGPU, 2003).

9. See Michael Holquist, "Bazarov and Sečenov: the Role of Scientific Metaphor in *Fathers and Sons*," *Russian Literature*, 1984, 14 (4): 359–74.

10. For a detailed examination of Tolstoy's attitudes toward science and medicine, see Grigorii Kulizhnikov, *L. N. Tolstoi i meditsina* (Moscow: Sfera, 1999).

11. See N. P. Vagner, *Povesti, skazki i rasskazy Kota-Murlyki* (St. Petersburg: Stasiulevich, 1887–1899), 7 vols.; I. V. Gete, *Faust*, transl. N. A. Kholodkovskii (St. Petersburg: Suvorin, 1890), vol. 1; and Dzhon Mil'ton, *Poterianyi rai*, transl. N. A. Kholodkovskii (St. Petersburg: Suvorin, 1911), Vol. 1–2.

12. For instance, in 1891, in a leading "thick" journal, *Herald of Europe*, Elie Metchnikoff published a lengthy article tellingly entitled "The law of life: Regarding certain works of the count Leo Tolstoy." Metchnikoff offered a scathing assessment of Tolstoy's views on science, challenging the writer's claim that science was at best useless in search for answers to the existential questions. See I. Mechnikov, "Zakon zhizni. Po povodu nekotorykh proizvedenii gr. L. Tolstogo," *Vestnik Evropy*, 1891, 5: 228–60. Tolstoy's responses to Metchnikoff's assessment, however, were collected and published only after both the writer's and the scientist's death; see L. N. Tolstoi, *O nauke* (Moscow: n.p., 1917). Although written with the goal of analyzing similarities in Metchnikoff's and Tolstoy's attitudes toward death and old age (as an example of more general Russian "modern" attitudes), Stephen Lovell's article ("Finitude at the Fin de Siècle: Il'ia Mechnikov and Lev Tolstoy on Death and Life," *Russian Review*, 2004, 63 (2): 296–316) inadvertently provides a vivid description of the challenge issued by a leading experimental biologist and a response to this challenge by a leading Russian writer of the time.

13. Literature on the history of popular science is vast; for useful overviews, see Terry Shinn and Richard Whitley, eds., *Expository Science: Forms and Functions of Popularization* (Sociology of the Sciences Yearbook 9) (Dordrecht: Reidel, 1985); Peter Broks, *Understanding Popular Science* (Maidenhead: Open University Press, 2006); Bernard Lightman, *Victorian Popularizers of*

Science (Chicago: University of Chicago Press, 2007); Peter J. Bowler, *Science for All* (Chicago: University of Chicago Press, 2009); and Faidra Papanelopoulou, Agustí Nieto-Galan, and Enrique Perdriguero, eds., *Popularizing science and technology in the European periphery, 1800–2000* (Ashgate, 2009). For a history of popular science in Russia, see Andrews, *Science for the Masses*.

14. Cited from G. W. F. Hegel, *Encyclopaedia of the Philosophical Sciences*, Part One: The Shorter Logic, Introduction: § 9. Speculative Logic, http://www.marxists.org/reference/archive/hegel/works/sl/slintro.htm#SL9, retrieved on March 21, 2010.

15. L. E. Obolenskii, "Biologi o zhenskom voprose," *Russkaia mysl'*, 1893, 2: 61–78, here–61.

16. See P. Bakhmet'ev, "Retsept dozhit' do XXI veka," *Estestvoznanie i geografiia*, 1901, 8: 1–13.

17. He further expanded such speculations in his article on "Biological analogies in crystals." See P. Bakhmet'ev, "Biologicheskie analogii u kristallov," *Estestvoznanie i geografiia*, 1906, 1: 1–16.

18. P. I. Bakhmet'ev, "Sedalishche dushi," *Nauchnoe obozrenie*, 1902, 9: 1–6; 10: 21–39.

19. E. Metchnikoff, *Études sur la nature humaine: Essai de philosophie optimiste* (Paris: Masson & C-ie, 1903); idem, [I. I. Mechnikov], "Etiudy o prirode cheloveka," *Nauchnoe slovo*, 1903, 4: 37–49; 5: 38–53; 6: 5–18; 7: 5–17; 8: 31–58; 9: 5–34; 10: 5–36; idem, *The Nature of Man: Studies in Optimistic Philosophy* (New York: Putnam, 1903); idem, *Studien über die Natur des Menschen: Eine optimistische Philosphie* (Leipzig: Verlag von Veit, 1904).

20. I. I. Mechnikov, *Etiudy o prirode cheloveka* (Moscow: Nauchnoe slovo, 1908), 3d ed., 220.

21. On Pavlov as an epitome of "objective" science, see Todes, *Pavlov's Physiology Factory*.

22. Pavlov used the phrase *nashe psikhicheskoe soderzhanie*, literally "our psychical substance." An English translation of the speech available at the Nobel Foundation website translates this phrase as "our psychical constitution"; see "Ivan Pavlov—Nobel Lecture: Physiology of Digestion," at http://www.nobelprize.org/nobel_prizes/medicine/laureates/1904/pavlov-lecture.html; I think, however, that "our psychical nature" or "the nature of our psyche" corresponds more closely to what Pavlov actually meant in his speech.

23. Ivan Pavlov, "Nobelevskaia rech'," http://ppip.su/arhiv_gl/2008_3_4/nomer/nom11.php; retrieved on March 21, 2010.

24. There are numerous examples of Russian biologists' attempts to answer the "perpetual" existential questions by using their particular scientific expertise. See, for instance, a voluminous treatise on "biological theories and questions of life" written by the founder of Russian experimental zoopsychology Vladimir Vagner, *Biologicheskie teorii i voprosy zhizni* (St. Petersburg: Obrazovanie, 1910). In this book, among many other subjects, Vagner also takes issue with Tolstoy's critique of science.

25. There are dozens of publications attesting to this trend. For an illuminating and quite extensive contemporary synopsis of such publications, specifically on the subject of "eternal life," see E. E. Fournier d'Albe, *New Light on Immortality* (London: Longman, Greens, and Co., 1908).

26. Useful introductions to the literature on the cultural impact of the Great War can be found in Daniel Pick, *War Machine: the Rationalization of Slaughter in the Modern Age* (New Haven: Yale University Press, 1993); and Modris Eksteins, *Rites of Spring: the Great War and the Birth of the Modern Age* (Toronto: Lester & Orpen Dennys, 1994).

27. There are numerous examples of scientists' claiming that mantle in the 1920s; see, for instance, the physicist Robert A. Millikan's *Science and Life* (Boston: Jordan & More Press, 1924), especially Chapter 4, "Science and Society," 65–85; or the psychologist William McDougall's *Janus: the Conquest of War, a Psychological Inquiry* (London: Kegan Paul, Trench, Trubner & Co., 1927).

28. On the history of the series, see a special issue of *Interdisciplinary Science Reviews*, 2009, 34 (1).

29. E. E. Fournier d'Albe, *Quo Vadimus? Some Glimpses of the Future* (London: Kegan Paul, Trench, Trubner & Co., 1925).

30. See H. S. Jennings, *Prometheus, or Biology and the Advancement of Man* (London: Kegan Paul, Trench, Trubner & Co., 1925); and J. D. Bernal, *The World, the Flesh, and the Devil: An Enquiry into the Future of the Three Enemies of the Rational Soul* (London: Kegan Paul, Trench, Trubner & Co., 1929). In addition, the series also included F. C. S. Shiller's *Tantalus, or the Future of Man* (London: Kegan Paul, Trench, Trubner & Co., 1925); Garet Garrett's *Ouroboros, or The Mechanical Extension of Mankind* (London: Kegan Paul, Trench, Trubner & Co., 1926); H. F. Carlill's *Socrates, or The Emancipation of Mankind* (London: Kegan Paul, Trench, Trubner & Co., 1927); and Ronald Campbell MacFie's *Metanthropos, or The Body of the Future* (London: Kegan Paul, Trench, Trubner & Co., 1928), all of which discussed the possible futures offered by various techniques and methods of experimental biomedical sciences.

31. Of course, the rapid advances of everyday technology (electric lamps and motors, telephone, telegraph, radio, cinema, automobile, zeppelin and airplane, phonograph, etc.), as well as more specialized and exotic technologies (such as chemical fertilizers, vaccinations, and x-rays), played a huge role in the rise of science's cultural authority in the decades around the turn of the twentieth century.

32. See, for instance, Lenin's famous lecture "On the state," delivered at the height of the Civil War in 1919 but published only a decade later. An English translation is available at http://www.marxists.org/archive/lenin/works/1919/jul/11.htm.

33. Recall the "ray of life" discovered by Persikov in electric light, but absent in sunlight.

34. For details on Bogdanov and his ideas, see Krementsov, *A Martian Stranded on Earth*.

35 L. Trotskii, *Literatura i revoliutsiia* (Moscow: Krasnaia nov', 1923), 195–97.

36. Of course, the attempts to explain scientifically human creativity and genius—ranging from phrenology to brain anatomy to eugenics—had a long history, but, as in many other endeavors, the coincidence of the Bolshevik Revolution with the experimental revolution in the life sciences gave such studies a new impetus and a new flavor in Soviet Russia.

37. V. V. Savich, "Popytka uiasneniia protsessa tvorchestva s tochki zreniia reflektornogo akta," *KN*, 1922, 4: 207–23. The same article had appeared a year earlier in a specialized journal, *Herald of the Lesgaft Petrograd Scientific Institute*, which of course had a much more limited circulation; see idem, "Popytka uiasneniia protsessa tvorchestva s tochki zreniia reflektornogo akta," *Izvestiia Petrogradskogo nauchnogo instituta im. P. F. Lesgafta*, 1921, 4: 199–212. In July 1922, Savich also delivered a long report entitled "An analysis of [the Russian] revolutionary process from a physiologist's viewpoint," at a meeting of the Petrograd Biological Society. See "Zasedaniia biologicheskogo obshchestva," *Poslednie novosti*, July 3, 1922, 3. Alas, I was unable to find a text of this report. Savich later elaborated and summarized his views in a monograph entitled "Foundations of human behavior. An analysis of human behavior from the viewpoint of the physiology of the central nervous system and internal secretions"; see V. V. Savich, *Osnovy povedeniia cheloveka. Analiz povedeniia cheloveka s tochki zreniia fiziologii tsentral'noi nervnoi sistemy i vnutrennei sekretsii* (Leningrad: Priboi, 1924).

38. See G. V. Segalin, ed., *Klinicheskii arkhiv genial'nosti i odarennosti (evropatologii), posviashchennyi voprosam patologii genial'noodarennoi lichnosti, a takzhe voprosam patologii tvorchestva* (Sverdlovsk, 1925–29).

39. On the activities of this group, see Irina Sirotkina, *Diagnosing Literary Genius* (Baltimore: Johns Hopkins University Press, 2001). On the history of "pathography" as a genre, see idem, "Patografiia kak zhanr: kriticheskoe issledovanie," *Meditsinskaia psikhologiia v Rossii: elektronnyi nauchnyi zhurnal*, 2011, 2 at http://medpsy.ru, accessed July 3, 2011.

40. Compare this preoccupation to Louis Berman's *The Glands Regulating Personality*, which provided similar "pathographies" of "genius-endowed" historical figures. Berman's analysis, however, included Napoleon Bonaparte, Charles Darwin, Friedrich Nietzsche, Julius Caesar, Florence Nightingale, and only one writer, Oscar Wilde.

41. On the role of the media in publicizing "medical breakthroughs" in the United States in the late nineteenth and early twentieth centuries, see John H. Warner, "Medicine, Media, and the Dramaturgy of Biomedical Research: Historical Perspectives," in Peter Snyder, Linda Mayes, and Dennis Spencer, eds., *Science and the Media: Delgado's Brave Bulls and the Ethics of Scientific Disclosure* (Elsevier, 2009), 13–25.

42. A. Kun, "Ob ugadyvanii chuzhikh myslei," *VM*, January 9, 1928, 2; italics are added.

43. I. I. Slitsan, "Stranitsa iz knigi zhizni i smerti," *Nasha gazeta*, December 9, 1928, 6.

44. M. K., "Opyty omolozheniia za granitsei i v Moskve," *VM*, February 16, 1926, 1.

45. See, for instance, Frances Bernstein's examination of the role of popular science in shaping sexual behaviors in revolutionary Russia in *The Dictatorship of Sex*, or Tricia Starks's examination of its role in promoting personal hygiene in *The Body Soviet: Hygiene, Propaganda, and the Revolutionary State* (Madison: University of Wisconsin Press, 2008).

46. In the 1920s, nearly all popular-science magazines had special sections devoted to readers' letters and regularly published responses that both answered individual inquiries and addressed subjects of particular interest for large groups of readers, such as rejuvenation.

47. See ARAN, f. 450, op. 4, d. 25, ll. 16–17rev, 22–22 rev.

48. For a general discussion of the visual representations of "medical breakthroughs" in popular publications, see Hansen, *Picturing Medical Progress*.

49. K. N. Krzhishkovskii, *Fiziologiia polovoi zhizni* (Leningrad: Priboi, 1926), 120.

50. See, for instance, Robert L. Forward, *Indistinguishable from Magic* (Baen, 1995).

51. In this respect, a comparison between *science fiction* and *fantasy* is particularly illuminating. In the West, fantasy as a genre that brought magic (and the supernatural more generally) back into literary fiction began to develop rapidly during the 1920s, in parallel and in competition with science fiction. In the Soviet Union, however, magic became a taboo. In *Master and Margarita* Bulgakov pointedly parodied this taboo in the famous scene with a séance of "black magic and its exposure" performed by Woland's retinue at the Moscow Variety Theater. Expectedly, fantasy found practically no following in Soviet literature until the dissolution of the Soviet Union in 1991, when it made a spectacular appearance in the writings of Nik Perumov. One could perhaps argue that the rise in the popularity of fantasy in the West during the 1960s and its continuing appeal to the widest audiences in recent decades (as witnessed, for instance by the extraordinary success of Rick Riordan's *Percy Jackson* and J. K. Rowling's *Harry Potter* stories), reflects a profound disillusionment in, and fear of, science in the age of nuclear, chemical, and bacteriological weapons, man-made disasters like Chernobyl and Fukushima, and global warming.

52. Bubus, "Obez'iany i chelovechestvo," *VM*, February 23, 1925, 3.

53. At about the same time J. B. S. Haldane leveled similar criticism at H. G. Wells's novels for adhering to the notions of "old" biology, as Mark B. Adams has noted in his pioneering study of "visionary biology." See Adams, "Last Judgment," 457–91.

54. Bulgakov's attention to detail is noted by all students of his writings. There are practically no "small" details in his works. Bulgakov carefully selected every word, every situation, every name, and every prop, in short—every detail, in his writings.

55. On the role of Voronskii and his journal in 1920s Soviet literature, see Maguire, *Red Virgin Soil*; for English translations of Voronskii's selected works, see Aleksandr Voronskii, *Art as the Cognition of Life. Selected Writings, 1911–1936*, transl. and ed. by Frederick S. Choate (Oak Park, MI: Mehring Books, 1998); for a recent detailed biography of Voronskii, see E. A. Dinershtein, *A. K. Voronskii: V poiskakh zhivoi vody* (Moscow: ROSSPEN, 2001).

56. For a similar interpretation of the interrelation of science and literature, see a series of essays written in the early 1920s by Mikhail Gershenzon, a prominent literary critic and theorist: M. Gershenzon, *Videnie poeta* (Moscow: n. p., 1920) and idem, *Gol'fstrem* (Moscow: Shipovnik, 1922). During the same period, several avant-garde painters, notably Kazimir Malevich, expressed similar sentiments regarding interrelations between science and the fine arts. On Malevich's stance, see, for instance, Larissa A. Zhadova, *Malevich: Suprematism and Revolution in Russian Art, 1910–1930* (London, New York: Thames & Hudson, 1982). For a recent discussion of fiction's cognitive elements from the viewpoint of epistemology, see Catherine Elgin, *Considered Judgment* (Princeton: Princeton University Press, 1996).

57. A. Voronskii, *Iskusstvo i zhizn'* (Moscow: Krug, 1924), 10, 28.

58. A. Voronskii, *Iskusstvo videt' mir* (Moscow: Sovetskii pisatel', 1987), 10–11.

59. L. Vasil'ev, "Piatnadtsat' tezisov normal'noi ideologii," *Bessmertie*, 1922, 1: 3–4.

60. See "Budushchee," *30 dnei*, 1927, 11: 76–92.

61. F. Gladkov, "Cherez 100 let," *30 dnei*, 1927, 11: 80.

62. See, for instance, N. Dolin, "Krov' mira," *Mir prikliuchenii*, 1927, 6: 2–12; and V. D. Nikol'skii, "Luchi zhizni," *ibid*, 1927, 10: 6–20.

63. For a fine-grained description of a similar situation in Western literature, see Rosslyn D. Haynes, *From Faust to Strangelove: Representations of the Scientist in Western Literature* (Baltimore: Johns Hopkins University Press, 1994).

64. Interestingly, in his very detailed and insightful analysis of the typology of various personages in Bulgakov's writings, Iablokov does not even mention either Persikov or Preobrazhenskii; see Iablokov, *Khudozhestvennyi mir Mikhaila Bulgakova*, 260–307.

65. On the image of "mad" and "evil" scientists in Western culture, see Christopher P. Toumley, "The Moral Character of Mad Scientists: A Cultural Critique of Science," *Science, Technology, and Values*, 1992, 17 (4): 411–37.

66. A. Voronskii, *Literaturnye zapisi* (Moscow: Krug, 1926), 68–69.

67. Only in the 1940s did the issues of creating and maintaining large scientific institutions begin to attract writers' attention. Veniamin Kaverin (Zilber) was one of the first to address these issues in his 1946 novel *Open Book*. Kaverin had intimate knowledge of the Soviet science system and its workings through his older brother, Lev Zilber, Russia's leading medical virologist and immunologist, and Zilber's wife, Zinaida Ermol'eva, Russia's foremost expert on antibiotics.

68. For details on the Soviet system of science patronage and its evolution, see Krementsov, *Stalinist Science*.

69. See, for instance, M. I. Barsukov, ed., *Ocherki istorii zdravookhraneniia SSSR (1917–1956)* (Moscow: Medgiz, 1957).

70. For a brief history of IEB, see Mark B. Adams, "Science, Ideology, and Structure: The Kol'tsov Institute, 1900–1970," in Linda L. Lubrano and Susan G. Solomon, eds., *The Social*

Context of Soviet Science (Boulder, CO: Westview Press, 1980), 173–204; for an analysis of Kol'tsov's institutional strategies, see V. V. Babkov, "N. K. Kol'tsov: Bor'ba za avtonomiiu nauki i poiski podderzhki vlasti," *VIET*, 1989, 3: 2–19.

71. See K. Rossiianov, "Experimental Stations and the Emergence of the Soviet Genetics Community, 1917–1925," *Journal of the Swedish Seed Association*, 1997, 107: 213–20.

72. See GARF, f. A-482, op. 1, d. 34, ll. 346–47.

73. For a short biography of Tarasevich, see L. I. Grabovskaia, *Lev Aleksandrovich Tarasevich, 1868–1927* (Moscow: Meditsina, 1970).

74. In the next decade, this practice would become a characteristic feature of the Soviet scientific culture.

75. On Bogdanov and proletarian science, see Krementsov, *A Martian Stranded on Earth*.

76. See Krementsov, *A Martian Stranded on Earth*.

77. Indeed, according to Julian Huxley, a leading British experimental biologist of the interwar period, one of the main functions of any religion always had been and still is to provide answers to the "big" questions. Huxley has argued that because, from the mid-nineteenth century on, science had gradually assumed that function, science must eventually become humanity's only religion. See Julian Huxley, *Religion Without Revelation* (New York, London: Harper & Brothers, 1927).

78. Literature on the history of interactions between science and religion is enormous. Useful introductions to the subject can be found in Gary B. Ferngren, ed., *Science and Religion: A Historical Introduction* (Baltimore: Johns Hopkins University Press, 2002); Richard G. Olson, *Science and Religion, 1450–1900. From Copernicus to Darwin* (Baltimore: Johns Hopkins University Press, 2004); and Peter Harrison, ed., *The Cambridge Companion to Science and Religion* (New York: Cambridge University Press, 2010).

79. The role of religion, especially Orthodox Christianity, in the country's cultural life is a large topic that has attracted much attention during the last two decades in Russia. See, for instance, the chronicles of special international conferences on the subject held from 1994 to 2003 in St. Petersburg: "Khroniki mezhdunarodnykh nauchnukh konferentsii 'Pravoslavie i russkaia kul'tura' (1994–2003)," in *Khristianstvo i russkaia literatura* (St. Petersburg: Nauka, 2006), vol. 5, 598–718. Unfortunately, science was not part of this wide-reaching enterprise. The interrelations of science and religion in Russia have so far attracted very limited attention. Soviet-era works have focused predominantly on "the struggle of science for materialism" and "against religion." See, for example, A. A. Maksimov, *Ocherki po istorii bor'by za materialistm v russkom estestvoznanii* (Moscow: Gospolitizdat, 1947); or A. I. Kuroedov, *Bor'ba vydaiushchikhsia uchenykh za materialism v biologii* (Moscow: MGU, 1966). Western students of the subject have concentrated largely on theories of evolution and theories of mind in their relation to religion in Russia. See, for instance, Vucinich, *Darwinism in Russian Thought*, 240–48. Only in the last few years have some historians begun to explore the science-religion nexus in any depth. See, for example, Alexei V. Nesteruk, "The problem of faith and scientific knowledge in Russian religious thought of the nineteenth-twentieth centuries," in Jitse M. van der Meer and Scott Mandelbrote, eds., *Nature and Scripture in the Abrahamic Religions: 1700-Present* (Leiden: Koninklijke Brill NV, 2008), 371–404; Loren Graham and Jean-Michel Kantor, *Naming Infinity: A True Story of Religious Mysticism and Mathematical Creativity* (Cambridge, MA: Belknap Press of Harvard University Press, 2010); and Efthymios Nicolaidis, *Science and Eastern Orthodoxy* (Baltimore: John Hopkins University Press, 2011), 140–50. For a more general discussion of the intersections among science,

religion, and Bolshevik ideology, see Michael David-Fox, "Religion, Science, and Political Religion in the Soviet Context," *Modern Intellectual History*, 2011, 8 (2): 471–84.

80. For a detailed analysis of the clergy as a particular subset of the Russian intelligentsia, see Laurie Manchester, *Holy Fathers, Secular Sons: Clergy, Intelligentsia, and the Modern Self in Revolutionary Russia* (DeKalb: Northern Illinois University Press, 2008); for a sketch of Moscow professors' attitudes toward religion, see N. N. Niks, *Moskovskaia professura vo vtoroi polovine XIX-nachale XX veka* (Moscow: Novyi Khronograf, 2008), 211–30.

81. Furthermore, the so-called "God's Law" (*Zakon Bozhii*) taught by the orthodox priest was an obligatory subject in all secondary schools (gymnasiums) in the Russian Empire.

82. On the role of the church in the state apparatus in Russia, see S. L. Firsov, *Pravoslavnaia tserkov' i gosudarstvo v poslednee desiatiletie sushchestvovaniia samoderzhaviia v Rossii* (St. Petersburg: Izdatel'stvo RKhGI, 1996); S. I. Alekseeva, *Sviateishii sinod v sisteme vysshikh i tsentral'nykh uchrezhdenii poreformennoi Rossii 1856–1904 gg.* (St. Petersburg: Nauka, 2003); and John D. Basil, *Church and State in Late Imperial Russia: Critics of the Synodal System of Church Government (1861–1914)* (Minneapolis: University of Minnesota Press, 2005). Of course, the Church was no monolith; for a recent analysis of different trends within the Russian clergy, see Jennifer Hedda, *His Kingdom Come: Orthodox Pastorship and Social Activism in Revolutionary Russia* (DeKalb: Northern Illinois University Press, 2008).

83. For a detailed examination of the interactions of the Russian biomedical community and the censorship in the second half of the nineteenth century, see Daniel P. Todes, "Biological Psychology and the Tsarist Censor: The Dilemma of Scientific Development," *Bulletin of the History of Medicine*, 1984, 58: 529–44. For the censorship's reaction to and actions against Haeckel's work, see A. Maksimov, "Gekkel' i ego 'Mirovye zagadki,'" in E. Gekkel', *Mirovye zagadki* (Moscow: OGIZ, 1935), 5–58. Although in the last twenty years the history of censorship in Russia has become a lively field of historical research, with hundreds of publications ranging from short notes to solid monographs, examinations of the interrelation of Russian science and censorship during both tsarist and Soviet periods have attracted very limited attention. See the proceedings of several conferences held in St. Petersburg, M. B. Konashev, ed., *Svoboda nauchnoi informatsii i okhrana gosudarstvennoi tainy* (Leningrad: BAN, 1991); idem, ed., *Tsenzura v tsarskoi Rossii i Sovetskom Soiuze* (Moscow: Rudomino, 1995); idem, ed., *Tsenzura i dostup k informatsii: istoriia i sovremennost'* (St. Petersburg: Russian National Library, 2005), as well as five volumes compiled by Mikhail Konashev and issued during the last ten years under the general title *Censorship in Russia: History and the Present*, M. Konashev, ed., *Tsenzura v Rossii: Istoriia i sovremennost'* (St. Petersburg: RNB, 2001–2011), vol. 1–5.

84. A survey of such major theological journals published in Russia at the time as *Theological Herald* (Bogoslovskii vestnik) and *Faith and Reason* (Vera i razum) demonstrates this trend convincingly. See also N. M. Solov'ev, *"Nauchnyi" ateizm. Sbornik statei o professorakh Gekkele, Mechnikove i Timiriazeve* (Moscow: Tvorcheskaia mysl', 1915).

85. The two works were Erich Wasmann, *Instinct und Intelligenz im Thierreich: ein kritischer Beitrag zur modernen Thierpsychologie* (Freiburg im Breisgau: Herder, 1897); and idem, *Vergleichende Studien über das Seelenleben der Ameisen und der höhern Thiere* (Freiburg im Breisgau: Herder, 1898).

86. The letter is cited in Kavaeev's introduction to the volume. See V. Karavaev, "Psikhologicheskie vzgliady V. Vagnera," in E. Wasmann, *Itogi sravnitel'noi psikhologii* (Kiev: Tipografiia Universiteta sv. Vladimira, 1906), vii.

87. See, for example, V. A. Vagner, *Biologicheskie osnovaniia sravnitel'noi psikhologii* (St. Petersburg: Vol'f, 1910), vol. 1; and idem, "Fiziologiia i biologiia v reshenii psikhologicheskikh problem," in V. A. Vagner, *Biopsikhologiia* (St. Petersburg: Vol'f, 1914), 1–37. Compare these reviews with the one written by a professor of the Moscow Theological Academy, S. S. Glagolev, "Vzgliad Vasmanna na proiskhozhdenie cheloveka," *Bogoslovskii vestnik*, 1911 (April): 621–42; (May): 1–16; (July–August): 417–50; (October): 233–64.

88. See A. Lozinskii, "Segal', S. L. Voprosy zhizni s tochki zreniia estestvoznaniia (Novocherkassk, 1906)," *Vrachebnaia gazeta*, 1907, 11: 339.

89. Of course, certain scientists did try to engage in a dialogue with theology and theologians, but strictly on their own terms, searching for "a substitute for religion suitable for the individuals with a consistent natural-scientific viewpoint," A. I. Iarotskii, *Tsennost' religii s biologicheskoi tochki zreniia* (Iur'ev: Mattisen, 1915), 58–73; here—i.

90. For instance, Ivan Pavlov, who vehemently opposed any theistic or idealistic interpretations in his science, protested the Bolsheviks' repressive policies against the Church and its clergy. For a detailed analysis of Pavlov's personal attitude and public stance toward religion and atheism; see Daniel P. Todes, *Ivan Pavlov* (New York: Oxford University Press, 2014, forthcoming). See also a similar assessment of the Bolsheviks' antireligious campaign by the director of the Narkomzdrav Institute of Biophysics Petr Lazarev found in the secret-police archives by Galina Savina, "Napisano v podvalakh OGPU. P. Lazarev, 'Moe politicheskoe kredo,' " *Vestnik RAN*, 1995, 65(6): 452–60.

91. See N. K. Kol'tsov, "Uluchshenie chelovecheskoi porody," *Russkii evgenicheskii zhurnal*, 1922, 1: 3–27.

92. Kol'tsov obviously used the word "religion" in the sense we currently use the word "ideology," comparing eugenics not only to Islam and Christianity, but also to nationalism and socialism.

93. For historical analyses of Russian eugenics, see Adams, "Eugenics in Russia"; and Krementsov, "From 'Beastly Philosophy' to Medical Genetics."

94. In a telling episode in Bulgakov's first novel *The White Guard*, God himself complains to one of the characters: "Don't even talk to me about priests. I myself can't figure out what to do with them. That is, there are no such fools as our priests anywhere in the world. I'll let you on a secret: they are a disgrace [to me], not priests." Mikhail Bulgakov, *Belaia gvardiia*, in idem, *Sobranie sochinenii v 5-ti tomakh* (Moscow: Khudozhestvennaia literatura, 1989–1990), vol. 1: 237.

95. On "The Dog's Heart" as an anti-scientistic pamphlet, see E. G. Stepanian, *O Mikhaile Bulgakove i "sobach'em serdtse"* (Moscow: Oklik, 2009).

96. See A. Ukhtomskii, *Intuitsiia sovesti* (St. Petersburg: Peterburgskii pisatel', 1996). For a brief account of Ukhtomskii's life and work in English, see A. D. Nozdrachev and L. A. Samoilova, "Academician Prince A. A. Ukhtomskii and his Scientific Heritage," *Human Physiology*, 2000, 26(6): 736–46; for a partial examination of the interrelation of Ukhtomskii's scientific and religious views, see Iu. V. Stolbun, *Psikhicheskaya dominanta Akademika A. A. Ukhtomskogo* (Tver—Iaroslavl: Liliia Print, 2002). Of course, there is no rule without an exception. Thus, Valentin Voino-Iasenetskii (1877–1961), a prominent surgeon and experimentalist, not only received a Stalin prize for his research, but was also an archbishop of the Russian Orthodox Church, even though he was arrested and imprisoned several times; see M. A. Popovskii, *Zhizn' i zhitie sviatitelia Luki Voino-Iasenetskogo, arkhiepiskopa i khirurga* (Sankt-Peterburg: Satis-Derzhava, 2002).

97. For analyses of the role of biological knowledge in such debates, see Denis R. Alexander and Ronald L. Numbers, eds., *Biology and Ideology from Descartes to Dawkins* (Chicago: University

of Chicago Press, 2010); for the role of the media in the transformation of scientific knowledge into a cultural resource in US contexts, see Peter Snyder, Linda Mayes, and Dennis Spencer, eds., *Science and the Media: Delgado's Brave Bulls and the Ethics of Scientific Disclosure* (Elsevier, 2009).

98. L. Fleck, *Genesis and Development of a Scientific Fact*, transl. F. Bradley and T. J. Trenn; eds., T. J. Trenn and R. K. Merton (Chicago: University of Chicago Press, 1979), 122–23.

99. See, for instance, an examination of "Darwinism" as a cultural resource in Nikolai Krementsov, "Marxism, Darwinism, and Genetics in Soviet Union," in Alexander and Numbers, eds., *Biology and Ideology*, 215–46; or an analysis of "Lysenkoism" as a cultural resource in William deJong-Lambert and Nikolai Krementsov, "On Labels and Issues: The Lysenko Controversy and the Cold War," *Journal of the History of Biology*, 2012, 45 (3): 373–88.

100. A popular misconception that the name "Frankenstein" refers to a "monster" created by "science" (and not to aspiring "scientist" Viktor Frankenstein, who created the "monster" in Mary Shelley's classic) exemplifies the misrepresentation of actual facts and the perpetuation of distorted information in the process of generating a particular cultural resource. For a fine-grained examination of Frankenstein as a cultural icon and the role of Hollywood movies in molding its modern perception, see Susan E. Lederer, *Frankenstein: Penetrating the Secrets of Nature* (New Brunswick, NJ: Rutgers University Press, 2002), 31–52.

EPILOGUE

1. See Gr. Grebnev, "Arktaniia," *Pioner*, 1937, 10: 64–86; 11: 42–66; 12: 73–97. Six months later the novel came out in book format with a first print run of 25,000 copies; see Grigorii Grebnev, *Arktaniia* (Moscow-Leningrad: Izdatel'stvo detskoi literatury, 1938).

2. Grebnev, *Arktaniia*, 31.

3. See Robert C. Tucker, *Stalin in Power: The Revolution from Above, 1928–1941* (New York: Norton, 1990); also Sheila Fitzpatrick, ed. *Cultural Revolution in Russia, 1928–1931* (Bloomington: Indiana University Press, 1978).

4. For details, see Krementsov, *Stalinist Science.*

5. *Sotsialisticheskaia rekonstruktsiia i nauka* (1931–36).

6. Grebnev, *Arktaniia*, 36.

7. Alexander Beliaev, "Sozdadim sovetskuiu nauchnuiu fantastiku," *Detskaia literatura*, 1938, 15–16: 1–8.

8. For details of this episode, see Eric Naiman, "Injecting Communism."

9. See http://www.prnewswire.com/news-releases/steven-seagal-asks-for-russian-pm-vladimir-putins-support-in-immortality-research-121948323.html; accessed May 20, 2011. The complete text of the letter is available at http://www.scribd.com/doc/55559143/Letter-to-Putin; accessed May 20, 2011.

10. For details on the group membership and its manifesto, see http://2045.com and http://2045.com/ideology.

11. See, for instance, a journalistic account of recent events in this area in the United States, Stephen S. Hall, *Merchants of Immortality: Chasing the Dream of Human Life Extension* (Boston: Houghton Mifflin Harcourt, 2003).

Index